COMMUNICATION NETWORKS

COMMUNICATION NETWORKS

Principles and Practice

Sumit Kasera
Nishit Narang
Sumita Narang

McGraw-Hill
New York   Chicago   San Francisco   Lisbon   London   Madrid
Mexico City   Milan   New Delhi   San Juan   Seoul
Singapore   Sydney   Toronto

Cataloging-in-Publication Data is on file with the Library of Congress.

2 3 4 5 6 7 8 9 0   DOC/DOC   0 1 3 2 1 0 9 8 7 6

ISBN-13: 978-0-07-147656-0
ISBN-10: 0-07-147656-3

This book was first published in India in 2005 by Tata McGraw-Hill.

The sponsoring editor for this book was Stephen S. Chapman and the production supervisor was Richard C. Ruzycka. The art director for the cover was Brian Boucher.

Printed and bound by RR Donnelley.

This book was printed on acid-free paper.

McGraw-Hill books are available at special quantity discounts to use as premiums and sales promotions, or for use in corporate training programs. For more information, please write to the Director of Special Sales, McGraw-Hill Professional, Two Penn Plaza, New York, NY 10121-2298. Or contact your local bookstore.

PREFACE

Origin

The origin of this book lies in the first edition of the book *ATM Networks: Concepts and Protocols* by Sumit Kasera. *ATM Networks* comprises both the basic concepts of networking as well as the details of ATM networks. The organization of the book made many reviewers opine that the target audience was unclear, largely because it incorporated, in one book, material that was too technical for beginners and material that was too basic for experts. It was clear therefore, that the basic concepts of networking demanded another form.

The options were not many. They could be deleted from the book, and hence, lost forever. Alternatively, they could be posted on the internet, in what would inevitably be a rather raw form. The third choice was to put in some effort on the contents, to give it a meaningful shape in the form of a book. To decide on which would be the best option, a quick survey was conducted among the readers of the book: Gurpreet Singh, Yogesh Garg, Paras Shah, and L. Sreenivasan were those who provided valuable comments. They were unanimous that the basic concepts should not be lost and that breaking the book into two separate books was a better idea by far. This verdict led to the birth of the second edition of *ATM Networks* focusing only on ATM technology; and a second book— a new book— on *Communication Networks: Principles and Practice*.

To fulfill the latter objective, the three of us came together, to co-author this project. Our task was made easier because we are colleagues at that great organization, Flextronics Software Systems, which has a vast knowledge base in the field of networking and communication.

Scope

Without any doubt there are a plethora of books in the field of data communications, networking, and computer networks. And without any doubt, many of them are written by great academicians and are extremely good books. We have had the opportunities to read and learn from these books. Given this, it was important for us to ask ourselves whether there was a genuine need for another book. If yes, what purpose would this book fulfill?

One clear need (or demand) for most readers is to have a book that is simple and fairly easy to read. However, as with the first edition of *ATM Networks*, most good books are comprehensive and cover many topics that people might not like to read. This is especially true for first time readers. Thus, a clear need emerges for a book that is easy to read, simple to understand, without jargon, and covering most of the important topics of communication networks (if not all). It is this need that our new book tries to fulfill. This book has a fairly simple language, covers only key concepts, does not attempt to cover networking protocols in any detail (except as examples to illustrate a point) and is not very thick. We hope therefore, that reading the book from cover to cover will not task our readers for time.

To this end, the authors have adopted many novel techniques. First of all, the focus is on concepts, not on technologies. Further, the chapters are not based on the OSI reference model, thus providing a greater flexibility to the organization of the contents. The book is planned such that each chapter first provides details of a particular aspect of networking (e.g. Routing) and then explains it further by using examples of a contemporary technology (like TCP/IP, ATM or even 3G Network). This approach allows the reader to learn the concepts and further helps him with actual realization in a popular technology. For more details, there are several good books on each individual technology that are mentioned in the Further Reading section. To summarize, this book tries to build a base for understanding the basic concepts of communication networks.

Many categories of people will find this book useful. These include students of basic courses on computer networks or data communications. Students taking advance courses will find it handy for revising basic concepts of networking. This book will also be of value to fresh telecom professionals coming from non computer-science background and lacking relevant experience.

Here, without any iota of doubt, it may be said that this new book does not diminish the utility of the existing good books. It only makes the reading of other books easier. It does not aspire to replace them.

About the Book

Most books on computer networks are typically based on the OSI reference model. Thus, they cover the seven layers or building blocks of communication networks. This OSI model is a fairly old concept and most of its principles have undergone modifications. Moving away from the OSI model, this book looks at several key concepts of networking and explains them with relevant examples. The concepts are divided in three Parts (see Fig. P.1).

Part 1 provides an introduction to the building blocks of communication networks. In this part, Chapter 1 provides an overview of connection-oriented and connectionless services and explains why these concepts are important. Chapter 2 provides an overview of protocol layering principles, including the OSI reference model. Chapter 3 then elaborates on the concept of transfer modes (including circuit switching and packet switching). This concept is tightly linked to the nature of network services (i.e. connection-oriented and connectionless services). Chapter 4 discusses some of the aspects related to network topology and extent.

Part 2 uses the building blocks of networks as discussed in Part 1 to elaborate upon seven core concepts of networking. Chapter 5 discusses physical layer aspects like transmission and multiple access. Chapter 6 looks at data link layer functionality and discusses issues like access resolution, sliding window protocol, flow control and error control. Chapter 7 looks at the means for connecting different networks together, a process referred to as bridging. Chapter 8 looks at switching issues and switch architectures. Chapter 9 elaborates upon the types of addressing and the relevance of addressing. Chapters 10 and 11 talk

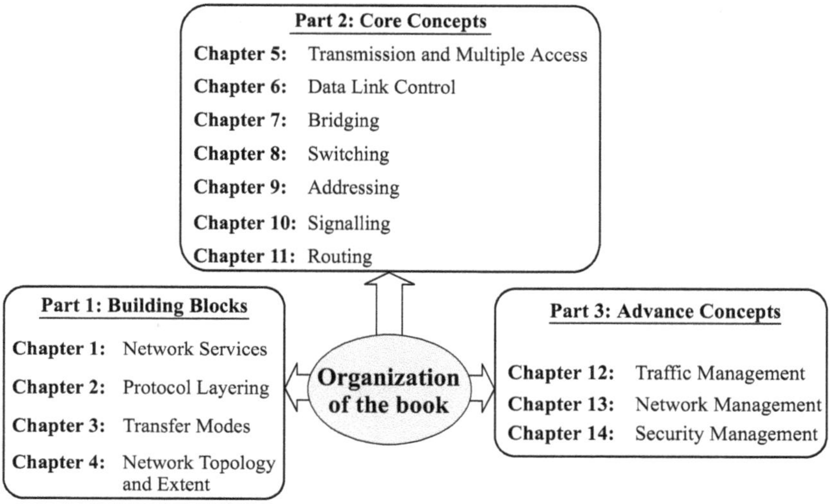

Part 2: Core Concepts

Chapter 5:	Transmission and Multiple Access
Chapter 6:	Data Link Control
Chapter 7:	Bridging
Chapter 8:	Switching
Chapter 9:	Addressing
Chapter 10:	Signalling
Chapter 11:	Routing

Part 1: Building Blocks

Chapter 1:	Network Services
Chapter 2:	Protocol Layering
Chapter 3:	Transfer Modes
Chapter 4:	Network Topology and Extent

Organization of the book

Part 3: Advance Concepts

Chapter 12:	Traffic Management
Chapter 13:	Network Management
Chapter 14:	Security Management

Fig. P.1 *Organization of the book*

about signalling and routing. Typically, addressing, signalling, and routing go together (and are typically applicable at the network layer).

Part 3 takes the discussion forward and touches upon three more concepts. The concept of traffic and traffic management is discussed in Chapter 12. The network management aspects are covered in Chapter 13 and security related issues and solutions are discussed in Chapter 14.

Web Site

To better interact with readers after the publication of this book, the authors have created a web site *http://nwbook.tripod.com* and a mirror web site managed by the publisher   *http://www.tatamcgrawhill.com/digital_solutions/ sumitnishit*. These web sites offers the following:

- Preface
- Table of Contents
- Errata
- Feedback and Review Comments
- References
- Other Related Material

Readers are encouraged to visit the web site and use the available material.

Suggestions

Your comments, feedback, and constructive criticism are valuable to us; so, please feel free to drop an email at *nwbook@lycos.com*. We would be glad to incorporate your comments in the subsequent editions of the book.

SUMIT KASERA
NISHIT NARANG
SUMITA NARANG

CONTENTS

Part 2

Core Concepts of Communication Networks

Part 3

Advanced Concepts of Communication Networks

ABBREVIATIONS

1G	First Generation
2G	Second Generation
3G	Third Generation
3GPP	Third Generation Partnership Project
AAL	ATM Adaptation Layer
AAL1	ATM Adaptation Layer type 1
AAL2	ATM Adaptation Layer type 2
AAL3/4	ATM Adaptation Layer type 3/4
AAL5	ATM Adaptation Layer type 5
ABM	Asynchronous Balanced Mode
ABR	Available Bit Rate/Area Border Router
ACM	Address Complete Message
ADCCP	Advanced Data Communication Control Procedure
ADR	Average Data Rate
AES	Advanced Encryption Standard
AFI	Address Family Indicator
AH	Authentication Header
AK	Anonymity Key
AKA	Authentication and Key Agreement
AN	Access Network
ANM	Answer Message
ANSI	American National Standards Institute
API	Application Programming Interface
ARM	Asynchronous Response Mode
ARP	Address Resolution Protocol
ARPA	Advanced Research Projects Agency
AS	Autonomous System
ASBR	Autonomous System Border Router
ASCII	American Standard Code for Information Interchange

ASK	Amplitude Shift Keying
ASN.1	Abstract Syntax Notation One
ATM	Asynchronous Transfer Mode
AuC	Authentication Centre
AUTN	Authentication Token
AVP	Attribute Value Pair
BD	Burst Duration
BER	Bit Error Rate
BG	Border Gateway
BGP	Border Gateway Protocol
BGT	Broadcast and Group Translator
B-ISDN	Broadband ISDN
bps	Bits Per Second
BPSK	Binary Phase Shift Keying
BRI	Basic Rate Interface
BSC	Base Station Controller
BSS	Base Station Sub-system
BT	Burst Tolerance
BTS	Base Transceiver Station
CAC	Connection Admission and Control
CAM	Content Addressable Memory
CAN	Campus Area Network
CBR	Constant Bit Rate
CC	Country Code
CDM	Code Division Multiplexing
CDMA	Code Division Multiple Access
CDV	Cell Delay Variation
CDVT	Cell Delay Variation Tolerance
CER	Cell Error Ratio
CIDR	Classless Inter-Domain Routing
CK	Cipher Key
CLP	Cell Loss Priority
CLR	Cell Loss Ratio
CMIP	Common Information Management Protocol
CMOT	Common Management over TCP/IP
CMR	Cell Mis-insertion Ratio
cps	Cycles Per Second
CPU	Central Processing Unit
CRC	Cyclic Redundancy Check
CS	Circuit Switched
CSMA	Carrier Sense Multiple Access

CSMA/CD	Carrier Sense Multiple Access with Collision Detection
CSU	Channel Service Unit
CTD	Cell Transfer Delay
dB	Decibels
DCF	Data Communication Function
DCN	Data Communication Network
DES	Data Encryption Standard
DHCP	Dynamic Host Configuration Protocol
DLCI	Data Link Connection Identifier
DLE	Data Link Escape
DNS	Domain Name System
DoD	Department of Defence
DSAP	Destination Service Access Point
DSU	Data Service Unit
DTE	Digital Terminal Equipment
DVMRP	Distance Vector Multi-cast Routing Protocol
ECC	Elliptic Curve Cryptography
EDGE	Enhanced Data Rates for Global Evolution
EFCI	Explicit Forward Congestion Indication
EGP	Exterior Gateway Protocol
EIGRP	Enhanced Interior Gateway Routing Protocol
EIR	Equipment Identity Register
EPD	Early Packet Discard
ESP	Encapsulating Security Payload
ETSI	European Telecommunications Standards Institute
ETX	End of Text
FCAPS	Fault, Configuration, Accounting, Performance and Security Management
FDD	Frequency Division Duplex
FDDI	Fiber Distributed Data Interface
FDM	Frequency Division Multiplex
FDMA	Frequency Division Multiple Access
FEC	Forward Error Correction
FIFO	First In First Out
FSK	Frequency Shift Keying
FSM	Finite State Machine
FTP	File Transfer Protocol
GCRA	Generic Cell Rate Algorithm
GFC	Generic Flow Control

GPRS	General Packet Radio Service
GSM	Global System for Mobile Communication
HDLC	High Level Data Link Control
HDLCP	High Level Data Link Control Procedure
HE	Home Environment
HEC	Header Error Control/Header Error Check
HEMP	High Level Entity Management Protocol
HEMS	High Level Entity Management System
HLR	Home Location Register
HOL	Head Of Line
HTTP	Hyper Text Transfer Protocol
IAB	Internet Activities Board
IAM	Initial Address Message
IC	Integrated Circuit
ICI	Interface Control Information
ICMP	Internet Control Message Protocol
ID	Identifier
IDU	Interface Data Unit
IE	Information Element
IETF	Internet Engineering Task Force
IGRP	Interior Gateway Routing Protocol
IISP	Interim Inter-Switch Protocol
IK	Integrity Key
IKE	Internet Key Exchange
ILC	Identifier-Length-Contents
ILCE	Identifier-Length-Contents-End_of_contents
IM	Input Module
IMEI	International Mobile Equipment Identity
IMP	Interface Message Protocol
IMS	IP Multimedia Sub-system
IMSI	International Mobile Subscriber Identity
INTERNIC	Internet Network Information Center
IP	Internet Protocol/Initial Permutation
IPv4	Internet Protocol Version 4
IPv6	Internet Protocol Version 6
IPSec	IP Security
IPX	Internetwork Packet Exchange
ISD	International Subscriber Dialling
ISDN	Integrated Services Digital Network
IS-IS	Intermediate System to Intermediate System
ISO	International Organization for Standardization

ISP	Internet Service Provider
ISUP	ISDN User Part
ITU	International Telecommunication Union
IV	Initialization Vector
IWF	Inter-Working Function
KDC	Key Distribution Center
LAN	Local Area Network
LAP	Link Access Procedure
LAPB	Link Access Procedure-Balanced
LAPD	Link Access Procedure for D-channel
LAPM	Link Access Procedure for Modems
LLA	Logical Layered Architecture
LLC	Logical Link Control
LRU	Least Recently Used
LSA	Link State Advertisement
MAC	Medium Access Control/Message Authentication Code
MADR	Maximum Allowable Data Rate
MAN	Metropolitan Area Network
MAP	Mobile Application Part
MAPsec	MAP security
MBS	Maximum Burst Size
MCR	Minimum Cell Rate
MD	Mediation Device
MD5	Message Digest 5
ME	Mobile Equipment
MF	Mediation Function
MIB	Management Information Base
MIPS	Million Instructions Per Second
MMI	Man-Machine Interface
MOSPF	Multicast Open Shortest Path First
MPLS	Multi-Protocol Label Switching
MS	Mobile Station
MSC	Mobile Switching Center
MSISDN	Mobile Subscriber ISDN Number
MTP	Message Transfer Part
MTU	Maximum Transfer Unit
NAT	Network Address Translation
NCP	Network Control Protocol
NDC	National Destination Code

NDS	Network Domain Security
NE	Network Element/Network Entity
NEF	Network Element Function
NFS	Network File Server
NIC	Network Interface Card
NMS	Network Management Station
NNI	Network-Node Interface/Network-Network Interface
NPC	Network Parameter Control
nrt-VBR	Non real-time VBR
NRM	Normal Response Mode
NRZ	Non-Return to Zero
NSAP	Network Service Access Point
NSF	National Science Foundation
NSN	National Significant Number
O&M	Operations and Maintenance
OAM	Operation, Administration and Maintenance
OS	Operations Systems
OSF	Operations Systems Function
OSI	Open Systems Interconnection
OSI-RM	OSI Reference Model
OSPF	Open Shortest Path First
PAN	Personal Area Network
PCI	Protocol Control Information
PCM	Pulse Code Modulation
PCR	Peak Cell Rate
PDA	Personal Digital Assistant
PDH	Plesiochronous Digital Hierarchy
PDU	Protocol Data Unit
PIM	Parallel Iterative Matching/Protocol Independent Multi-cast
PLMN	Public Land Mobile Network
PNNI	Private Network Node Interface
POTS	Plain Old Telephone System
PPD	Partial Packet Discard
PPP	Point-to-Point Protocol
PRI	Primary Rate Interface
PS	Packet Switched
PSK	Phase Shift Keying
PSTN	Public Switched Telephone Network
PVC	Permanent Virtual Circuit

QA	Q Adaptor
QAF	Q Adaptor Function
QoS	Quality of Service
QPSK	Quadrature Phase Shift Keying
RAM	Random Access Memory
RAN	Radio Access Network
RAND	Random Number
RARP	Reverse Address Resolution Protocol
RED	Random Early Discard
RFC	Request for Comments
RIP	Routing Information Protocol
RM	Resource Management
RPC	Remote Procedure Call
RSA	Rivest, Shamir and Adleman
RSVP	Resource Reservation Protocol
RTP	Real-time Transport Protocol
rt-VBR	Real-time VBR
RZ	Return to Zero
SA	Security Association
SAP	Service Access Point
SAPI	Service Access Point Identifier
SAR	Segmentation and Reassembly
SCR	Sustainable Cell Rate
SCTP	Stream Control Transmission Protocol
SDH	Synchronous Digital Hierarchy
SDLC	Synchronous Data Link Control
SDU	Service Data Unit
SECBR	Severely Errored Cell Block Ratio
SGSN	Serving GPRS Support Node
SHA1	Secure Hash Algorithm 1
SIM	Subscriber Identity Module
SLA	Service Level Agreement
SMI	Structure of Management Information
SMS	Short Message Service
SMTP	Simple Mail Transfer Protocol
SN	Subscriber Number
SNA	System Network Architecture
SNMP	Simple Network Management Protocol
SNMP AE	SNMP Application Entity
SNR	Signal to Noise Ratio
SONET	Synchronous Optical Network

SPF	Shortest Path First
SPR	Special Purpose Register
SQN	Sequence Number
SS7	Signaling System No. 7
SSAP	Source Service Access Point
STD	Subscriber Trunk Dialling
STDM	Statistical Time Division Multiplexing
STM	Synchronous Transfer Mode
STX	Start of Text
SVC	Switched Virtual Circuit
TAT	Theoretical Arrival Time
TCP	Transmission Control Protocol
TDD	Time Division Duplex
TDM	Time Division Multiplexing
TDMA	Time Division Multiple Access
TFTP	Trivial File Transfer Call
TLV	Type Length Value
TMN	Telecommunication Management Network
TMSI	Temporary Mobile Subscriber Identity
TP	Transport Protocol
TS	Technical Specification
UBR	Unspecified Bit Rate
UDP	User Datagram Protocol
UE	User Equipment
UMTS	Universal Mobile Telecommunications System
UMTS AKA	UMTS Authentication and Key Agreement
UNI	User-Network Interface
UPC	Usage Parameter Control
URL	Uniform Resource Locator
USA	United States of America
USIM	Universal Subscriber Identity Module
USSR	United Soviet Socialist Republic
VBR	Variable Bit Rate
VC	Virtual Circuit/Virtual Channel
VCC	Virtual Channel Connection
VCI	Virtual Circuit Identifier/Virtual Channel Identifier
VCI	Virtual Channel Identifier
VLR	Visitor Location Register
VLSI	Very Large Scale Integration
VoIP	Voice Over IP

VPC	Virtual Path Connection
VPI	Virtual Path Identifier
VPN	Virtual Private Network
WAN	Wide Area Network
WCDMA	Wideband Code Division Multiple Access
WDM	Wavelength Division Multiplexing
WF	Workstation Function
WS	Workstation
WWW	World Wide Web
XOR	Exclusive- Or
XRES	Expected Response

(from left) Sumit Kasera, Nishit Narang, Sumita Narang

AUTHORS' PROFILES

Sumit Kasera is Senior Technical Leader at Flextronics Software Systems (formerly Hughes Software Systems), India. He has a B. Tech degree in Computer Science and Engineering from IIT, Kharagpur, India, and an M.S degree in Software Systems from Birla Institute of Technology and Science, Pilani, India. His current area of interest is software development for GSM, GPRS, and EDGE access network. Sumit is also experienced in software development for networking protocols like ATM, TCP/IP and 3G UMTS. He has been an active participant in various technical forums like ITU-T, ATM Forum and 3GPP, where he has presented papers and conducted seminars; and also participated in the review of 3GPP specifications.

Sumit is the author of two books—*ATM Networks: Concepts and Protocols* and *3G Networks: Architecture, Protocols and Procedures.*

Nishit Narang is Senior Technical Leader at Flextronics Software Systems, India. He has a B. Tech. degree in Computer Science and Engineering from IIT, Delhi, India, and an M.S degree in Software Systems from Birla Institute of Technology and Science, Pilani, India. Nishit's current areas of interest include software development for GSM, GPRS and 3G networks. He has wide experience in software development for networking protocols (like ATM and TCP/IP), network convergence, and routing protocols over satellite-based networks. He has been an active participant in various technical forums like IEEE ACM, ATM Forum and 3GPP and has presented papers and conducted technical seminars at these forums.

Nishit has authored a book titled *3G Networks: Architecture, Protocols and Procedures.*

Sumita Narang is a Senior Software Engineer at Flextronics Software Systems. She has a B. Tech. degree from Delhi Institute of Technology, Delhi, India and an M.S degree in Software Systems from Birla Institute of Technology and Science, Pilani, India. Sumita has been associated with Flextronics Software Systems for over three years. Her areas of interest include building applications for Element Management systems and satellite based broadband networks. Sumita has also been actively participating in various projects related to wireless technologies (GSM, GPRS).

COMMUNICATION
NETWORKS

- Network Services

- Protocol Layering

- Transfer Modes

- Network Topology and Extent

Part 1

BUILDING BLOCKS OF COMMUNICATION NETWORKS

In the modern world, a life without communication networks is inconceivable. Be it the railway booking counter, or automated banking, the ubiquitous Internet, or the global wireless telephony systems—communication networks or simply, networks, have found a wide variety of uses. In business organizations, networks are used for client—server applications, for connecting geographically-dispersed sites, for office automation, for audio/video conferencing and for many other miscellaneous purposes. Individuals use networks for a multitude of reasons like for exchanging personal information (via telephonic conversation or via e-mails), for surfing the internal and for personal transactions. Business organizations in the aviation and railways sectors maintain global databases through well-connected networks. This has considerably eased the process of ticket reservation and cancellation. In the banking sector, networks have made multi-city accounts and automated teller machines possible. The list of possible applications of communication networks is endless, and this list is bound to increase in the days to come.

What is it in a network that makes its use so pervasive? It is their ability to transcend geographical barriers that has made communications networks so omnipresent. This is, by far, one of the most important aspects of communications networks. Besides this, networks are also an efficient means to share resources. For example, in a corporate environment, few licenses for key applications are shared among a large number of users. Precious hardware resources like network printers and servers are also shared among a group of people. These are possible only because of communication networks.

A *communication network* can be viewed as an interconnection of communicating entities. A communicating entity is defined as a stand-alone entity involved in the process of communication. Personal computers, laptops, telephones, pagers and mobile phones are all examples of communicating entities. A communication network may involve computers, but this does not necessarily mean that it is synonymous with a computer network. Strictly speaking, a computer network is merely one of the forms of communication networks. A 'computer network', essentially means an interconnection of computers. In contrast, the term 'communication network' is used in a much broader sense. Today, when the world is heading towards integrated networks, there is a drastic change in the way one perceives a computer network. Telephones, intelligent television sets and multimedia desktops could be connected via the same network. Thus, it is better to refer to modern-day network as a communication network, rather than just a computer network.

As mentioned above, a communication network is an interconnection of communicating entities. This definition, however, is a very simple way of looking at things. The nitty-gritty of a complex network cannot be encompassed by the single word 'interconnection'. Therefore, let us go beyond this simple definition and find out the concepts and elements involved in defining a communication network. Consider the following questions:

- What are the broad categories of services offered by a network?
- How are network services actually provided? What set of software or hardware is required to realize these services?
- How is information packaged and transferred across the network?
- Is each entity in a network connected with every other entity of the network? Is there a limit on the geographical extent of the network?

The questions mentioned above touch upon certain key aspects in the field of networking. Analyzing these aspects and finding their engineering solutions is very important. This part of the book is an attempt to understand the concepts that will help the reader in building the basic foundations of communication networks. For this purpose, the following concepts are defined:

- **Network Services:** This refers to the classification of network services into connection-oriented and connectionless services. This topic is covered in Chapter 1.
- **Protocol Layering:** This refers to a layered architecture that simplifies the process of development and inter-operability of network entities. This topic is covered in Chapter 2.
- **Transfer Mode:** This refers to the means of packaging, sending and receiving information on the network. This topic is covered in Chapter 3.
- **Network Topology and Extent:** This refers to the physical layout and interconnection structure of a network. It also includes the classification of networks into Local Area Networks (LANs) and Wide Area Networks (WANs). This topic is covered in Chapter 4.

NETWORK SERVICES

1.1 INTRODUCTION

The primary task of a network is to provide the means to transfer user information from one network entity to another. Depending upon the way information is transferred, the services offered by a network are classified into two distinct categories, namely *connection-oriented service* and *connectionless service*. Choosing one of the two services has a marked effect on the capabilities of a network. Specifically, factors like the quality of service provided to the users, the range of applications supported, and the extent of resource utilization in the network are dependent upon whether the network provides a connection-oriented service or a connectionless service. This chapter explains these two distinct categories of network services.

1.2 CONNECTION-ORIENTED SERVICE

Connection-oriented service, as the name suggests, entails establishing a *dedicated connection* between communicating entities before data is exchanged. The connection establishment may be at the *physical* or the *logical* level and involves some form of signalling. The exact mechanism depends upon the transfer mode (discussed in Chapter 3). Irrespective of the nature of the mechanism, the connection establishment is typically accompanied by some form of resource reservation (e.g. reserving link bandwidth or buffers). This is followed by the exchange of user data. After this, the connection is cleared and all resources for the connection are freed. A typical example of a connection-oriented service is the telephone network. A dedicated connection is first set up after which the conversation takes place. After the conversation, the connection is disconnected and the reserved resources are freed.

A connection-oriented service is *generally* characterized by the following:

- A connection is first established (through signalling), which is usually accompanied by some form of resource reservation. After data transfer is over, the connection is released.
- The services are generally reliable in the sense that the data loss is either minimal or there are mechanisms to re-transmit lost data.
- The transfer mode for a connection-oriented service is circuit-switching or virtual-circuit-switching (see Chapter 3).
- For a packet-based connection-oriented service, the packets are delivered sequentially (i.e. the packets for a connection follow the same path).
- Further, packets do not carry the full address of the destination. Due to this, the header size is small and per packet overhead of the scheme is minimal.
- The scheme is suitable for long and steady transmissions.

Some of these concepts will become clear in the course of this book.

1.3 CONNECTIONLESS SERVICE

Unlike a connection-oriented service, a connectionless service does not require a connection to be established before the exchange of data. Information is transferred by using *independent data units*, wherein each data unit bears the complete destination address. Thus, a connectionless service is analogous to the way a postal mail is delivered. Each mail carries a postal address. When a mail reaches a post office, the postal address written on the mail is used to forward it to the next post office. This process is continued till the mail reaches its destination. Thus, without setting an explicit communication path or connection, the information reaches the destination end-system.

A connectionless service is *generally* characterized by the following:

- A connection does not need to be established before data exchange. Thus, there is hardly any prior resource reservation.
- The scheme provides unreliable or best-effort services in the sense that lost or dropped packets are not re-transmitted.
- The transfer mode for a connectionless service is routing (see Chapter 3).
- Packets may or may not arrive sequentially (i.e. data packets may or may not follow the same path).
- Packets carry the full address of the destination to enable routing. Due to this, the header size is large and the per packet overhead of the scheme can be significant.
- The scheme is suitable for bursty transmissions.

Why is it that the characteristics are generally true and not always true? The reasons are manifold. First and foremost, the characteristics are inextricably linked to the transfer modes and the protocol under consideration. For example, the connections in Asynchronous Transfer Mode (ATM) are virtual in nature and do not provide any re-transmissions; whereas the Transmission Control Protocol (TCP) connections are logical and provide re-transmission of lost packets. At an abstract level, both are said to provide a connection-oriented service, but their detailed characteristics are not the same. A related issue is that with the transmission media becoming more reliable, some functionality of old protocol stacks does not exist any longer. Thus, it is not necessary for a connection-oriented service to provide re-transmission of lost packets (e.g. ATM virtual connections). Secondly, the characteristics are subjective in nature. What is a reliable service for an application may be unreliable for another. For example, even if a few bits are dropped in a telephonic conversation, the effect is not reflected in the quality of the talk. The same does not hold for data transfer, wherein even a single-bit error can render the whole data useless. Thus, it is not easy to define reliability very objectively. Lastly, there are cases wherein even a connectionless protocol like Internet Protocol (IP) uses a connection-oriented service to provide better performance by using protocols like Resource Reservation Protocol (RSVP) and Multi-Protocol Label Switching (MPLS). This makes the differences between connection-oriented and connectionless service more ambiguous.

1.4 CONNECTION-ORIENTED VERSUS CONNECTIONLESS SERVICE

Most comparisons between two networking technologies finally boil down to a debate between connection-oriented and connectionless service. This debate centers around a few key issues, namely *resource utilization*, *service guarantees*, *state information*, and *call-setup latency* (or *delay*).

In a connection-oriented service, since the resources are reserved at call-setup time, a user is quite likely to get his share of the bandwidth. Thus, either the connection set-up request is rejected straightaway or, if accepted, the network provides adequate resources to satisfy the needs of a user application. This is why a connection-oriented service is said to ensure some form of service guarantees to the user (it may be noted that the exact service guarantees depend upon the nature of the connection-oriented service offered by the network and the extent of statistical sharing as discussed below). The advantage of resource reservation leads to shortcomings on several other fronts.

Firstly, resource reservation leads to sub-optimal utilization of network resources. This is because the actual use of dedicated resources is related to the amount of information generated by the source. If the source generates information sporadically, the network resources are also used intermittently. During the remaining time, the resources are unused, thereby resulting in a wastage of network resources. One important aspect pertaining to resource utilization in a connection-oriented service is the extent of statistical sharing. If more connections are statistically shared than the network can support, there is degradation in service guarantees. If fewer connections are supported, there is under-utilization of network resources. Thus, while deciding the number of user connections to be supported, it is important to take into consideration the amount of resources in the network and the requirements of user applications.

Secondly, in certain situations, call-setup and call-clearing delays can be time-consuming and may take more time than the actual time taken for data transfer. Thus, a connection-oriented service is not preferable for short-lived connections.

Thirdly, a connection-oriented service requires the network to maintain a large amount of state information. For example, using virtual-circuit in networks, the state information is maintained in the form of virtual circuit tables. This state information is directly proportional to the number of connections supported. This fact retards the scalability of connection-oriented services.

A connectionless service does away with the three drawbacks mentioned above. Firstly, as no resource is reserved, the resource utilization is directly linked to the current load, which should ideally be the case. Secondly, as there is no call-setup, there is no call-setup latency. Thirdly, since there are no connections, no per-connection state information is maintained. Here, it may be noted that a connectionless services also needs to maintain some form of routing tables to route a packet to the destination. But this information is global and not on a per-connection basis.

However, all this does not make connectionless service an ideal choice. One of its most important shortcomings is its failure to provide service guarantees. Although some argue that by providing links with very high bandwidths, this problem can be solved, this is not true. As link speeds increase, the bandwidth requirements of user applications also increase. Thus, there never arises a situation wherein the bandwidth is so plentiful that the need for reserve resources can be done away with. This is a very serious drawback of protocols following the connectionless service paradigm.

The second important drawback of a connectionless service pertains to routing infrastructure bottlenecks. Since a router must maintain an entry for each host in the world (or at least each network or group of networks in the world), the size of routing tables is directly related to the number of hosts or networks. As the number of hosts or networks increases exponentially, there is a concomitant increase in the size of routing tables and routing table look-up. All this considerably slows down the maximum speed at which routers can forward packets.

1.5 EXAMPLES

Till now, the two broad categories of service, viz. connection-oriented and connectionless service were explained in this chapter. This section looks at some of the important protocols and categorizes them in one of the two categories. These protocols are listed in Table 1.1 along with the service type and a brief description.

Table 1.1   *Examples of Connection-oriented (CO)/Connectionless (CL) Protocols*

Name	Type	Description
PSTN	CO	Provides connection-oriented service through dedicated physical connections.
Frame Relay	CO	Provides connection-oriented service through virtual-circuit connections.
ATM	CO	Provides connection-oriented service through virtual-circuit connections.
IP	CL	Provides connectionless unreliable service using packet routing.
TCP	CO	Provides connection-oriented reliable service over unreliable IP protocol.
UDP	CL	Provides connectionless unreliable service over IP protocol.

As shall be further elaborated in Chapter 3, the service type is tightly linked to the transfer modes. Given that the transfer modes can be circuit switching as in Public Switched Telephone Network (PSTN), virtual-circuit switching as in (ATM and Frame Relay) or routing as in (IP), the service type of PSTN, ATM, Frame Relay and IP are defined accordingly. This categorization is applicable to lower layer protocols. For higher layer protocols, the nature of the service is linked to the scope of error control (i.e. re-transmission of lost packets). Based on the error control functionality, Transmission Control Protocol (TCP) and user Datagram Protocol (UDP) are classified as connection-oriented and connectionless, respectively. There can be a host of other protocols but they are not listed here.

1.6 CONCLUSION

This chapter looked at two broad categories of network service, viz. connection-oriented and connectionless service. This broad categorization defines the basic capabilities of network protocols in terms of service guarantees, scalability, and the extent of resource utilization. Given this, the comparison of any two protocols belonging to different categories boils down to the difference between connection-oriented and connectionless service. Some of these aspects are linked to transfer modes and the topic is revisited in Chapter 3.

REVIEW QUESTIONS

Q 1. *What are the typical characteristics of connection-oriented and connectionless service? What are the merits and demerits of each?*

Q 2. *Why is the categorization of network service into a connection-oriented or a connectionless service so important?*

FURTHER READING

This chapter introduced the concept of network services. For additional reference to this topic, the reader is referred to [Gen A. Tanenbaum], [Gen W. Stallings], [Gen S. Keshav] and [Gen J. Kurose].

2

PROTOCOL LAYERING

2.1 INTRODUCTION

One of the most fundamental concepts underlying all modern-day communication is the concept of *protocol*. A protocol is a set of rules and frame formats that govern the way two entities communicate. Since all networking devices are dumb in the sense that they only understand fixed commands, a protocol provides commands to send and receive information. For example, if a sender sends a stream of 0s and 1s, the receiver should be able to unambiguously interpret the information sent. This is possible only if the receiver knows that, say, the first 24 bits of the information stream are the address bits, the next 12 bits are the control bits, and so on and so forth. This knowledge is obtained via a protocol. A protocol consists of two parts, viz. *syntax* and *semantics*. The syntax provides the means to interpret an incomprehensible data stream consisting of 0s and 1s. For example, the information stream is divided into the control field, address field, data field and so on by using the syntax of the protocol. The semantics of the protocol provides the actions to be taken on the basis of the received information. For example, one of the values of the control field may be a re-transmission request, while another value may be an acknowledgement for data already sent. The semantics helps in deciding the future course of action according to the rules associated with each possible value of the received data.

The protocols in communication systems are not organized as a single monolithic piece. Rather, they are organized as a set of *layers*, wherein each layer has a well-defined function. This design approach, wherein modules are distributed into layers, results in a *layered architecture*. In order to build a layered architecture, there must be a consensus on the assignment of tasks at different layers. In order to formalize a layered architecture and to arrive at a consensus over functional distribution, the International Organization for Standardization (ISO) developed the *Open Systems Interconnection-Reference Model (OSI-RM)*. This chapter introduces the concept of layered architecture, and explains the OSI-RM. It then uses the example of the TCP/IP protocol suite to elaborate upon the layering concepts.

2.2 LAYERED ARCHITECTURE

In order to appreciate the need for layered architecture, consider the problem of building a system that will perform 'N' tasks. How does one go about designing this system? The design process will have two steps. First, one has to identify the modules that will build the system; second, one has to define the interaction between different modules. Now, suppose that the system has been built and is to be upgraded (i.e. some functionality of the system has to be altered), how does one go about making this change? There is a strong possibility that in order to accommodate this change, the entire system may have to be re-designed. Even when the changes are localized, it is quite possible that in the absence of well-defined interfaces, changes made in the one module have to be reflected in all the neighbouring modules.

The above argument necessitates that a well designed system should have two things—*well-defined modules* and *well-defined interfaces*. The layered architecture is based on this logic. *Layered architecture* is a design approach in which each layer provides a well-defined service to the layer above it and uses the services of the layer below it. A related phrase called 'layered abstraction' is also used to describe this approach. The term 'abstraction' implies that the implementation detail of a layer is hidden to the layers above and below it.

Layered abstraction offers several advantages. First, by using layered abstraction, each layer can be built independent of other layers. The scope of the work gets more focused (i.e. providing the functionality associated with the layer); each solution does not necessitate solving the whole problem. Moreover, if the services and interfaces provided to a higher layer are kept the same, altering the implementation of a layer does not require a change in the other parts of the system. Besides this, layered abstraction facilitates re-use. For example, though different applications may serve different purposes, they may still require the same services from the underlying layers, i.e. reliable data transfer. Thus, a single layer or a set of layers providing this function can be used by different application layers, even though they themselves provide different functionality. In summary, the layered abstraction approach thus divides a complex task into several sub-tasks and assigns one or more sub-tasks to each layer.

Even though the layered approach has its inherent advantages, there are some disadvantages as well. The most important among these is that the approach may not necessarily be optimal. Thus, there may be occasions wherein a sub-task is repeated in more than one layer. Moreover, clear identification of the functions of each layer may not be easy. The layered approach is also inflexible in the sense that it is not possible to easily change the basic layered structure even though a layer may have outlived its utility. Lastly, even though it is desirable that the internal variables of a layer be hidden from the other layers, there is a need for a layer to access internal variables of other layers. For example, if the physical layer can know (through higher layer's internal variables) that the higher layer does not have any data to transmit, it might not send the scheduled interrupts, thereby reducing load on message queues. However, accessing variables of other layers violates the spirit of layered abstraction.

2.3 SERVICES, INTERFACES, PRIMITIVES AND SERVICE ACCESS POINTS

The previous sub-section presented the motivation for having a layered abstraction. This section discusses the central concepts around which layer abstraction works. Specifically, the following concepts need to be elucidated:

- Concept of Services

- Concept of Interfaces, Primitives and Service Access Points (SAPs)
- Concept of Peers and Peer-to-Peer protocols
- Concept of Interface Data Units (IDUs), Service Data Units (SDUs) and Protocol Data Units (PDUs).

When the layered approach is being used, each layer has a well-defined task to perform. For performing its assigned task, each layer takes the help of the services provided to it by the layer below it. In turn, the layer provides services to the layer above it. Thus, each layer is a *service provider* for the layer above it, and a *service user* for the layer below it. The two exceptions to this are the lowest layer, which only provides services to the next higher layer, and the highest layer, which only uses the services of its lower layer.

The services provided by a layer form the *service definition* of that layer as per the protocol defining it. Note that the service definition of a layer merely states the tasks to be performed by it. The service definition does not include any mention of how the task has to be performed. (This is deliberately done so that the means used to achieve a given end are left open for innovation and competitiveness). The service definition also does not reveal as to how the upper layer avails itself of the services provided by the lower layer. This information is provided by means of interfaces and primitives.

An *interface* is the boundary between two adjacent layers. The interface definition states how a layer accesses the services of the layer below it. This definition is in the form of *primitives*. Primitives can be viewed as the operations available to a service user to access the services of the service provider. For example, consider a layer that provides the service of reliable data transfer. Such a layer will provide primitives of the form SEND_DATA(control_info, data) and RECV_DATA (control_info, data).

All primitives are defined in terms of four basic primitives, which are defined below:

1. **REQUEST:** The service user requests the service provider for a particular service.
2. **INDICATION:** The service provider informs the service user about the occurrence of an event/trigger (e.g., receipt of a message).

3. **RESPONSE:** The service user responds after getting an indication from the service provider.

4. **CONFIRMATION:** The service provider gives an indication to the service user that the requested service has been confirmed.

Figure 2.1 illustrates the four basic service primitives between layer (N+1) and layer (N) entity. As per this figure, layer (N) entity refers to the entity at layer 'N'. In a single system, only one layer (N) entity can reside. However, a layer (N) can have more than one protocol running on it. The layer (N+1) entity is the service user of layer (N) entity, and layer (N) entity is the service provider for layer (N+1) entity. The layer (N+1) entity communicates with the layer (N) entity via a *Service Access Point (SAP)*. SAPs are the connection entities that adjacent layers use to communicate with each other. Each SAP has its unique address. For a layer (N+1) entity to communicate with another layer (N+1) entity, the former must have the layer 'N' SAP address of the latter. (In postal systems, the postal addresses are the SAP addresses. Thus, you need your friend's postal address to write to him!)

Two layer (N+1) entities residing on different machines are called peers and the layer (N+1) protocol between two layer (N+1) entities is referred to as *peer-to-peer protocol*. The concept of peers is very important to understand the way data is carried from one entity to another. In simple terms, peers can be viewed as entities talking in the same language. Just as communication between two persons can take place only in a language understood by both of them, similarly, two

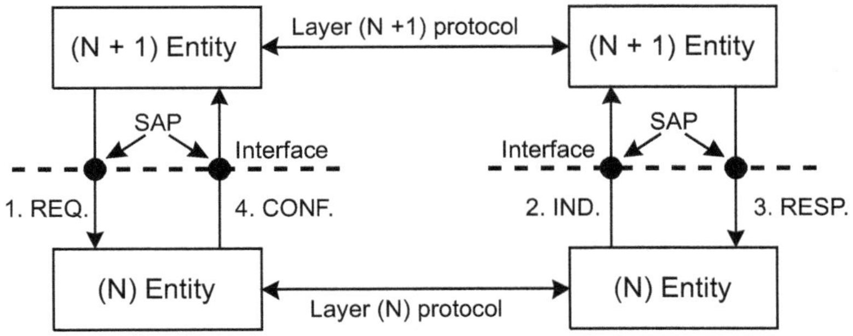

Fig. 2.1 *Interface, primitives and Service Access Points (SAPs)*

communicating entities can communicate only when they talk in the same language. This language that is understood by both communicating entities is called the peer-to-peer protocol.

For the purpose of communication, peers use the services provided by the layers below them. Now consider how two layer (N+1) entities communicate with each other. One of the layer (N+1) entities forms an *Interface Data Unit (IDU)* and passes this to the layer below it. An IDU is the basic unit of information exchanged between two adjacent layers through an SAP (see Fig. 2.2). An IDU consists of two parts—the *Service Data Unit (SDU)* and the *Interface Control Information (ICI)*. The SDU is the information (or the payload) part that layer (N+1) wishes to send to its peer (i.e. another layer (N+1) entity).

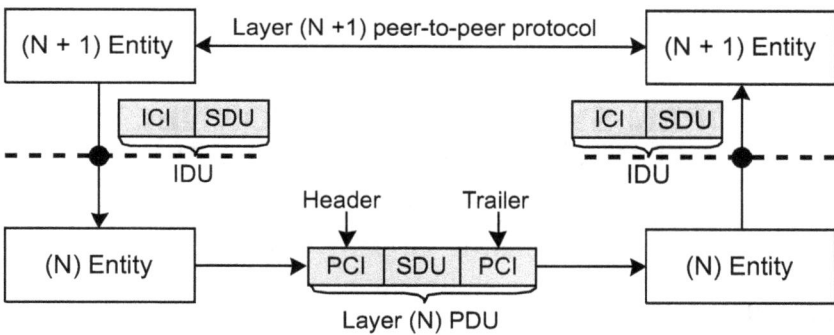

ICI: Interface Control Information   PCI: Protocol Control Information
IDU: Interface Data Unit   PDU: Protocol Data Unit
SDU: Service Data Unit

Fig. 2.2 *Data communication using layered architecture*

ICI contains the size of the payload, the interface on which the SDU is to be sent and other relevant information. ICI is required by the layer (N) to take appropriate action on the SDU sent by the layer (N+1). After receiving an IDU, layer (N) adds some control information—called the *Protocol Control Information (PCI)* to the SDU at the beginning and/or at the end and sends this to the layer below it. As example, PCI may contain addressing information, length, checksum and other control information. Adding PCI to received SDU is called *encapsulation*. This is done by using the ICI information received from the higher layer. The SDU, along with the PCI, is called a *Protocol Data Unit (PDU)*. The control information is added because this is required by the peer layer (N) entity to process the other part of the message (i.e. the SDU). The PDU thus formed is sent to the lower layer (i.e. layer (N-1) entity) and the entire process is repeated at all layers of the sending side. At the receiving end, the reverse process takes place. Layer (N) entity strips off the PCI and hands over the IDU consisting of ICI and SDU to the layer (N+1) entity. The process of stripping off the PCI to extract SDU is referred to as *decapsulation*. The encapsulation and decapsulation form the basis of how protocols in a layered architecture operate (i.e. the sending layer encapsulates, a process which is done by all layers of the sending side, and the receiving layer decapsulates, a process which is done by all layers of the receiving side).

2.4 OPEN SYSTEMS INTERCONNECTION-REFERENCE MODEL (OSI-RM)

In the late 1970s and the early 1980s, a model was developed by the International Standards Organization (ISO) to facilitate communication between multi-vendor systems. In the absence of well-defined protocols, it was almost impossible for systems of different vendors to inter-operate. Realizing the need to have internationally accepted communication protocols, the ISO developed the Open Systems Interconnection-Reference Model (OSI-RM) in 1983. The OSI model only lays down guidelines for the development of protocol stacks for open systems. In other words, the OSI reference model defines a layered architecture for developing protocol stacks and protocol software. It is not concerned with the implementation aspects of protocols and software.

One need to provide a clarification before discussing the specifics of the OSI model. Three terms, namely *reference model*, *protocol stack*, and *protocol software*, are sometimes used interchangeably, which is not quite incorrect. The following paragraphs explain the exact connotations of these three terms.

A reference model (in the current context) refers to a layered architecture that lists down the layers required to build a system, along with a set of well-defined functions associated with each layer. Each layer is expected to perform its function. However, there is no compulsion for it to do the same. Thus, a reference model merely acts as a guide for the development of protocol stacks.

Protocol stacks are developed by using the reference models. As the name suggests, a protocol stack is a stack of protocols, with one or more protocols residing in each layer. While developing a new networking technology, a reference model is defined first. This model is then used to identify the protocols that will be required at each layer. Once all the protocols have been identified and developed, a new protocol stack gets defined.

Developing a reference model and developing a protocol stack need two totally different approaches. In the former, the focus is on capturing all the required functionality in one layer or the other. The focus is not as much on details as it is on completeness. In contrast, developing a new protocol stack requires the confrontation of issues related to optimization, simplicity and efficiency. In essence, a reference model gives a broad outline of where things lie, whereas a protocol stack gives a complete description of the different layers. It may be noted, however, that developing a new reference model is not mandatory for the development of a new protocol stack. A networking technology may adopt an existing reference model or develop a new one. In order to avoid duplication of effort, the OSI model was developed with the hindsight that it would act as a reference model for all networking technologies. However, because of the serous limitations of the OSI model, many new reference models have been developed since then.

If the concept of a protocol stack is well-understood, the definition of protocol software becomes straightforward. Protocol software refers to the actual implementation of a particular protocol stack. Development of protocol software entails choosing a programming language for development, designing graphical user interfaces, defining system level interfaces, etc.

2.5 DEFINITION OF OPEN SYSTEM

According to the OSI standards, the definition of an open system is quite difficult to comprehend. To make matters simple, an *open system* is defined as a system for which:

- Standards/recommendations are fully published.
- Systems are open for communication with other systems.
- Communication takes place using internationally accepted protocols.
- Multi-vendor solutions are inter-operable.
- Standards are developed by independent and neutral bodies.

In essence, OSI is a seven-layered model and is based on the concept of layered abstraction. Each layer provides services to the layer above it (except the highest layer), and uses the services of the layer below it (except the lowest layer). The following sub-section elaborates the seven layers of the OSI model.

2.6 SEVEN-LAYERED OSI-RM

The OSI reference model is a *seven-layered* model. The seven layers are the *application layer, presentation layer, session layer, transport layer, network layer, data link layer* and *physical layer*. Figure 2.3 shows the model of communication in the OSI environment, based on the seven layers. As shown in the figure, the lower three layers operate *link-by-link*, while the remaining four layers (the transport layer upwards) operate *end-to-end*.

Link-by-link (also called hop-by-hop) operation means that peering takes place between two entities connected by a physical link. For example, the network layer being link-to-link, implies that at each intermediate node, the network layer packets are processed. Further forwarding takes place using the Protocol Control Information (PCI) (for example addressing information) contained in the network layer PDU. Similarly, the data link layer being link-to-link, implies that at each intermediate node, a data link layer frame is processed. This is not true for the transport layer PDU, which is only processed at the destination node. This is why a transport layer protocol

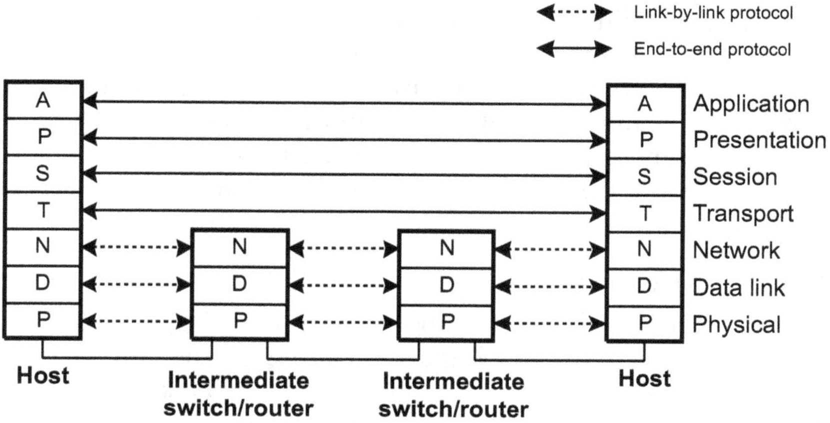

Fig. 2.3   *Seven-layered OSI model*

is an end-to-end protocol, implying that peering takes place only between end-systems, and not between intermediate switches/routers.

In summary, the lower layers of the OSI model (i.e. network layer, data link layer and physical layer) operate at the link-to-link level, while the higher layers (i.e. the application layer, presentation layer, session layer and transport layer) operate end-to-end. Thus, intermediate entities like routers only look up to the network layer packets. The higher layer units that are considered as payload for these intermediate entities transparently forward the higher layer data units to the next node. The seven layers of the OSI model are indivdually discussed in the following sub section.

2.6.1 Physical Layer

The physical layer is the lowest layer of the OSI reference model. Its primary function is to provide the *transmission of bit streams* over a physical transmission medium. Here, the physical transmission medium refers to an actual physical medium (e.g. copper cables), or to a wireless medium (e.g. satellite link).

A physical layer standard specifies the electrical, mechanical, functional and procedural behavior of a physical layer interface. Typically, this includes specifying the type of signal (electrical or optical), the voltage level of '0' and '1', the line coding rules (return to zero (RZ), non-return to zero (NRZ),

Manchester coding and so on), the modulation techniques, the number of pins in a connector and the function of each pin. The physical layer is also responsible for activating, maintaining and de-activating physical connections, sequential delivery of bits, clock synchronization and fault notification.

Examples of physical layer specification include X.21/ X.21bis, V.35 and RS-232-C.

2.6.2 Data Link Layer

The data link layer resides above the physical layer. Its most important function is to impart meaning to the information streams transferred over the physical layer. This is required because at the physical layer, all information is merely a stream of zeroes and ones. By itself, the physical layer is devoid of any intelligence and is incapable of attaching any meaning or structure to the carried data. It is the function of the data link layer to impart meaning to these bit streams. This is done by exchanging information via data link layer PDUs (called frames) and by providing the means to identify the start and end of frames.

Besides performing the above function, the data link layer also carries out *error detection and control*. Generally, the physical transmission media are prone to error and do not ensure transparent data transfer. A bit sent as '0' can be misread as '1' at the receiving end, a process called *data corruption*. Besides data corruption, there can also be data loss due to timing errors. The data link layer prevents data corruption and data loss by re-transmitting the corrupted packets. Data link also provides *flow control*, a mechanism whereby a fast sender is throttled from flooding a slow receiver.

Other functions of the data link layer include segmenting the data given by the upper network layer into frames, transmission of frames, and delivery of re-assembled frames to the network layer (after removing the data-link layer information).

Examples of data link layer standards include High-level Data Link Control (HDLC) and Link Access Procedure-Balanced (LAP-B).

Data link protocols reside over both *shared* and *non-shared media*. In case of a shared medium, a host requires a Media Access Control (MAC) layer protocol to arbitrate and provide the means for it to access the shared medium. Ethernet, Token Ring and Token Bus are examples of MAC layer protocol. In case of a non-shared medium, a point-to-point link is set up between the communicating entities. The Point-to-Point Protocol (PPP) used for Internet connectivity via a dial-up connection, is an example of a data link protocol for a non-shared medium.

2.6.3 Network Layer

The network layer sits between the data link and the transport layer. To the higher layers, network layer is the equivalent of an end-to-end data pipe (see Fig. 2.4). The higher layers are transparent to the existence of the lower layers or to the existence of numerous intermediate nodes traversed to reach the destination.

In order to give the impression of an end-to-end data pipe to its users, the network layer performs its two most important functions *addressing* and *routing*. Each network layer protocol entity has its unique network layer address. For example, hosts using Internet Protocol (IP) have an IP address. The network address, irrespectiv of its type, is used by a network layer entity to communicate with its peer entity. This communication requires that the network layer protocol data units (called packets) be routed over a number of intermediate entities. An intermediate entity has its own physical, data link and network layer (see Fig. 2.3).

If required, the network layer also provides, the means for *segmenting* the upper transport layer data into packets, *sending* of these packets, and *re-assembly* of these packets at the destination end.

Internet Protocol (IP) and the Internetwork Packet Exchange (IPX) protocol are two examples of network layer protocol.

2.6.4 Transport Layer

The transport layer is a true *end-to-end layer*, in the sense that at this layer connections are established between the source and the destination. In all the three layers discussed earlier, the connections are established on a hop-by-hop (i.e. link-by-link) basis. The intermediate entities like the switches and routers only facilitate the communication between a source and destination. Thus, they are the means and not the ends. Only at the transport layer and above are the connections truly end-to-end (see Fig. 2.4).

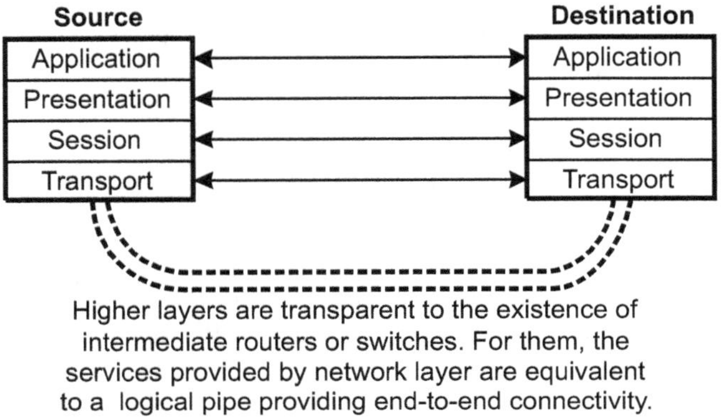

Higher layers are transparent to the existence of intermediate routers or switches. For them, the services provided by network layer are equivalent to a logical pipe providing end-to-end connectivity.

Fig. 2.4 *Network layer's equivalent*

The transport layer is mainly concerned with *interconnecting session layer* entities. The session layer entity can even specify the class of service it wants. In order to provide the requested service, the transport layer maps this requested service to a particular network layer capability. End-to-end flow *control, error detection/recovery* and *segmentation and re-assembly,* are some other important functions of the transport layer.

Transmission Control Protocol (TCP) and Transport Protocol (TP) classes 0–4 are a few examples of transport layer protocol.

2.6.5 Session Layer

The session layer provides the means to *establish, manage* and *release session-connections.* The session-connections are connections between peer entities used for the orderly exchange of data. Session-connections are established by the session layer when it receives a request for the same from the presentation layer. During the establishment process, the connection parameters are also negotiated. During the release of a connection, it is ensured that there is no data loss. Data loss usually occurs when there is no synchronization between peer entities (i.e. one of them aborts the connection while the data sent by the other entity is already in transit). In order to prevent this possibility, the peer session layers negotiate the release of a connection.

The session layer manages sessions by providing services like token management, synchronization, resynchronization, exception reporting, etc.

X.225 is an example of a session layer protocol.

2.6.6 Presentation Layer

The presentation layer is mainly concerned with the *representation of data* exchanged between two communicating entities. This layer defines the syntax of the data exchanged. It also helps in the negotiation of the abstract syntax through which data is transferred. It is important to distinguish between the syntax and the semantics of data. Syntax defines the means of representing data. Semantics, on the other hand, refers to the meaning of data. The presentation layer is only concerned with the syntax of data, and not with the semantics.

While sending data to the session layer, the presentation layer accepts data (usually in the form of characters and integers) from the application layer and converts it into a standard format (e.g. American Standard Code for Information Inter change (ASCII) format). The reverse occurs when the presentation layer receives data from the session layer. Here, data is converted from the standard format to characters and integers, and passed on to the application layer.

Abstract Syntax Notation One (ASN.1) is a popular language used to represent and exchange data.

2.6.7 Application Layer

The application layer is the topmost layer of the OSI reference model. This layer supports various *application processes* like file transfer, and message handling. The application processes, in turn, act as an interface between the end-user and the protocol stack.

Examples of application layer protocols include X.400 (inter-personal messaging system), X.500 (directory management), and T.411 (office document architecture). Virtual terminal services, database support and network management are some other examples of services provided by the application layer.

2.7 SHORTCOMINGS OF THE OSI-RM REFERENCE MODEL

It has been more than two decades since the OSI reference model was formalized. Even though the layered architecture remains an integral philosophy behind any communication framework, the model suffers from certain shortcomings. This section highlights some of these drawbacks.

In the early 1980s, when the OSI model was being finalized, it was highly inspired by the System Network Architecture (SNA). The SNA was the propriety, seven-layered model of the IBM company. At that time, IBM wielded so much clout in the computing and related industries that no one had the audacity to question the efficacy of their model. The SNA model, which was almost 10 years older to the OSI one, acted as a convenient starting point for the development of the latter. The net effect was a model that was, at best, a shadow of the SNA model.

The OSI model is not only a shadow of the SNA model, but it is also flawed. The session layer is of little use for most of the applications. Similarly, the presentation layer is almost non-existent in most of the protocol stacks. Both the layers, the session layer and the presentation layer, are remnants of the mainframe/terminal communications era, and are not applicable today. Most technologists are of the view that a *five-layer model*, without the session and presentation layers, is a much better choice (see Fig. 2.5). Whatever functionality is required at these two layers, can be clubbed with the application layer.

However, developing a modified version of the OSI reference model also does not solve all problems. Many network functions such as error control and flow control, are duplicated in almost every layer (in particular, in the data link layer and the transport layer). This results in a lot of redundant functions, which consume precious processing time. At a time when the world is moving towards high-speed networks, such redundancy is not permitted. Moreover, the

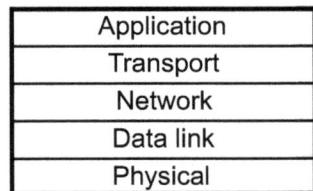

Application
Transport
Network
Data link
Physical

Fig. 2.5 *Modified five-layered OSI reference model*

issues that plagued the communication systems two decades back are not applicable today. Today things are radically different. The advent of smart and powerful desktops, and high-speed and error-free links has altered the face of present-day networks. Smart desktops have meant that network functionality is shifted from the network to the desktops. This has reduced the functions of the hitherto heavily burdened protocol stacks. The high-speed, error-free links have meant that link-by-link error control is no longer a necessity. This has led to the near disappearance of the data link layer and the development of new models that cannot be easily mapped to the OSI.

Another important development is the clear segregation between *user plane* (for transmission of the user's data) and *control plane* (for transmission of signalling information). Such distinctions are not very evident in the OSI model.

The above arguments seem to suggest that the OSI reference model has lived its life, and a new model needs to be developed. Although this is true to some extent, keeping the criticism of the OSI model aside, it still elucidates the concept of layered abstraction and encapsulation that are central to understanding any protocol stack. Hence, it is a good starting point for any course of networking. Moreover, through its shortcomings, one gets to know the look and feel of the new-generation protocol stacks.

2.8 EXAMPLE: TCP/IP PROTOCOL SUITE

In this section, the layering concepts of the OSI model are explained by using the TCP/IP protocol suite. TCP stands for Transmission Control Protocol and IP for Internet Protocol. These two are among the most popular transport and network layer protocols, respectively. These two protocols, along with a host of other protocols, form the TCP/IP protocol suite. The following sub-sections elaborate various aspects of the TCP/IP protocol suite.

2.8.1 Historical Background

The TCP/IP network suite (or the Internet), as one knows it today, was the result of some pioneering work done in the late

1960s and early 1970s by the developers of the ARPANET. ARPANET derived its name from Advanced Research Projects Agency (ARPA), an organization of the USA's Department of Defence (DoD). The ARPANET project was the progeny of the Cold War between the USA's and the then USSR. The USA government wanted a network that could withstand the destructive might of another nuclear war. The development of the ARPANET was a step in this direction.

By 1983, the Transmission Control Protocol/Internet Protocol (TCP/IP) protocols were developed, with all the machines connected to the ARPANET supporting these two protocols, replacing the earlier used Network Control Protocol (NCP) and the Interface Message Protocol (IMP). Another important development around this time was the ARPA's decision to allow the University of California at Berkeley to integrate the TCP/IP code with the existing Berkeley UNIX. This led to the development of the socket interface and many other networking applications. Overall, these developments spurred the widespread deployment of the TCP/IP protocol suite.

Another US organization, the National Science Foundation (NSF), closely followed the developments of the ARPANET. In the mid-1980s, it built the NSFNET, which was intended to serve as a backbone, connecting the important research and scientific centers of the US. By 1990, the backbone link, which started as a 56 kbps link, was upgraded to a speed of 1.5 Mbps. The interconnection of the ARPANET and the NSFNET, and a host of other developments from then onwords, meant an explosive growth of the Internet. Today, there are thousands and thousands of networks, with millions of hosts. And this number keeps doubling almost every nine months.

2.8.2 TCP/IP Reference Model

The TCP/IP follows a four-layered reference model, as shown in Fig. 2.6. It is more or less similar to the modified five-layered OSI model presented earlier. The only difference between the two is the clubbing of the physical and the data link layers in the TCP/IP reference model to form the link layer (or the Intranet layer). This difference needs some explanation. In a markedly different approach, the TCP/IP protocols were developed first and then fitted into a reference model. In contrast, the OSI

model was developed first and
then came the OSI protocols.
Thus, there is no formal definition
of a TCP/IP reference model.
What Fig. 2.6 depicts is a reference model intuitively derived
from the protocol suite. Also note
that the TCP/IP protocol suite is

| Application layer |
| Transport layer |
| Internet layer |
| Link layer (undefined) |

Fig. 2.6   *Four-layered TCP/IP reference model*

intended to be independent of the data-link-layer technology.
Thus, it can reside over any possible combination of data-link
and physical layer. This explains why the last layer in the reference model is left undefined.

On the basis of this reference model, Fig. 2.7 illustrates the
TCP/IP protocol stack. As briefly mentioned, the protocols of
the physical layer and the data link layer are outside the realm
of the TCP/IP protocol suite. Thus, various options are possible
at the physical layer (e.g. twisted pair, SONET and T1/E1) and
at the data link layer (Ethernet, ATM and Frame relay).

The Internet Protocol (IP), along with the Internet Control
Message Protocol (ICMP), the Address Resolution Protocol
(ARP), Reverse Address Resolution Protocol (RARP), and
routing protocols like Routing Information Protocol (RIP) and
the Open Shortest Path First (OSPF) form what is known as the
Internet layer.

The twin protocols of Transmission Control Protocol (TCP)
and User Datagram Protocol (UDP) form the *transport layer.*

At the application layer, there are a host of *application layer*

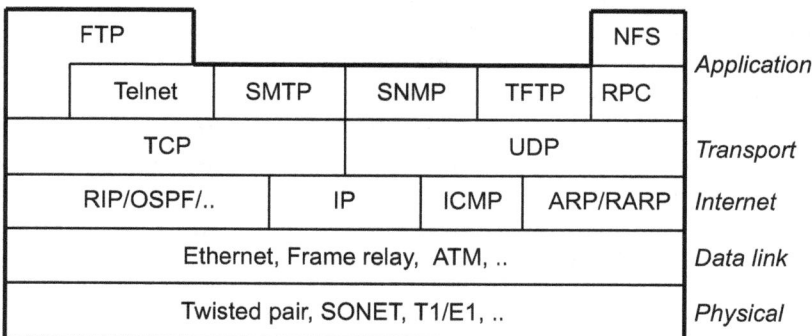

FTP					NFS	*Application*
Telnet	SMTP	SNMP	TFTP	RPC		
TCP			UDP			*Transport*
RIP/OSPF/..	IP	ICMP	ARP/RARP			*Internet*
Ethernet, Frame relay, ATM, ..						*Data link*
Twisted pair, SONET, T1/E1, ..						*Physical*

Fig. 2.7   *TCP/IP protocol stack*

protocols like the File Transfer Protocol (FTP), Telnet, Simple Mail Transfer Protocol (SMTP) and Simple Network Management Protocol (SNMP), among others. A thorough discussion of all these protocols is beyond the scope of this book. Here, only the three important protocols, viz IP, TCP and UDP are discussed. Some of the other protocols are also discussed in the subsequent chapters of this book through examples.

It may be noted that in Fig. 2.7 though some protocols (e.g. ICMP, RIP, OSPF) are shown at the Internet layer, they use the services of IP/TCP/UDP. Thus, strictly speaking, these protocols should have been placed above the particular layer whose services they are using. But doing this complicates the matter because routing protocols are neither application layer protocols nor transport layer protocols. This complication arises primarily because, as mentioned earlier, the OSI model fails to distinguish between the user plane and the control plane. When a single plane is used to depict the relative positions of the various protocols, such ambiguities arise. To conclude, it may be assumed that Fig. 2.7 is correct according to the broad functional division of the OSI, keeping in mind that the service user/service provider concept is not being adhered to.

2.8.3 Internet Protocol (IP)

The *Internet Protocol (IP)* is a network layer protocol described in [RFC 791]. The primary function of IP is to *route IP datagrams* from the source to the destination. IP datagrams are IP protocol data units exchanged between two IP layer entities. Each IP datagram is independently routed from the source to the destination. In order to carry out routing, each IP datagram carries the complete address of the source and the destination. The source and destination addresses are specified in the form of 32-bit addresses called the IP addresses. Here, it may be noted that the IP under discussion is also referred to as *IP version 4 (IPv4)*. An advanced version of Internet Protocol, referred to as *IP version 6 (IPv6)* has been defined in [RFC 2460] that provides a 128-bit address space. For the current discussion, only the basic IPv4 version is considered.

Although the *datagram forwarding* (or *routing*) process is explained in Chapter 3, a brief description is provided here. A source first sends an IP datagram to its nearest router. Using the destination IP address, this router then forwards the datagram to the next router in the direction of the destination. This process continues till the datagram finally reaches its destination.

IP does not ensure a guaranteed delivery; it just provides a connectionless, *best-effort* service. Best-effort service implies that the IP tries its best to deliver a datagram, without making any commitment of doing so. An IP datagram, on its way to its destination, can be dropped or misrouted, for which the IP is not responsible. It is left to the higher transport layer protocols (like TCP) to ensure reliable delivery. The best-effort nature of the IP makes the IP services unpredictable in nature.

Besides providing a connectionless service to the transport layer, IP provides network layer addressing, and fragmentation and re-assembly procedures. Thus, IP is comparable to a typical network layer protocol of the OSI model.

2.8.4 Transmission Control Protocol (TCP)

Transmission Control Protocol (TCP) is a transport layer protocol described in [RFC 793]. It sits over the connectionless IP and provides end-to-end connection-oriented service. To many, the fact that a connection-oriented protocol (TCP) sits over a connectionless protocol (IP) and still provides *reliable service* is quite confusing. The following text explains how this is possible.

As mentioned, IP provides best-effort service; i.e. IP can misroute datagrams or even drop them. In order to ensure that the end-to-end connection between the two communicating TCP entities is reliable, a logical connection is first established between them. Then, each segment (i.e. TCP protocol data unit) transmitted by the sender, is acknowledged by the receiver. To be precise, each transmitted byte is acknowledged by the receiving entity. On occasions when datagrams carrying the TCP segments are dropped or misrouted, the original segment is re-transmitted. For this purpose, a timer is maintained by the sender to ensure that if a response is not received within a reasonable time, the segments are re-transmitted. This explains how a connection-oriented layer sitting over a connectionless protocol can still provide reliability. Thus, TCP maintains a logical connectivity, and uses sequence numbering and packet re-transmissions to ensure reliable and sequential delivery even in the wake of dropped or misrouted datagrams. Apart from providing end-to-end connectivity, TCP provides flow control, error control and congestion control.

Just like IPv4 has an advanced protocol in the form of IPv6, there is an improved form of TCP called the *Stream Control Transmission Protocol (SCTP)*, which provides certain enhancements over TCP (e.g. removal of head-of-line blocking by providing more than one path for data transfer). The SCTP protocol is not considered in the present discussion.

2.8.5 User Datagram Protocol (UDP)

User Datagram Protocol (UDP) is a connectionless transport layer protocol described in [RFC 768]. When TCP was being built, it was felt that the connection-oriented nature of TCP was unsuitable for certain applications because of the following reasons:

- The numerous features (like flow control, sequencing and acknowledged transfer) provided by TCP made it *relatively slow.*
- Some of the features like *flow control were inapplicable* for real-time data transfer.
- For delay-sensitive applications, it was desirable that the *processing overheads were kept minimal.*

All these issues led to the development of the UDP.

UDP sits over the IP layer and uses its services. Since UDP is connectionless, it does not provide reliable delivery. In fact, UDP does very little that IP cannot do. The only reason for having a skeletal transport layer protocol like UDP is that it supports port numbers like the TCP to identify application layer processes.

UDP is mainly used for real-time applications, delay-sensitive applications and multi-cast applications. Real-time Transport Protocol (RTP), Trivial File Transfer Protocol (TFTP), Remote Procedure Call (RPC), and many other protocols use UDP as their transport layer protocol.

2.8.6 Layering Concepts in TCP/IP

As shown in Fig. 2.7, there are a host of protocols in the TCP/IP protocol stack. At the top of the stack are a number of application layer protocols like Telnet, FTP, SNMP and SMTP, among others. All the application layer protocols use some transport layer protocol (e.g. TCP or UDP). At the network layer, there is not much choice and the transport layer protocols use IP for data transfer.

A basic rule that governs TCP/IP and in general, any layered architecture, is the use of multiplexing identifiers, which enable the receiver to distribute PDUs of different protocols. Different layers use different names for the multiplexing identifiers.

At the transport layer, both TCP and UDP allow a number of higher-layer protocols/applications to run over it. In order to do this, it is necessary for a packet to specify the application process for which the segment is due. The destination IP address only identifies the destination machine and is not sufficient to specify the application type. For this purpose, TCP and UDP use the concept of *port-numbers*. The port numbers can be viewed as transport-level SAP that identify an interface between TCP/UDP and a particular higher-layer protocol. Certain port numbers are reserved for well-defined services. For TCP, the port number 21 is reserved for file transfer (FTP), 23 for remote login (telnet), and 79 for finger. Thus, if a user wants to perform file transfer with a server, he uses the IP address of the server and uses 21 as the destination port number. Like TCP, UDP has well-defined port numbers such as Trivial File Transfer Protocol (port 69) and Network Time Protocol (port 123).

The port number, along with the IP layer address, forms what is referred to as a *socket*. A pair of socket defines a TCP/UDP connection. For example, a TCP connection between machines with IP addresses IP(a) and IP(b), and using port numbers 6001 and 21 respectively, is defined as <IP(a), 6001, IP(b), 21>.

As many protocols use services of TCP and UDP, the IP layer too provides services to many service users including TCP and UDP. This is done by using the protocol identifier field that is a part of the IP header.

Figure 2.8 depicts the de-multiplexing of frames in TCP/IP protocol stack operates. When the IP layer receives a packet, then after some processing, it uses the protocol field in the datagram header to forward the datagram to one of the upper-layer protocols (like TCP or UDP). If the datagram's protocol field specifies the TCP, the datagram is forwarded to the TCP module and vice versa. At the TCP level, the port numbers in the TCP segment are used to forward this to a particular application layer process. Similar is the case for de-multiplexing at the UDP layer.

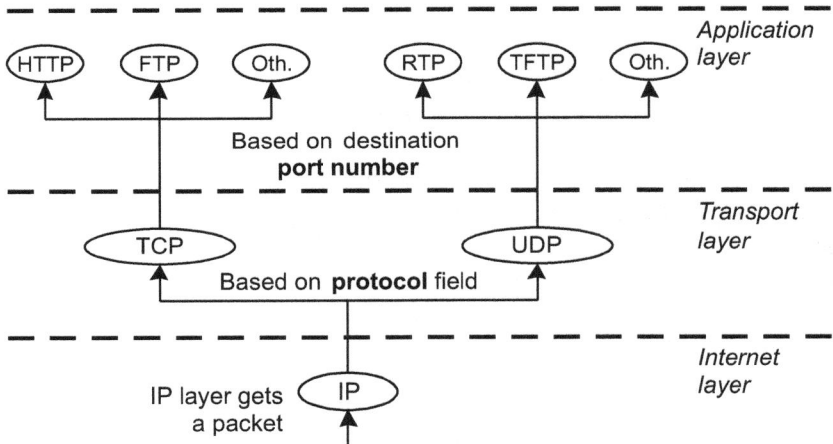

Fig. 2.8 *De-multiplexing of frames in TCP/IP protocol stack*

The previous paragraph only elaborates a general approach for de-multiplexing frames. For a specific example, consider Fig. 2.9, which depicts how IP datagrams and TCP segments are decapsulated. Consider an IP datagram, which arrives at a particular IP entity. The IP datagram header with protocol field = 6 indicates that the payload corresponds to a TCP segment. Thus, the IP header is removed to give a TCP segment. The TCP segment header with port number = 80 indicates that the data is HTTP data. Thus, the TCP header is removed to give HTTP data. This is how received datagrams/segments are de-multiplexed and payload is decapsulated.

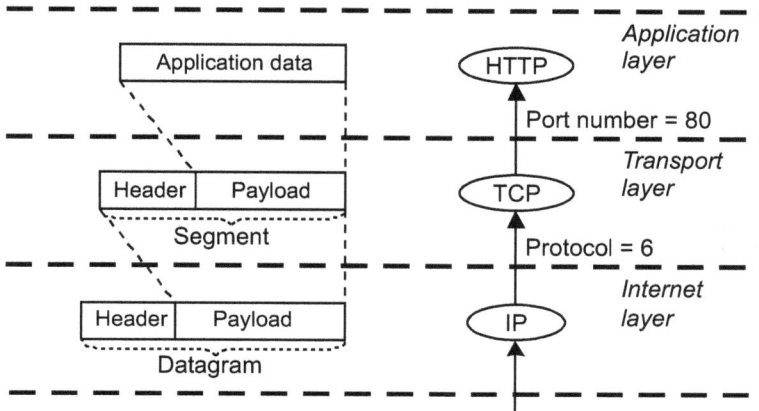

Fig. 2.9   *Decapsulation of a TCP segment belonging to HTTP protocol*

2.9 CONCLUSION

This chapter examined the concept of layered abstraction, which forms the basis for the protocol stack design. Various concepts like service-user/service-provider, interfaces, SAPs, encapsulation/decapsulation and multiplexing/de-multiplexing were explained in this chapter. The OSI reference model was then explained which defines a seven-layered structure. Finally, a simpler five-layered model of TCP/IP protocol stack was presented as an example to understand various aspects of layered abstraction.

REVIEW QUESTIONS

Q 1. Distinguish between Interface data unit (IDU), Service data unit (SDU) and Protocol data unit (PDU) with the help of an example.

Q 2. Why are seven layers defined in an OSI model? What could be the minimum number of layers that could provide similar functionality to that in the OSI model?

Q 3. What are the main drawbacks of the OSI reference model? Compare the OSI reference model with the reference model used in TCP/IP.

Q 4. How can a reliable protocol like TCP reside over an unreliable datagram service like IP?

Q 5. What are port numbers? Why are reserved ports required? Give examples of some reserved port numbers.

FURTHER READING

For details of the topics covered in this chapter, a plethora of options are available. [Gen A. Tanenbaum] and [Gen W. Stallings] remain two very popular texts. Apart from this, few other books [Gen S. Keshav] and [Gen J. Kurose] also provide different ways to understand protocol layering concepts. For the case study of TCP/IP, [TCP/IP R. Stevens] and [TCP/IP D. Comer] are two popular texts.

3

TRANSFER MODES

3.1 INTRODUCTION

The fundamental function of a network is to transfer user information from one entity to other. In order to understand how the transfer of information takes place, it is important to understand the notion of *transfer mode*, which is the central theme of this chapter. Transfer mode refers to the techniques used to transmit, switch and multiplex information. In other words, transfer mode is a means of packaging, sending and receiving information on the network. In current literature, interpretation of the term 'transfer mode' is subject to wide variations. According to some authors, Asynchronous Transfer Mode (ATM) and Synchronous Transfer Mode (STM) are the only two possible transfer modes. For others, circuit-switching and packet-switching are the two basic transfer modes. Still others consider telegraphy and telephony also to be variants of transfer mode.

According to this book, there are various ways of classifying transfer mode. If one looks at the way information is packaged, the two basic techniques are *circuit switching* as embodied by telecommunication networks and *packet switching* as embodied by frame relay networks and the Internet. In circuit-switching, a dedicated circuit (or a channel) is established from source to destination and the information is sent as a bit stream through the fixed bandwidth channel. This technique is used primarily to carry voice. In packet switching, user information is carried in variable-sized or fixed-sized packets. These packets are also called frames, cells or datagrams. Each packet has a header and a payload. The header carries address information and is used to make switching decisions and to forward the packet to the next hop. The payload carries the actual user information. This technique is used predominantly to carry data.

If one looks at the way information is transferred and resources reserved, there can be three different techniques, viz. *circuit switching*, *virtual circuit switching* and *routing*. In circuit switching, the information is carried over a dedicated connection and resources are reserved for the duration of the connection. In virtual circuit switching, virtual circuits are established and the packets are forwarded by using the virtual circuit numbers. A variant of this technique is *cell switching* in which fixed-sized small cells are used instead of variable-sized frames. In routing, each packet header carries a well-defined destination address. This destination address field is used at each intermediate hop to forward the packet. This form of forwarding is also referred to as *datagram forwarding*.

The following sections discribe the circuit switching, routing and virtual circuit switching techniques, each of which can be viewed as a distinct transfer mode. Cell switching is discussed under the section on virtual circuit switching.

3.2 CIRCUIT SWITCHING

Circuit switching is used primarily in telecommunication networks to carry voice. In circuit switching, a *dedicated connection* is established between two end-systems. The connection may be a physical connection using copper cables, or it may be a wireless link using radio waves or satellite links. Irrespective of the nature of the connection, a dedicated non-sharable channel is reserved between the source and destination end-systems for the entire duration of the connection. The dedicated connection is established using a process called *call-setup*. Call-setup implies finding a path from the source to the destination and establishing a connection between them. Data transfer takes place only after call-setup. After the data transfer is over, the call is cleared, and all resources reserved for the call are freed. In essence, information transfer in a circuit switched network is a three-step process call-setup, data transfer and call clearing.

A typical switch has 'm' incoming links and 'n' outgoing links. The number of incoming links may or may not be equal to the number of outgoing links. However, if they are equal, the switch is said to be a *symmetric*. Each incoming link receives a frame that contains 'N' slots. The frame is switched by using a switching table or a translation table. Each time slot on each frame is switched to a new time slot on an outgoing link, as per the translation table. No header information is necessary to perform switching. The position of the slot implicitly identifies the source and destination end-systems. This information is obtained when the connection is established. It is at the time of connection establishment that the entries in the translation table are filled. The entries are removed when the connection is cleared.

The circuit switching technique primarily relies on *Time Division Multiplexing* (TDM) to carry user information from one node to another. Time Division Multiplexing is a technique in which the individual input channels are multiplexed onto a single physical channel. The multiplexed channel is divided into *frames* and each frame is further sub-divided into *slots*. Each input channel gets a fixed time slot in each frame. This ensures fairness because input channels equally share the time slots. However, this method is wasteful because an allocated slot is wasted if the input channel has nothing to send for that slot.

Figure 3.1 illustrates the TDM technique in which four input channels are multiplexed onto a single output channel. In each frame, a slot is reserved for each of the input channels *A, B, C* and *D*. Irrespective of whether the input channel has valid data

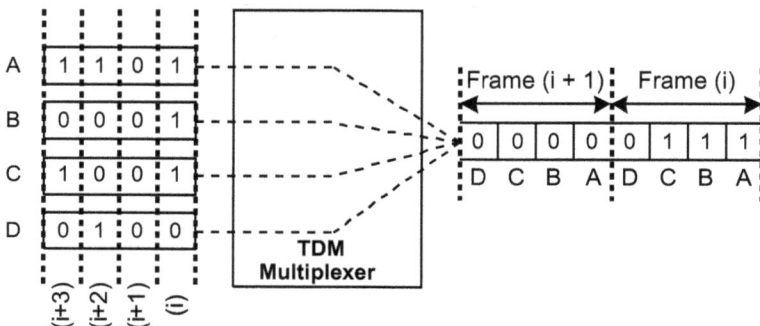

Fig. 3.1 *Time division multiplexing*

or not, the output channel reserves the slot for the input channels. Since the allocation of time slots is static, this wastage is unavoidable. Additional intelligence needs to be incorporated into the multiplexers to avoid this wastage. *Statistical Time Division Multiplexing* (STDM) is one of the ways for reducing this wastage.

In the STDM technique, no slot is reserved for an incoming channel. Rather, a slot is assigned as and when user data is received. In order to allow the proper sharing of bandwidth, resource reservation takes places through the use of statistical methods. These methods entail calculation of the probability distribution functions of all the incoming connections and prediction of the requirements of each of them. However, *a priori* calculation of these distribution functions is not easy and hence, only a rough estimate is made. The request for a new connection is accepted only if there are ample resources throughout the path of the connection. In case the network cannot support new connections, the connection request is rejected. This ensures that each multiplexer is able to support all the incoming connections. The STDM is also referred to as 'asynchronous TDM'. The Asynchronous Transfer Mode is an example of STDM and is further detailed in Section 3.6.

3.3 ROUTING

In contrast to circuit switching, routing is a *hop-by-hop datagram forwarding* technique in which the destination address field of the packet header is used to make forwarding decisions. The packets are typically large in size (~1000 bytes) and are called *datagrams*. In this technique, there is no concept of connection and each datagram is routed to the destination independent of other datagrams in a connectionless manner. *Datagram forwarding*, *datagram switching* and *layer 3 forwarding* are some of the other terms used for routing. Among all the terms, *routing* is the most commonly used one. Hence, the nodes forwarding the datagrams are referred to as routers.

As an example of routing, consider Fig. 3.2, which depicts how datagrams are routed from host A1 towards host A3. A1 forwards the packets destined for A3 to its nearest router (i.e. router R1). When a packet from A1 reaches R1, R1 uses the

Routing table at R2

Destination	Outgoing If.
A3	R2–R4
A2	R2–R3
A6	R2–R3
...	...

A	Host (with address A)
(R)	Router
H P	Datagram with Header (H) and Payload (P)

Routing table at R1

Destination	Outgoing If.
A3	R1–R2
A2	R1–R3
A6	R1–R3
...	...

Routing table at R4

Destination	Outgoing If.
A3	R4–A3
A2	R4–R2
A6	R4–R2
...	...

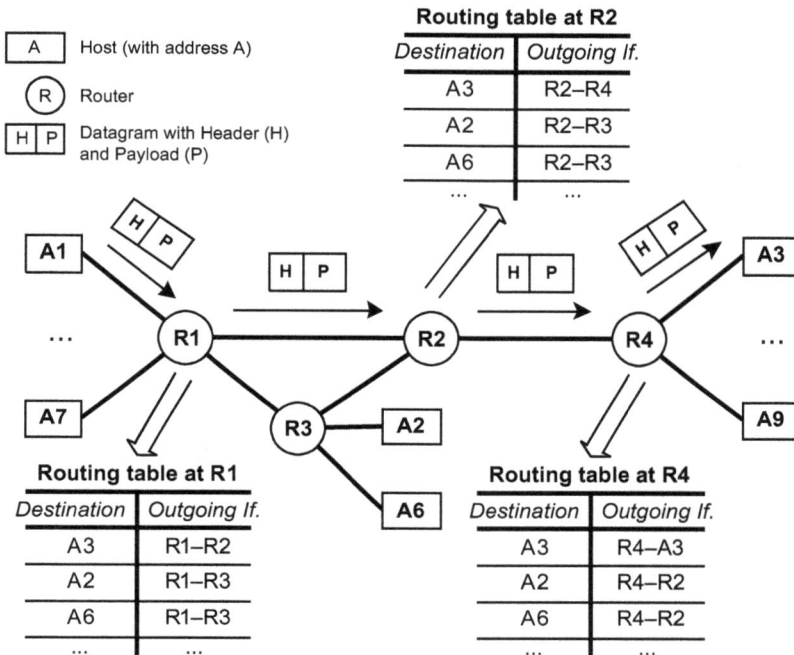

Fig. 3.2   *Example depicting the mechanism of routing*

destination address field (A3 in this case) to forward it further. For routing purposes, every router maintains a routing table with each entry having an association between a destination address and the link to which it is to be forwarded (i.e. outgoing interface on which the received packet is to be forwarded). This entry is used to make routing decisions. In case no entry is found, datagrams are forwarded to a default route. Since R1 has an entry for destination address A3, it forwards the datagram on the link R1–R2. When R2 receives a datagram, it uses its own routing table to forward the datagram further. In this way, each intermediate router forwards the datagrams hop-by-hop, using the destination address field until the datagram reaches the destination host.

Datagram routing lays the foundation of Internet architecture and its popularity seems to be rising unabated with the exponential growth of the Internet. However, this growing popularity causes a lot of problems related to the routing table look-up. As the number of hosts increases, there is almost a linear growth in the number of entries that a router must maintain. As the number of entries increases, so does the processing time associated with finding the outgoing link for a datagram, and forwarding it. This has led to new forwarding paradigms like *Multi-Protocol Label Switching* (*MPLS*), in which the concepts of virtual circuit switching and datagram routing are merged, resulting in a faster and scalable routing architecture. Virtual circuit switching is detailed in the next section.

⚡ 3.4 VIRTUAL CIRCUIT SWITCHING

Two techniques have been discussed so far. In circuit switching, there is a dedicated connection which is used to transfer information. In routing, there is no concept of connection nor is there any resource reservation; each packet carries the complete destination and a router routes the packet in a best-effort fashion. Virtual circuit switching *lies between* these two extremities. In this technique, information is transferred by using *virtual circuits*. Virtual circuits are logical channels (or circuits) between two end-systems that are identified by small identifiers or labels. For the sake of discussion, let us call this label *Virtual Circuit Identifier (VCI)*.

In virtual circuit switching, each packet that is transmitted by the sender carries a VCI. When this packet reaches the next hop, the VCI field of the packet header is used to forward it further. Before forwarding, this field is overwritten with a new label. This process of overwriting an incoming label with a new label is called *virtual circuit switching* or *label swapping*.

As in circuit switching, the connection in virtual circuit switching goes through three phases *call-setup*, *data transfer* and *call clearing*. The virtual circuit is established in the call-setup phase. In order to understand how a call is setup, consider a simple example in which three connections are established A-E (VC1), D-C (VC2), and A-E (VC3) (see Fig. 3.3). For the first connection, A sends a message to B, requesting it to establish a connection with E. B returns a VCI value (34) and forwards the call to E. A, upon receiving the VCI value 34, fills it in its VC table. Meanwhile, the message forwarded by B reaches E. E accepts the request and returns a VCI (47) to B.

VC table at A

Incoming		Outgoing	
Link	VCI	Link	VCI
-	-	AB	34	← For VC1
-	-	AB	30	← For VC3

VC table at B

Incoming		Outgoing	
Link	VCI	Link	VCI
For VC1 →AB	34	BE	47
For VC2 →DB	39	BC	43
For VC3 →AB	30	BE	30

VC table at E

Incoming		Outgoing	
Link	VCI	Link	VCI
For VC1 → BE	47	-	-
For VC3 → BE	30	-	-

- · — · · VC1
- · · · · · · · VC2
- · — — — VC3
- ◯ Switch

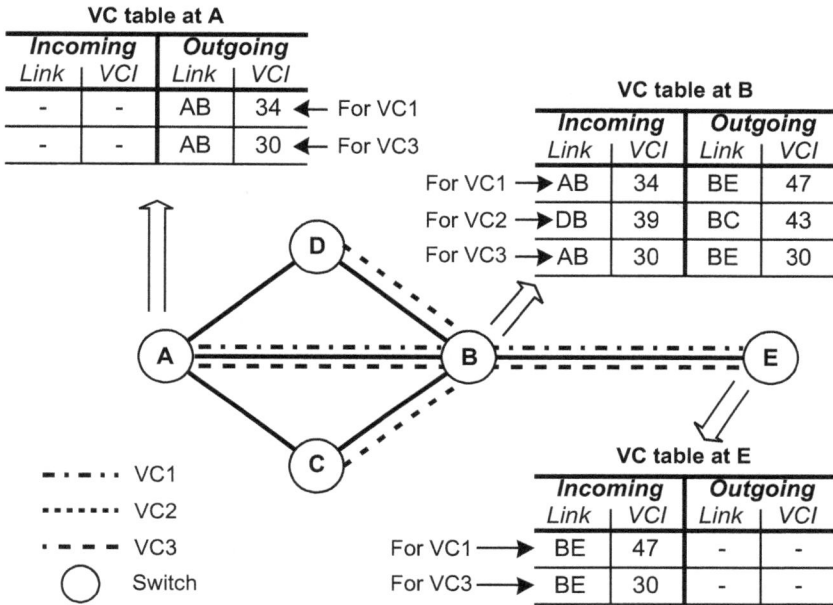

Fig. 3.3 *Example depicting establishment of virtual circuits*

E also enters the returned VCI in its VC table. B, upon receiving the message from E, maps the VCI 34 with VCI 47. This is how VC1 is established.

As shown in the figure, the VC table has two important fields an *incoming VCI* and an *outgoing VCI*. Both these collectively identify a part of the virtual connection. The total connection is identified by the set of incoming and outgoing VCIs present in all intermediate nodes, including the source and the destination. For example, VC1 has two distinct parts, link AB and link BE. Link AB is identified by VCI 34 and link BE by VCI 47. Therefore, {(AB, 34) (BE, 47)} collectively identifies the virtual circuit VC1.

After VC1 is established, the virtual circuit tables at A, B and E contain only the first row entries. After the virtual connections VC2 and VC3 are made, the complete virtual circuit tables for nodes A, B and E contain entries as shown in Fig. 3.3. Note that the dashed column entries in the VC table indicate that the node is either an ingress node or an egress node.

This is as far as establishing virtual circuits is concerned. If the data transfer between A and E for the first connection is now considered, then each packet sent from A for this virtual circuit will have a VCI 34 written in its header field. When this packet reaches node B, the virtual circuit table at node B is searched for the tuple (AB, 34). Here, AB refers to the link on which the packet is received, and 34 refers to the VCI value contained in the received packet. After a row entry is found, the VCI value in the packet is swapped with the value written in the outgoing VCI column of that row (i.e. VCI 34 is replaced with VCI 47). The packet is then forwarded on the outgoing link as per the entry in the virtual circuit table (i.e. BE). This is how label swapping and packet forwarding take place.

One important thing to note about VCIs is that they are *locally unique* and *not globally unique*. For example, consider VC1 in Fig. 3.3. The value of VCI in link AB is not the same as VCI in link BE, i.e. the same value of VCI is not maintained end-to-end. However, this does not imply that a VCI value used in one portion of the link cannot be used in other part. For example, for VC3, the VCI value for both the AB and BE links is 30. This is acceptable as long as two virtual circuits on the same incoming link have different VCIs. The Sharing of VCIs across different links is allowed.

The discussions on virtual circuit switching and routing beem to indicate that little separates datagram routing from virtual circuit switching. The few differences centre around the following aspects:

1. The global uniqueness of complete addresses in datagram routing versus local uniqueness of small VCIs in virtual circuit switching, and
2. Packet routing over a connection-less network versus label swapping over a virtual circuit.

Section 3.5 highlights the other differences between these two techniques.

The virtual circuit concepts are not new to the data communication world. In fact, this concept has been in use in X.25 networks since the 1960s. The virtual circuits in X.25 are identified by a 12-bit Logical Channel Number. Among the contemporary technologies, frame relay employs virtual circuit concepts. In frame relay, the Data Link Connection Identifier

(DLCI) field identifies the virtual circuits. The latest technology to use virtual circuit switching is ATM. Virtual circuit switching in ATM is also referred to as *cell switching*. The most important difference between cells and frames is that ATM cells are small and have a fixed size, whereas frames are usually large and have variable sizes. Otherwise, conceptually, there is very little difference between virtual circuit switching in frame relay and cell switching in ATM. However, the important thing to note here is that this single conceptual difference results in a number of other dis-similarities (e.g. transmission delay and switching speeds).

3.5 COMPARISON OF TRANSFER MODES

The discussion on transfer modes is incomplete without an assessment of the relative strengths and weaknesses of these techniques. Table 3.1 lists the differences between four different transfer modes, viz. circuit switching, virtual circuit switching, cell switching and routing. To classify one of them as the best option is unjustified as different scenarios warrant different solutions.

Since all switching techniques, except datagram routing, are connection-oriented, most of the advantages and disadvantages of connection-oriented services are applicable to these switching techniques. The advantages of connection-oriented services include *low propagation delay, sequenced data transfer* and *reliable service*. On the downside are factors like latency associated with call-setup and *inefficient resource utilization* due to dedicated allocation. However, despite being connection-oriented, circuit switching, virtual circuit switching and cell switching do not have identical characteristics.

As mentioned earlier, circuit switching is mainly used to carry voice. Dedicated channels ensure that there is always enough bandwidth in the network to carry voice samples. A connection request for voice call can be refused, but once it is accepted, the voice transfer is of good quality. Although allocating dedicated channels provides better services and is thus preferable, it comes at the cost of sub-optimal resource utilization. This is because even though the resource for a call remains allocated for the entire duration, it is actually used when there is useful

Table 3.1 *Comparison of Transfer Modes*

Circuit Switching	Virtual Circuit Switching	Cell Switching	Routing
Connection-oriented; a dedicated connection is established before data transfer	Connection-oriented; a virtual circuit is established, before data transfer	Connection-oriented; a virtual circuit is established before data transfer	Connectionless; no connection is established
Sequential delivery	Sequential delivery	Sequential delivery	Out of order possible
No header overheads	Label overheads	Small cells cause higher overheads	Overheads of carrying complete address information
Fixed resource allocation	Statistical resource allocation	Statistical resource allocation	No resource reservation
Stream oriented	Packets or frames oriented; variable size packets or frames	Cell-oriented; small fixed size cells	Packet (datagrams) oriented; variable size packets
Highest reliability	Reliable	Reliable	Best-effort service
Low delay and jitter	Not very good delay/jitter behaviour	Controlled delay/jitter	Unpredictable delay/jitter
Mainly for voice traffic	Mainly for data traffic	For all kinds of traffic	Mainly for data traffic
Hardware-based switching	Hardware or software based switching	High-speed hardware based switching	Hardware or software-based switching
Very fast, but inefficient resource utilization	Relatively slow, but better resource utilization	Fast and optimal resource utilization	Slowest of all, but robust and flexible
Ex: Plain old telephone systems (POTS)	Ex: X.25 and Frame relay networks	Ex: ATM networks	Ex: TCP/IP networks (i.e. Internet)

data to transmit. For example, consider the case of a telephonic conversation in which a person normally talks for only about 40 per cent of the time. The remaining time, there is nothing to transmit, and hence the reserved channel carries no useful data. This is gross wastage of bandwidth. For data communication, a circuit-switched network is more wasteful because data transfer is essentially bursty in nature. There are sharp bursts followed by prolonged periods of inactivity. If resource is reserved for maximum burst, then for a major period (which is nearly 90-95), the link will be under-utilized or unutilized.

By virtue of being connectionless, datagram routing is ideal for carrying *bursty data traffic*. No resource is reserved and packets are forwarded as and when they arrive. This precludes the possibility of wasted bandwidth. However, this also means that during heavy traffic conditions, the quality deteriorates rapidly. There is no guarantee that adequate bandwidth will be available and that the packet will not be dropped. Also, because of the lack of bounds on the delay involved in the transfer of datagrams, this method is not very popular for voice.

Virtual circuit switching is a *compromise* between dedicated resource allocation and connectionless switching. By using *statistical multiplexing*, the resources are shared in such a manner that neither is the capacity under-used, nor does the quality of a connection suffer because of lack of resources. However, this form of switching assumes that the resource requirements of a connection can be predicted *a priori,* which is not always true. Moreover, unlike a router, a virtual circuit switch does not maintain information on all possible destinations. Informations or taining to only established virtual circuits is maintained at each switch. This reduces the size of the VC table and makes it look up faster.

Cell switching is the latest form of virtual circuit switching. In this technique, small fixed-sized cells are used instead of variable-sized packets. The small cell size ensures that the delay *is predictable* due to which the voice quality can be maintained. For carrying data, multiple cells are carried, instead of a large packet.

⚡3.6 EXAMPLE: ASYNCHRONOUS TRANSFER MODE (ATM)

In this section, the virtual circuit switching concept is explained by using the example of ATM. Formally, ATM is a transfer mode in which the information is organized into cells; it is asynchronous in the sense that the recurrence of cells containing information is not periodic. The above definition encompasses three basic terms, viz. *transfer mode*; *cell-based transfer* and asynchronous transfer. The following sub-sections describe upon each of these terms and also explain a few important aspects of ATM.

3.6.1 Transfer Mode

As discussed in this chapter, ATM is a transfer mode that fits in between the two extremes of circuit switching and packet switching because it uses a small-sized frame (53 bytes) and uses virtual circuits. By using the small-sized cell, ATM retains the speed of circuit switching while still offering the flexibility of packet switching. This is why ATM is also referred to as *fast packet switching technology.*

3.6.2 Cell-based Transfer

Information in ATM is organized into *cells*, which means that the lowest unit of information in ATM is a cell. A cell is a fixed size frame of 53 bytes, with 5 bytes of header and 48 bytes of payload. The header carries information required to switch cells, while the payload contains the actual information to be exchanged.

Figure 3.4 illustrates the concept of cell-based transfer. Information from various sources is multiplexed and segmented, resulting in a stream of ATM cells. Each cell is transmitted and received independent of other cells. A cell is identified by the labels carried in the header. Here, label refers to the *Virtual Channel Identifier (VCI)* and *Virtual Path Identifier (VPI)* fields. These fields collectively identify the virtual circuit to which a cell belongs. The two identifiers enable a two-level information flow where cells can either be switched using VPI alone or can be switched using both VPI and VCI. In summary, ATM is a cell-based technology, which employs virtual circuit concepts to forward information streams.

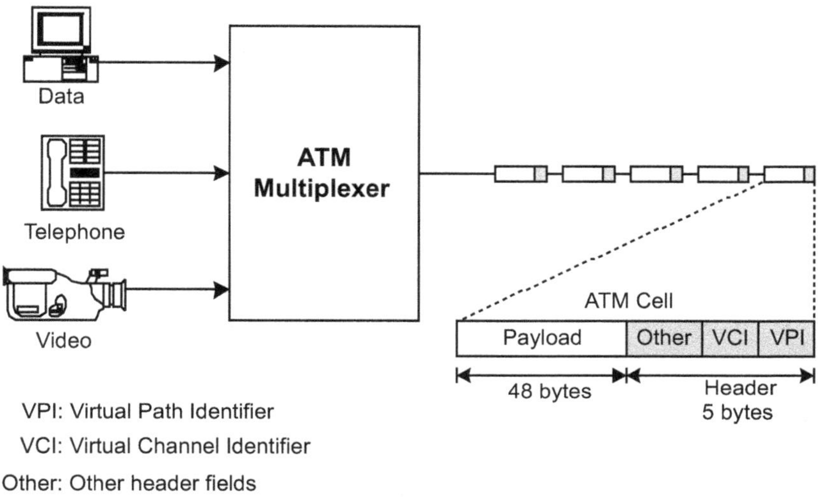
Data
Telephone
Video
ATM Multiplexer
ATM Cell
Payload
Other
VCI
VPI
48 bytes
Header 5 bytes
VPI: Virtual Path Identifier
VCI: Virtual Channel Identifier
Other: Other header fields

Fig. 3.4 *Cell-based transfer*

3.6.3 Asynchronous Transfer

ATM is an asynchronous transfer mode. There is considerable confusion regarding the term 'asynchronous' in ATM. While the definition provided earlier ('asynchronous in the sense that the recurrence of cells containing information is not periodic') clarifies the meaning of the term 'asynchronous', the connotation of 'asynchronous' is still not very clear.

Usually, the terms 'synchronous' and 'asynchronous' refer to the way data is transmitted. In the *synchronous mode*, the transmitter and receiver clocks are synchronized and frames are sent/received periodically. Time division multiplexing, wherein each time slot is reserved for a particular voice channel, and where frames recur at 125 μ sec interval, is a good example of synchronous transfer. In the *asynchronous mode*, timing information is derived from the data itself, and the transmitter is not compelled to send data periodically (though it is allowed to do so). The RS-232 protocol, where start/stop bits are used to indicate start/end of the transmission, is an example of asynchronous data transfer.

With this knowledge, where does the term 'asynchronous' fit in the definition of ATM? There is more confusion if Synchrounous Optical Network (SONET) is the underlying transmission medium for ATM because SONET operates synchronously. The answer lies in realizing that the term 'asynchronous' is used for the ATM, layer and not for the physical layer, that is, the multiplexing of cells on to the physical medium is asynchronous, and not the transmission of cells. Unlike TDM, in ATM, no slot is reserved for a logical channel and cells are transmitted as and when they arrive. This makes the transfer of cells for a particular channel non-periodic, which is why ATM is an asynchronous transfer mode.

Note that in circuit switching, each information unit is identified implicitly by the slot position in which it is carried. For example, data from Source A is carried in slot 1, that from Source B is carried in slot 2, and so on. Thus, whenever any source has nothing to send, the slot goes empty, thereby wasting bandwidth and reducing efficiency. However, this is not the case in ATM, because there is no coupling between the slot position and the source number. This decoupling results in a more efficient utilization of the link bandwidth because the link does have to carry empty slots. Also, ATM also does not the couple application data rate with underlying transport mechanism. To understand this, note that to carry voice, TDM partitions bandwidth into 64 Kbps channels. This is an example where the application data rate (64 Kbps) is tied to the underlying transport mechanism (64 Kbps channels). The effect of this coupling is that even if technological advancements permit voice transfer at lower data rates, the widespread deployment of transport infrastructure hinders the use of this advancement. ATM does not have this drawback as information is packaged into independent cells, and depending upon application requirements, the number of cells transferred can be altered. This makes ATM much more flexible. However, the flip side of the above strategy is that ATM requires each cell to explicitly carry information about its virtual channel. This information is carried as VPI/VCI fields in ATM cell header, accounting for an overhead of 10.4 per cent (5/48).

When both the strategies are compared, it is observed that the overhead of the cell header is more than compensated by better link utilization and a gain in flexibility. This is why the switching of cells in ATM takes place asynchronously.

3.6.4 Fixed and Small-sized Cell

In the late 1980s, when the ATM cell format was being discussed, two important issues were subjects of fierce debate. The first issue partained to the decision as to whether the ATM cell would be of a *fixed size* or of *variable size*. The second issue was related to finding the *right cell size*, provided it was decided to keep the size fixed.

At the time the decision was made, and even now, almost all packet-based technologies (including X.25, frame relay, and Ethernet) employ a variable-sized frame/packet. The actual length is indicated in one of the fields of the packet header. A variable-sized packet, as compared to a fixed-sized packet, offers two key benefits. First, since the packet size is variable, there is no bandwidth wastage and only the requisite amount of data is transmitted. This is not true for fixed-sized packets because irrespective of the size of the higher layer packet, the packet transmitted always remains constant. So, when the number of bytes to be sent is less than the (fixed) size of the packet, the packet contains padding bytes, which carry no useful information.

Second, if the packet size is fixed and if the higher layer PDU does not fit into one packet, an intermediate segmentation and re-assembly layer is required. For variable-sized packets, in contrast, this intermediate segmentation and re-assembly layer is not required because in the variable-sized frame-based technologies like Ethernet, an IP datagram fits directly into an Ethernet frame. This condition is true provided the datagram is less than the maximum payload that the Ethernet can carry. Since an Ethernet frame can carry up to 1500 bytes of payload, segmentation and re-assembly are generally not required. However, assuming a small cell size (say 53 bytes), ATM requires an intermediate layer that segments IP datagrams at the transmitter and re-assembles them at the receiver. If a very large cell size is chosen to prevent segmentation and re-assembly, a majority of the cells will carry partially filled cells. This leads to inefficient bandwidth usage, which is not acceptable. To conclude, for a variable-sized packet, there is no bandwidth wastage and no need for intermediate segmentation and re-assembly layer. For a fixed-sized packet, if the packet size is too small, a segmentation and re-assembly layer is required, and if the size is too large, there is lot of bandwidth wastage.

Despite the drawbacks, a fixed-sized cell was preferred over a variable-sized packet. Since ATM was designed for both voice and data communication, it was mandatory to keep transit delay within reasonable bounds. In this regard, a fixed-size cell, as compared to a variable-sized packet, offered many advantages like predictable and lesser switching delay and simpler memory management in switch buffers. Keeping variable size packets meant that each cell carried the length field. This introduced extra processing in intermediate switches because the length field had to be processed and appropriate size buffers had to allocated. Since guaranteeing bounded delay was a greater concern than minimizing overheads, a fixed-sized cell was chosen.

Once it was decided that the cell would be of a fixed size, choosing the right size became critical. For a large cell which was only partially filled, there was bandwidth wastage. A small cell size meant less propagation and switching delay. Moreover, a smaller-sized cell required less time to get filled. For example, to fill a 53 bytes cell using a source of 64 Kbps, 0.75 ms is required. If the cell size was increased, the packetization delay also increased proportionately.

In fact, the whole world was divided into two camps. The Europeans were in favour of 32 bytes because they thought that having smaller payloads would ensure transit delays below a threshold, thereby obviating the need for echo cancellers. The US and Japan were in favour of 64 bytes because this would have resulted in higher transmission efficiency. Ultimately, a compromise was arrived at and the cell size was fixed at 53 bytes, with a 48 bytes payload and a 5 bytes header.

In essence, the decision of relating to a variable-sized or fixed-sized cell, and choosing the cell size, was a battle between the telecommunication world and the data communication world. The former group was concerned about delay while the latter was concerned about transmission efficiency. In the end, the telecommunication world seems to have triumphed. This was expected because ATM was primarily designed by the telecommunication body ITU-T.

3.6.5 Connection-oriented Approach

ATM is a connection-oriented technology. The connection-oriented approach of ATM is characterized by the following:

- ATM uses *virtual connections* to switch cells from one node to another. The virtual connections are identified by VPI and VCI. The VPI/VCI fields are carried in the ATM cell headers and are called connection identifiers.
- Cells belonging to a virtual connection follow the *same path*. Thus, cell sequence is maintained implicitly.
- The bandwidth allocated to a virtual connection is assigned at the time of the connection set-up and is based on the requirements of the source and the available capacity.
- Each virtual connection is provided with a Quality of Service (QoS).
- The connections are *bi-directional* in nature. The bandwidth allocated in forward and backward directions may or may not be same (symmetric/asymmetric connection). The bandwidth in the backward direction can even be zero (rendering the connection unidirectional).
- The *same VPI/VCI* value is used across a link in both the directions.

The virtual connections in ATM are established either *statically* or *dynamically*. A static connection is one in which the end-points of the connection are defined in advance and the bandwidth is pre-determined. Such static connection is established through subscription and is referred to as *Permanent Virtual Circuit (PVC)*.

Dynamic connections, on the other hand, offer much more flexibility. They are established when there is data to be transferred and released when the data transfer is over. Such a dynamic connection is referred to as *Switched Virtual Circuit (SVC)*. The establishment and release of SVCs takes place by using signalling procedures.

Between the two, PVCs are preferred when there is relatively stable data flow between two end-points. SVCs are useful when the communication is for a very short duration only. Public carriers providing services over ATM networks generally offer PVC connections only. Of late, carriers have started offering SVC connections also.

3.6.6 Virtual Channels and Virtual Paths

As stated earlier, ATM uses VPI/VCI fields of a virtual connection to forward cells. At each intermediate node, VPI/VCI values are swapped and new values are filled. This process is also referred to as *cell relaying*, *cell switching* or *cell forwarding*. Figure 3.5 shows VPI/VCI swapping performed by an ATM switch. The VPI/VCI pair of an incoming cell is used as an index to search the VPI/VCI translation table. In the VPI/VCI translation table, a separate entry is maintained for each established virtual circuit. For example, the table in the figure maintains entries for two virtual circuits. The values in the table get filled when the connection is established. For a PVC connection, each switch between the source and the destination is configured by using some management tool. For an SVC connection, this connection takes place automatically via an exchange of signalling messages.

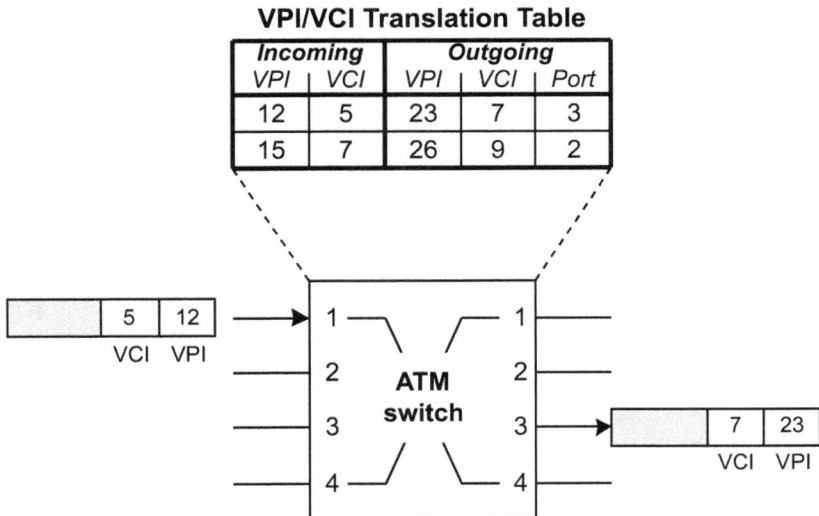

VPI/VCI Translation Table

| Incoming | | Outgoing | | |
VPI	VCI	VPI	VCI	Port
12	5	23	7	3
15	7	26	9	2

Fig. 3.5 *VPI/VCI translation in an ATM switch (cell switching)*

Through this table, the new VPI/VCI values are extracted and filled in the incoming cell's header. The entry in the table also provides the port number on which the cell is to be forwarded. For example, in the figure a cell with VPI = 12 and VCI = 5 arrives on port number 1. This VPI/VCI value is used to obtain the new VPI/VCI values, VPI = 23 and VCI = 7, and also the port number (3 in this case) on which the cell is to be forwarded.

The VPI/VCI swapping in ATM is not as simple as explained. Actually, the virtual circuits or connections in ATM are classified into two categories, namely *Virtual Channel Connection* (VCC) and *Virtual Path Connections* (*VPCs*). This establishes a two-level hierarchy in the forwarding of cells. In a VCC, switching is done by using the combined VPI/VCI value. In contrast, a VPC is switched by using the VPI values only.

It may not seem very obvious as to why there are two identifiers (VPI and VCI) in ATM at the very first place. After all, both X.25 and frame relay, which are virtual circuit-based technologies, use only a single label. However, there are many advantages of having a two-level hierarchy of identifiers. The first advantage is that a two-level hierarchy allows virtual paths to be distinguished on the basis of QoS. Thus, the resource may be allocated on the basis of virtual paths, and depending upon the nature of the connection request, a virtual channel from a particular virtual path can be granted. This reduces the workload on the switch and makes granting/rejecting connections easier.

Further, by having a single virtual path connection between networks A and B, the two networks can establish multiple virtual channels. This gives them the flexibility to add, modify and delete virtual channels without bringing this to the notice of the service provider. The service provider also does not bother about the number of channels established because the charging will be based upon the parameters of the virtual path connection.

Apart from the advantages mentioned above, VP switching (which is based upon VPI values only) is faster as the VCI value is not looked at. This facilitates the aggregation of similar flows (i.e. virtual connections) over a single virtual path.

3.6.7 Negligible Error and Flow Control

In ATM, significant processing has been pushed out of the network domain and very little error-control/flow-control is done within the network. Error correction is limited to the 1-byte Header Error Control (HEC) field, which is a part of the

5-byte cell header. The HEC protects the cell header only. There is no error correction for the payload. Moreover, even if cells are lost or discarded due to exceptional conditions, no retransmission takes place. It is left for the higher layers to retransmit the lost packets.

The decision to keep negligible error control/flow control is based on the following statements:

- Throughput is inversely related to the processing done within the network. To achieve high throughput, it is necessary to have minimal error control/flow control.
- Most of the real-time applications (like voice conversation and video conferencing) do not permit any kind of flow control. This is because a delayed real-time packet is totally useless.
- It is assumed that the majority of underlying transmission mediums for ATM will be fiber. Since error rates in fiber are low and since they seldom occur in bursts, it is sufficient that only the cell header be protected by a 1-byte Cyclic Redundancy Check (CRC) and that the cell payload be left unattended.
- Today the end-systems have much more computing power than what they had a decade back.

All the aforementioned points imply that a high-speed technology supporting real-time services must have minimal error and flow control.

3.7 CONCLUSION

This chapter provided an overview of the basic transfer modes. The comparison between different transfer modes can be reduced to the difference between connection-oriented and connectionless services. The trade-off here is between reduced resource utilization at better service guarantees, versus higher resource utilization at the cost of a possible loss in service quality. It is said that ATM offers the best of both worlds, i.e. connection-oriented statistical multiplexing resulting in better resource utilization along with small cell size resulting in the ability to carry voice and real-time traffic by providing a controlled delay and jitter response.

REVIEW QUESTIONS

Q 1. What are transfer modes? Why are they important in the context of networking techniques? What is the basis of classifying these transfer modes?

Q 2. What is circuit switching? What are its advantages and disadvantages?

Q 3. Why are both routing and virtual circuit switching referred to as packet switching? What are the essential differences between the two?

Q 4. What is the difference between virtual circuit switching and cell switching?

Q 5. Compare the four transfer modes, viz. circuit switching, virtual circuit switching, cell switching and routing on the following aspects, with proper reasoning and justification:

- *Nature of connection*
- *Resource reservation*
- *Resource utilization*
- *Reliability*
- *Ability to carry voice*
- *Header overheads*

Q 6. Discuss the important aspects of ATM.

FURTHER READING

For details of the topics covered in this chapter, a plethora of options are available. [Gen A. Tanenbaum] and [Gen W. Stallings] remain two very popular texts. Apart from this, a few other books including [Gen S. Keshav] and [Gen J. Kurose] also provide good insight. For the case study of ATM, a number of options are available including [ATM S. Kasera], [ATM W. Goralski], [ATM U. Black] and [ATM M. Prycker].

NETWORK TOPOLOGY AND EXTENT

4.1 INTRODUCTION

This chapter discusses some of the aspects related to network topology and extent. The *topology* of a network defines the physical interconnection of its constituent elements (i.e. topology refers to the layout of connected devices on a network). This chapter looks at the important topologies like bus, ring, star, tree and mesh.

A concept related to network topology is *network extent*. The network extent defines the geographical boundaries of networks and also leads to the classification of networks into Area Networks. Two important classes of area network are *Local Area Network (LAN)* and *Wide Area Network (WAN)*. A LAN is a network that is restricted to a small region (e.g. a building). In contrast, WANs cover large geographical distances. LAN, WAN and some other types of area networks are discussed in this chapter.

4.2 NETWORK TOPOLOGY

The topology of a network defines the *physical interconnection* of its constituent elements. In other words, the topology of a network refers to the way nodes (or hosts) are interconnected in a network. The interconnection may either be *real* or *logical*. Real interconnection refers to the way the elements are actually (i.e. physically) connected. In contrast, a logical interconnection refers to the way data is exchanged between the constituent elements. For example, 'N' nodes may be physically connected via a single piece of cable, but it is quite possible that the nodes exchange data in some cyclic order. Thus, real and logical interconnections may not necessarily be the same. Nonetheless, for the purpose of discussion, let us assume that the topologies refer to physical connection.

The most common topologies are bus, ring, mesh (partially connected and fully connected), star and tree. All these topologies have been illustrated in Fig. 4.1.

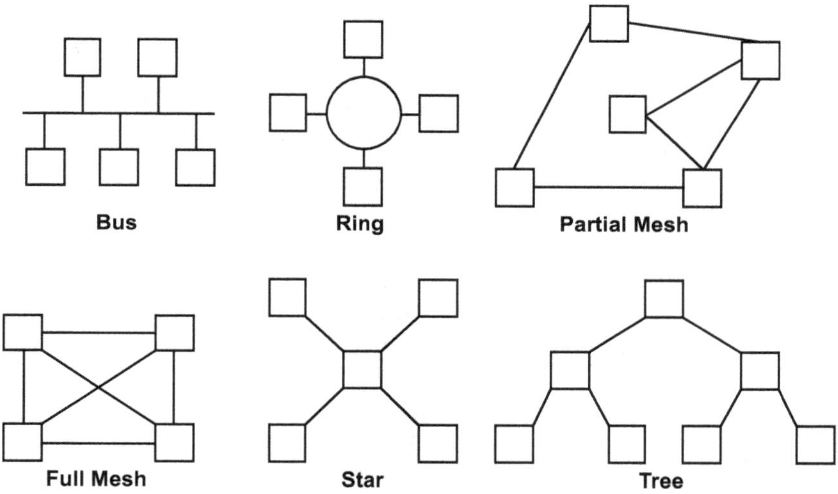

Bus   **Ring**   **Partial Mesh**

Full Mesh   **Star**   **Tree**

Fig. 4.1 *Network topologies*

In a *bus* interconnection, all the nodes are connected via a single bus. The nodes connect the bus by using an interface connector (e.g. T base connector). A node wanting to communicate with other node broadcasts the packet on the bus with the destination address. All nodes except the destined node ignore the message. Only the addressed destination accepts and processes the sent packet. The best example of bus-based topology is Ethernet, which provides a medium for a small set of nodes to share a communication medium. In case there is any fault in the shared medium, the entire network fails.

The *ring* interconnection has all the nodes connected in the form of a ring. All packets move either clockwise or anti-clockwise. Like the shared medium, any fault in the ring breaks the loop and has the potential to halt the operations of the entire network. The solution to this problems lies in creating a redundant ring or providing a loop back at the nodes between which the link has failed. Examples of network using ring topology are Token Ring and SONET.

In a *mesh* topology, each node is connected to two or more nodes. In a *partially connected mesh*, a node has two or more neighbours. For a fully connected mesh, there is a direct path between any two nodes of the network. The fact that there are two paths to reach another node provides robustness against single faults. The mesh topology is adopted in Internet routing wherein a set of routers route packets by using mesh topology.

In a *star* interconnection, each node is connected to a central node (also called the *hub* or the *switch*). Two non-central nodes in a star topology communicate with each other via the central node. This topology is preferable over the bus topology because a single failure does not bring down the entire network. As discussed later in this chapter, this switched topology provides higher throughput and performance and a good example of it is Fast Ethernet.

Finally, in a *tree* topology, multiple star connections form a topological tree. This topology is preferable when the network is to be scaled as it provides a hierarchical arrangement of the network.

Network topology affects a number of factors. Some of these factors include the number of hops required for data transmission (e.g. star topology requires two hops, while fully meshed topology requires one hop and so on), vulnerability to single point failures (e.g. star topology is vulnerable to the failure of central node) and cost (e.g. fully meshed topology is costlier to build than other topologies).

4.3 NETWORK EXTENT

Network topology is tightly coupled with the concept of network extent and the classification based on this leads to different types of *area networks*. The reason for this is that networks come in different shapes and sizes. Further, the extent of a network has a significant influence on the behavioural aspects of a network. For example, factors like maximum bandwidth, error rates, and ease of management are related to the network extent.

Two important classes of area network are Local Area Network (LAN) and Wide Area Network (WAN). A LAN is a network that is restricted to a small region. In contrast, WANs cover large geographical distances. The concepts of LANs and WANs are detailed in the following sections.

4.4 LOCAL AREA NETWORKS (LAN)

A *Local Area Network (LAN)* is a network that is restricted to a small region (see Fig. 4.2). LANs are widely used in business enterprises to connect personal computers, network servers, printers and other network entities. LANs are characterized by the following attributes:

- They are usually shared communication medium to which various nodes are attached. In order to grant access to a shared medium, access resolution protocol is necessary. Some LANs use a switched medium, which does not

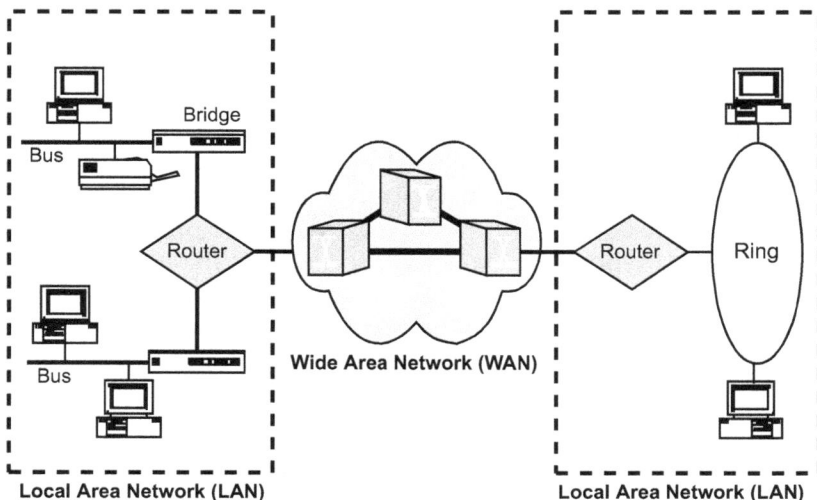

Fig. 4.2 *Local and wide area network*

require access resolution. Chapter 6 highlights the access resolution aspects.

- They have a diameter of the order of few kilometres.
- They are generally privately owned by an organization.
- Their bandwidth is considered to be free and hence bandwidth cost is not an important consideration. (Note that the cost is important only when the network is built. After that, irrespective of the bandwidth used, the fixed and the variable costs remains more or less the same. Thus, bandwidth is said to be free).
- Low-speed LANs provide bandwidth to the tune of 4 16 Mbps. Higher speed LANs can provide up to 100–1000 Mbps.
- Error rates are lower by an order of magnitude as compared to WAN counterparts.
- Ethernet, Fast Ethernet, Gigabit Ethernet, Token Ring, Fiber Distributed Data Interface (FDDI) and ATM are few examples of LAN technologies.

Among LAN technologies, the most prominent is the Ethernet Family, which includes Ethernet (10 Mbps), Fast Ethernet (100 Mbps) and Gigabit Ethernet (1 Gbps). Till recently, the bus-based Ethernet was considered adequate to satisfy the needs of most LAN applications. However, with the growth in the size of the networks and user requirements, there was a need to augment the existing network capacity and to provide more bandwidth to the users. This increased demand for bandwidth is tackled by migrating from *shared* LANs to *switched* LANs. In a shared LAN, multiple hosts share a common communication channel [see Fig. 4.3(a)]. Thus, for a 10 Mbps Ethernet LAN with 10 hosts, each host, on an average, gets 1 Mbps of bandwidth. However, this is true only if the LAN operates at 100 per cent capacity. This, unfortunately, is not the case because the *Carrier Sense Multiple Access with Collision Detection (CSMA/CD)* technique used in Ethernet reduces the effective throughput due to collisions. During heavy load conditions, the numerous collisions further reduce the throughput. Moreover, in a shared LAN, a frame is broadcast to all the hosts, which requires that hosts filter unwanted frames. Besides this, broadcasting frames to all hosts also makes the network traffic susceptible to security violations.

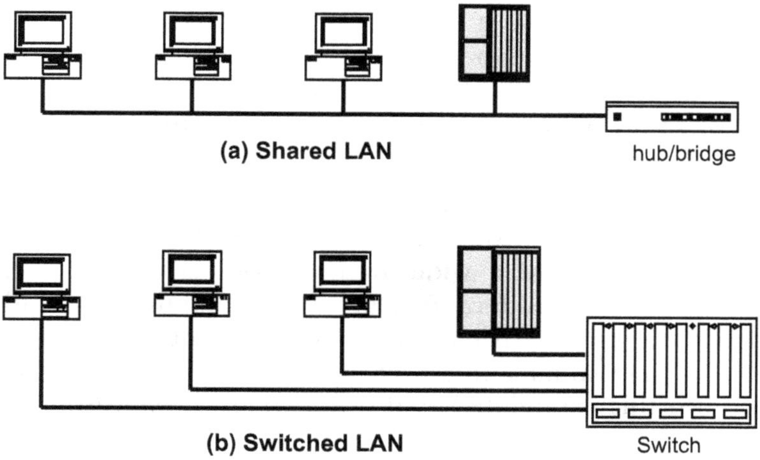

(a) Shared LAN   hub/bridge

(b) Switched LAN   Switch

Fig. 4.3 *Shared and switched LAN*

A switched LAN (e.g. Switched Ethernet) seeks to alleviate some of these problems by providing a dedicated bandwidth to each host. Each host in a switched LAN is directly connected to a switch (or an intelligent hub) and thus gets the entire channel bandwidth for itself (see Fig. 4.3(b)). Since the switch forwards frames according to the destination address, the filtering overheads and security concerns as in shared LAN, are eliminated. Moreover, fault diagnosis becomes easier because the switch can monitor each link individually. In essence, a switched LAN provides higher throughput and better manageability. However, this comes at a cost, because each host in the LAN has to get direct connectivity with the switch.

4.5 WIDE AREA NETWORK (WAN)

Unlike LAN, a *Wide Area Network (WAN)* covers a large geographical distance, spanning multiple states/countries. WANs are used primarily to connect dispersed sites. WANs are characterized by the following attributes:

- They have a diameter of the order of a few 1000 kilometers.
- They are seldom owned by one organization.
- Their bandwidth is very costly. Hence, bandwidth cost plays an important role in the decision-making process.

- Their bandwidth is of the order of 1–45 Mbps.
- Error rates are higher by an order of magnitude as compared to their LAN counterparts.
- X.25, frame relay and ATM are few examples of WAN technologies.

WAN technologies typically use point-to-point dedicated connections for data transfer (e.g. X.25, frame relay and ATM). Broadcasting data over a wide area network is not feasible. There are security considerations, geographical limitations and a host of other issues. In contrast, point-to-point connections are generally not suitable in a LAN environment. One possible exception to this is satellite-based WAN where broadcasting is possible.

Besides broadcasting, cost is also an important issue. Since bandwidth is free in a LAN environment, protocol efficiency is not a very critical issue. However, the same is not true in WAN, where the cost per byte of data transferred is important. This mandates that WAN protocols be simple and efficient.

Another significant difference between LANs and WANs is robustness. Since WANs serve a large number of organizations, they are required to be much more robust than LANs. Thus, it is likely that one finds redundant WAN links to cater to an emergency.

4.6 OTHER AREA NETWORKS

Apart from the two broad classes of area networks (LAN and WAN), literature provides references to many other forms of area networks. Some of the important terms are explained in this section.

The most popular among these terms is the Metropolitan *Area Network (MAN)*, which covers an area larger than a LAN but smaller than a WAN (e.g. covers a city). Some provide another category between LAN and MAN and refer to it as *Campus Area Network (CAN)*. As the name suggests, CAN refers to a connection of networks within a campus (e.g. connecting various departments within a college campus). Examples of MAN and CAN are Gigabit Ethernet, ATM and Fiber Distributed Data Interface (FDDI).

A very recent term is the *Personal Area Network (PAN)*. A PAN encompasses all communication devices [e.g. laptop, mobile phone, Personal Digital Assistant (PDA) and pagers] within the limits of someone's personal space. The communication is typically done by using a wireless protocol (e.g. Bluetooth). The PAN is relevant for a person at his/her home, car or even in the office.

4.7 EXAMPLES

This chapter described various types of network topologies and types of area networks. This section lists various protocols/ technologies along with the supported topologies and categorizes them into one of the three types of area networks (LAN, MAN and WAN). These protocols/technologies, along with the topology and supported area networks, are listed in Table 4.1.

Table 4.1 *Protocols/Technologies and Associated Topology and Area Network*

Name	Topology	Area Network	Description
Ethernet	bus, star	LAN	Depending upon whether LAN is shared or switched, a bus or star topology is followed.
Fast Ethernet	bus, star	LAN	- same as above -
Gigabit Ethernet	bus, star	LAN, MAN	- same as above -
Token Ring	ring	LAN	Provides LAN connectivity in ring topology.
FDDI	ring	LAN, MAN	Based on token ring, can connect large number of users.
SONET	ring	MAN, WAN	Based on ring topology; primarily used for high-speed data transfer.
X.25	mesh	WAN	Based on virtual circuit switching, provides low data rates.
Frame Relay	mesh	WAN	Based on virtual circuit switching, provides low data rates.
ATM	mesh	LAN, MAN, WAN	Like X.25 and frame relay, they are primarily point-to-point, but they can emulate the broadcast environment.

LAN-based technologies have traditionally been one of the two types, viz. Ethernet and Token ring. ATM was pushed in the LAN environment but has not been very successful. Among new technologies, the Gigabit Ethernet (and even 10 Gigabit Ethernet) are becoming popular in the LAN and MAN environment.

The WAN environment is dominated by virtual circuit technologies like X.25, frame relay and ATM. The new technologies like IP over SONET are also becoming popular in the WAN environment.

4.8 CONCLUSION

This chapter described various types of topologies including bus, ring, star, tree and mesh. The different types of area networks were also explained with a special focus on LANs and WANs. Finally, various examples were provided along with a mapping between the topology and the type of area network.

REVIEW QUESTIONS

Q 1. Why is there a need for various types of network topologies? Which of the topologies is most preferred in the LAN environment and why?

Q 2. Is broadcast possible in WAN? If not, why? What are the alternative topologies?

Q 3. Compare shared and switched LAN.

Q 4. What are the key differences between a LAN and a WAN?

FURTHER READING

For details of the topics covered in this chapter, a few options are available. [Gen A. Tanenbaum] and [Gen W. Stallings] are two popular texts.

- Transmission and Multiple Access

- Data Link Control

- Bridging

- Switching

- Addressing

- Signalling

- Routing

Part 2

CORE CONCEPTS
OF COMMUNICATION
NETWORKS

The previous part of the book focused on some of the high level aspects (or the building blocks) of communication networks. The essence of communication is the *communication protocol*, which is the language used by entities for communication. The protocols are typically layered to give a *layered architecture*. The layered architecture and the OSI reference model were discussed in the first part of the book. Apart from this, other concepts like transfer mode was also discussed.

This part of the book raises a few more questions and seeks their answers. In particular, it considers the following questions:

- How is information carried at the physical layer? Is it shared across a medium? Is there any contention for the physical access? If yes, how is such a contention resolved?
- How do two peer entities engage in a reliable communication? How are errors handled? How is it ensured that the sender and receiver are in a position to send and receive?
- How are various devices connected together? What are the typical means to connect two segments of a network?
- How does an entity forward or switch packets? What is the internal structure of a switch?
- How are entities identified in a network? What are the typical addressing mechanisms?

The aforementioned questions touch upon core concepts in the field of networking. This part of the book is an attempt to describe these concepts to provide readers further insight into the field of communication networks. For this purpose, the following concepts are defined:

- **Transmission and Multiple Access:** This refers to the sharing of a resource between multiple users and resolving access contention between them. This topic is covered in Chapter 5.
- **Data Link Control:** This includes concepts like framing, flow control and error control that are useful for two-link layer entities to engage in a proper communication. This topic is covered in Chapter 6.
- **Bridging:** This refers to the means of connecting two similar or different LANs. This topic is covered in Chapter 7.
- **Switching:** This refers to the transfer packets from input ports to the appropriate output ports. The concept of switching and important aspects of switch design is covered in Chapter 8.

- **Addressing:** This refers to the process of assigning unique numerical identifiers to entities for the purpose of communication. This topic is covered in Chapter 9.
- **Signalling:** This is used between the user and the network, or between two network elements, to exchange various control information and to establish, manage and release connections. This topic is covered in Chapter 10.
- **Routing:** This refers to the techniques used for forwarding packets (at the network layer) between two routers. This topic is covered in Chapter 11.

TRANSMISSION AND MULTIPLE ACCESS

5.1 INTRODUCTION

The primary task of a network is to provide the means to transfer user information from one network entity to another. User information can be transferred from one point in the network to another through different transmission formats, and over a variety of transmission media. A discussion on these transmission concepts is provided in section 5.2. Besides user information transfer for a single user, it is also essential for the network to define mechanisms to allow multiple users to simultaneously make use of the network services. This introduces the concept of *Multiple Access*, which allows multiple users to simultaneously access network resources. Concepts related to Multiple Access are described in detail in section 5.3. The concept of multiple access is explained with the help of two examples. First, the Frequency Division Multiple Access (FDMA)/Time Division Multiple Access (TDMA) scheme in the Global System for Mobile Communication (GSM) is explained in section 5.4. Thereafter, section 5.5 describes a more complex scheme of Wideband Code Division Multiple Access (WCDMA).

5.2 TRANSMISSION CONCEPTS

The transfer of information between different network entities is undoubtedly the most critical service offered by any network.

Offering an error-free information transfer service necessitates multiple functions implemented by the network (e.g. information routing) that should be functioning correctly. While the remainder of this book will touch upon all these functions in detail, this section discusses some of the transmission concepts related to the physical layer that form the basis for all further discussion. These concepts deal with the following aspects of information transmission:

- What is the nature of the information being transmitted?
- What is the actual physical media used for information transmission?
- What is the capacity of the physical media used to carry this information?
- In what format is the information actually carried over the physical media?

The subsequent sections discuss these concepts related to information transmission in detail.

5.2.1 Analog and Digital Transmission

An important concept in information transmission is related to the nature of the information to be transmitted, and the way it is handled at the physical layer. The information to be transmitted across the network may be either in *analog form* or *digital form*. Similarly, the physical media used to transmit this information may transfer the information either as *analog signals* or as *digital signals*.

Information in analog form corresponds to *continuous* information. In other words, the characteristic of this information is such that it takes continuous values in an interval of time. Acoustic data is one of the most familiar examples of analog information. In contrast to this is the concept of digital information. Information in digital form corresponds to *discrete* information. In other words, the characteristic of the digital information is such that it takes discrete values. An example of digital information is a sequence of integers, or a sequence of characters forming the text of a particular language.

Irrespective of the nature of the information, the physical media may choose to transmit the information either as a continuously varying signal (analog signal) or as a discrete-valued signal (digital signal). Analog signal is a continuously varying electromagnetic wave that can be carried over any physical media. On the other hand, a digital signal is a sequence of voltage pulses that can take only discrete values (e.g. +5 volts and 0 volts). Figure 5.1 depicts information being transmitted as analog signals and digital signals (binary signal, consisting of 1s and 0s).

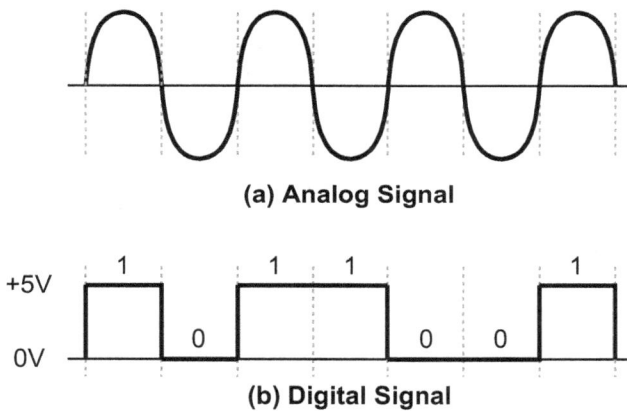

(a) Analog Signal

(b) Digital Signal

Fig. 5.1 *Analog signal and digital signal*

Information in analog form can be transmitted as digital signals, while information in digital form can be transmitted as analog Signals. This requires conversion from one form to another at the time of transmission on the physical media. The following sections discuss the conversion of information from digital form to analog form, and vice-versa.

5.2.2 Digital to Analog Conversion

Digital information can be transmitted over the physical media in the form of analog signals. Some of the commonly used mechanisms to convert digital information into analog signals are as follows:

- **Amplitude Shift Keying (ASK):** In this scheme, different discrete values of the digital data are represented by different unique amplitudes of the carrier frequency.

- **Frequency Shift Keying (FSK):** In this scheme, different discrete values of the digital data are represented by different unique frequencies. Thus, unlike ASK, which uses variations in the amplitude of the carrier frequency to depict different digital values, FSK uses different frequencies near the carrier frequency to depict the different digital values.
- **Phase Shift Keying (PSK):** Another commonly used mechanism to convert digital information into analog signals is Phase Shift Keying (PSK). In PSK, shifting the phase of the carrier signal differently represents the different discrete values of the digital data.

Figure 5.2 depicts the different schemes for conversion of digital information (as binary data) into analog signals. These are the three basic schemes and are also referred to as *digital modulation* techniques. There can be variations of these basic schemes. For example, in the PSK technique, the example depicted in Fig. 5.2 is a special case wherein the phase difference is 0 and π, and can also be referred to as *Binary Phase Shift Keying (BPSK)*. A more general case is the M-ary PSK where the phases are $2\pi m/M$, where m = 0, 1, ... (M-1). From this general case, there can be BPSK as discussed or other cases like *Quadrature Phase Shift Keying (QPSK)* in which there are four different phases, viz. 0, $\pi/2$, π and $3\pi/2$. This scheme has the advantage that for the same bandwidth, it can provide double the bit rate. To get three times the bit rate of basic case, we have 8-PSK where there are eight different phases. The 8-PSK scheme is used in Enhanced Data Rates for Global Evolution (EDGE) networks which can be viewed as an enhancement over the GSM network because of the better modulation scheme providing higher data rates.

It may appear that M-ary PSK should always be the preferred option because it provides higher bit-rates depending upon the value of M. However, the issue is actually related to the cost of making the de-modulation circuit. If the de-modulation scheme requires a carrier phase, it is a coherent scheme, otherwise it is an incoherent scheme. In the latter case, the scheme is sensitive to the power signal only. For example, in the ASK scheme, the signal is present or absent, while in the FSK scheme, the power is present at a given frequency or not. The cost of an incoherent

(a) Amplitude Shift Keying (ASK)

(b) Frequency Shift Keying (FSK)

(c) Phase Shift Keying (PSK)

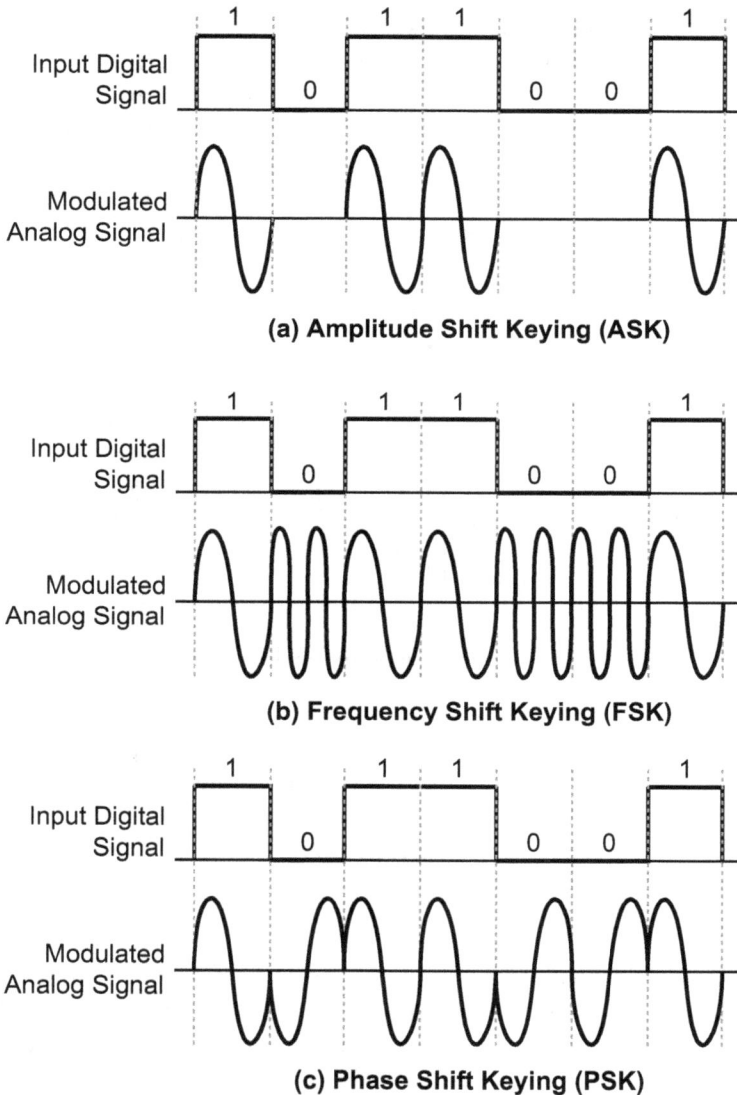

Fig. 5.2 *Conversion of digital information to analog signals*

de-modulation circuit is low, but it gives poor performance. In contrast, the coherent schemes provide higher performance at a greater cost. The design complexity and cost increase as a higher number of phases is used (i.e. M > 2).

5.2.3 Analog to Digital Conversion

Just like digital information can be transmitted over the physical media in the form of analog signals, similarly analog information can be transferred into digital form over the physical media. One of the most commonly used conversion mechanisms is called *Pulse Code Modulation (PCM)*. Figure 5.3 depicts the use of PCM for the conversion of an analog signal into digital form.

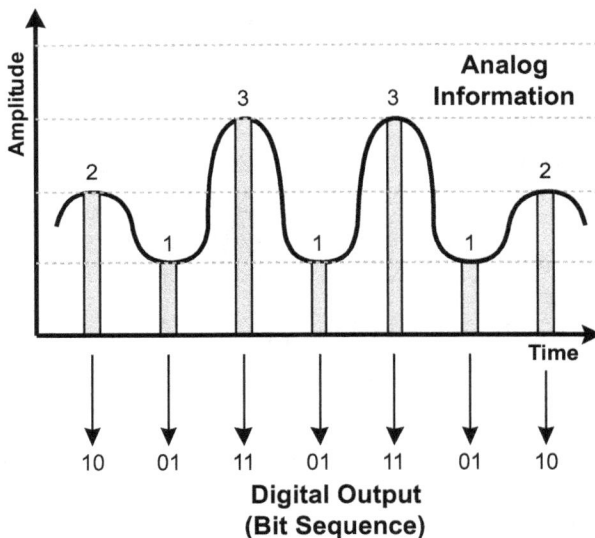

Fig. 5.3 *Pulse Code Modulation*

Pulse Code Modulation defines a scheme in which samples of the original analog signal are taken at regular intervals. The value (or amplitude) of the analog signal at these sampling instances is then converted into binary form. This sequence of values forms the output digital signal for the input analog information.

In this scheme the process of conversion of continuously varying analog information into discrete valued signal leads to some loss of information. In order to minimize loss of information, the sampling rate should be as high as possible, so that the original analog information is sampled more frequently. However, for the same input analog signal, increasing the sampling rate will increase the number of bits to be transmitted as part of the converted digital signal. Thus, a trade-off exists, wherein increasing the sampling rate reduces the loss of information, but at the same time it increases the amount of information that is required to be transmitted over the physical media. Here, the *Sampling Theorem* comes into play, which is as follows: *If the sampling of the analog information is performed at a rate higher than twice the highest signal frequency in the input analog information, then the digital samples thus obtained contain enough of the analog information to perform the reverse conversion at the receiver-end.* In other words, sampling at twice the rate of the highest signal frequency ensures that the original analog signal can be recovered by the receiver from the received digital information, by using a low-pass filter.

5.2.4 Bandwidth and Capacity

The previous section elucidated how information is transmitted over the physical media, in either digital form or analog form. This section explores the concepts related to the capacity of the physical media in carrying this information. Here, it is important to define the two related concepts: bandwidth and capacity.

Bandwidth is simply a measure of the difference between the highest and the lowest frequency that can be transmitted by using the physical communication media. It is expressed in Hertz (Hz in short), or Cycles Per Second (cps), and it is constrained by the nature of the physical media, and the transmitter. For example, analog signals transmitted over copper cables in telephone communication systems have a frequency range of 400 Hz (lowest frequency) to 3400 Hz (highest frequency). In this sense, the copper cables in telephone communication systems have a bandwidth of 3000 Hz (3400–400 Hz).

Capacity, on the other hand, is defined as the rate at which data can be transmitted over a physical media of known bandwidth. For digital transmission, this is measured in Bits per Second (bps). This is also sometimes commonly referred to as the Data Rate of the physical media or the communication channel. For a given physical media of finite bandwidth, the maximum data rate that is achievable is limited. H. Nyquist defined this limit on capacity by relating it to the bandwidth of the physical media, and the number of discrete levels used to carry the digital signal (binary digital signal has two discrete levels, 0 and 1). This relation is commonly known as the *Nyquist Theorem*.

Nyquist proved that if a signal is made to pass through a low-pass filter of bandwidth B, the filtered signal could be completely reconstructed by taking exactly $2B$ samples per second of the input signal. While it is possible to sample the input signal at a rate greater than $2B$ samples per second, it will not provide any additional benefit over a sampling rate of exactly $2B$ samples per second. This is because while sampling at a rate faster than $2B$ samples per second can recover the higher frequency components, these components would have already been filtered out when the input signal passed through the low-pass filter.

On the basis of these findings, Nyquist defined the maximum limit on the *Capacity* of a physical media of finite bandwidth as:

Capacity (or maximum data rate) = $2B * \log_2 N$ bits/sec,

where B is the bandwidth of the physical media, and N is the number of discrete levels in the digital signal.

While defining the maximum capacity of a physical media, the Nyquist Theorem considers ideal conditions, where the physical media is noiseless. Claude Shannon extended Nyquist's work further to define a more practical upper limit on the capacity of a physical media, limited also by the thermal noise present in the media. The amount of thermal noise present in a physical media is defined by the *Signal-to-Noise Ratio (SNR)*, and is defined as the ratio of the Signal Power (S) to the Noise Power (N). The unit of measurement of SNR is called decibels (dB), and is defined as $10*\log_{10}(S/N)$.

Shannon's law for the maximum data rate is thus defined as:

Capacity (or maximum data rate) = $B * \log_2 [1 + (S/N)]$ bits/sec,

where B is the bandwidth of the physical media, and S/N is the ratio of the Signal Power to the Noise Power in the physical media.

5.2.5 Transmission Media

The physical media used for information transmission are basically divided into two categories, viz. guided media and wireless media. Guided media can be defined as a transmission media wherein a visible physical cable runs between two or more network elements, thus connecting them together. The term 'guided' is used to denote that the information transfer takes place using a defined/guided path over the physical cable. Some of the commonly used guided media are:

- **Twisted Pair:** A twisted pair cable consists of two insulated copper wires, which are twisted around each other in a regular spiral pattern. The two cables in the twisted pair jointly work as a single communication link. These cables are used to transmit both analog and digital signals.
- **Coaxial Cable:** Coaxial cables, similar to twisted pair cables, consist of two wires. However, unlike in the twisted pair, coaxial cables have an arrangement in which one wire forms the outer cylindrical core of the cable, while the second wire forms the inner wire, inside the cylindrical core. Coaxial cables are most popular for the transmission of cable television over long distances. Like twisted pair cable, coaxial cables can also be used to transmit both analog and digital signals.
- **Optical Fiber:** Optical cables are cables manufactured by using glass and plastic. Unlike the twisted pair and coaxial cables, optical fibers do not have wires for carrying information signals. Instead, information is carried by using optical rays of different frequencies, which are carried within the optical cable.

Figure 5.4 depicts the guided media described above pictorially.

Unlike guided media, wireless media, on the other hand, is unguided media. In them, transmission and reception of information signals is done by using antennas. The physical medium of transmission is usually air. Antennas transmit and receive information signals as electromagnetic waves in the air. Some of the commonly used wireless media are:

- **Terrestrial Microwave:** Terrestrial microwave uses transmitter and receiver antennas, which are located at a substantial height above the ground level. A typical usage of terrestrial microwave is found in cellular communication systems.

- **Satellite Microwave:** Satellite microwave uses concepts similar to terrestrial microwave. However, unlike terrestrial microwave, satellite microwave makes use of satellites positioned at a much higher level above the ground. The satellites act as relay transmitters, which are used to relay signals between two transmitter receiver stations on the earth.

(a) **Twisted Pair**

(b) **Coaxial Cable**

(c) **Optical Cable**

Fig. 5.4 *Guided media*

- **Infra-red Communication:** Infra-red communication is a means of communication where transmitters and receivers use infra-red light to transmit information. Such communication systems are highly dependent on line-of-sight, i.e. the transmitter and receiver must be within the line of sight of each other for them to communicate with each other. Typical examples of infra-red communication include usage in remote control devices for electronic gadgets.

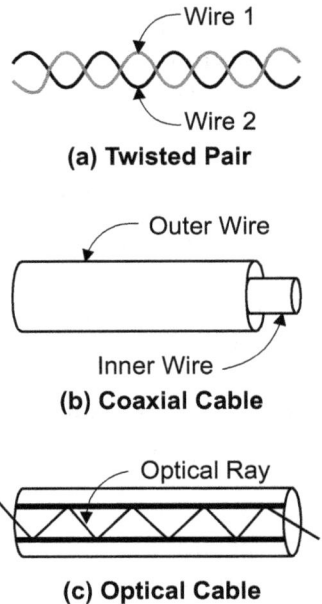

5.3 MULTIPLE ACCESS CONCEPTS

Multiplexing is defined as a mechanism to allow multiple streams of user information to share the same transmission medium. Multiplexing can exist at any layer in the OSI reference model, which works on the basic premise that a layer (N) entity can carry PDUs of multiple protocols sitting at layer (N+1). In other words, the layer (N) multiplexes different streams of information from multiple protocols sitting above it at layer (N+1). Frame types, protocol identifiers and port numbers are some of the commonly used identifiers for the purpose of multiplexing and de-multiplexing.

While multiplexing is not only a physical layer concept, the focus of this section is on multiplexing performed at the physical layer. Figure 5.5 depicts the basic schemata for multiplexing. At the transmitter side, N different input streams are multiplexed onto a single transmission channel. The single multiplexed channel can be a copper cable, optical fiber, or an unguided wireless media. At the receiver-end, the multiplexed channels are de-multiplexed to get back the N output streams.

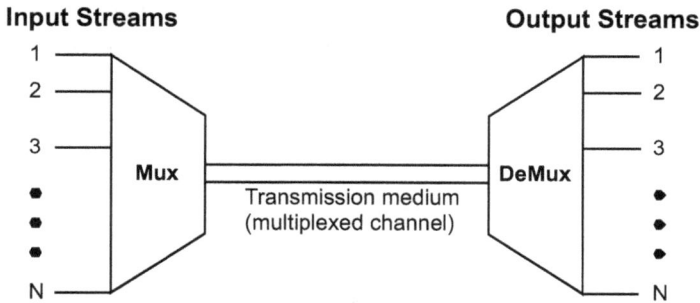

Input Streams   **Output Streams**

Fig. 5.5 *Basic schemata for multiplexing/de-multiplexing*

At the physical layer, multiplexing techniques are classified into four categories, namely Frequency Division Multiplexing (FDM), Time Division Multiplexing (TDM), Wavelength Division Multiplexing (WDM), and Code Division Multiplexing (CDM). The following sections describe these multiplexing techniques.

5.3.1 Frequency Division Multiplexing

Frequency Division Multiplexing (FDM) is a broadband analog technique in which multiple input streams are transmitted simultaneously by using different frequency bands. In this technique, the input signal is used to modulate a carrier frequency in a different frequency band. Thus, using different carrier frequencies, multiple input streams can be multiplexed onto a single channel. A typical example of FDM is radio transmission, wherein different radio stations broadcast signals at different frequency bands at the same time. The listener can then tune his or her radio set to a particular frequency channel to receive signals from a particular radio station. FDM has been very successful in radio broadcasts for transmitting multiple channels.

While it is not completely standardized, there does exist some agreement on the use of different frequencies for FDM. The most commonly used frequency range is the 12-channel group in the 60 to 108 Khz frequency band (see Fig. 5.6). While each channel requires a 3000 Hz bandwidth, each channel in the group is actually provided with a frequency of 4000 Hz. The additional 1000 Hz facilitates better channel separation. However, even with the additional bandwidth, the overlapping region between two contiguous channels does result in a phenomenon known as *Crosstalk*, which occurs when the signals of one channel interfere with those of the adjoining channel (see Fig. 5.6).

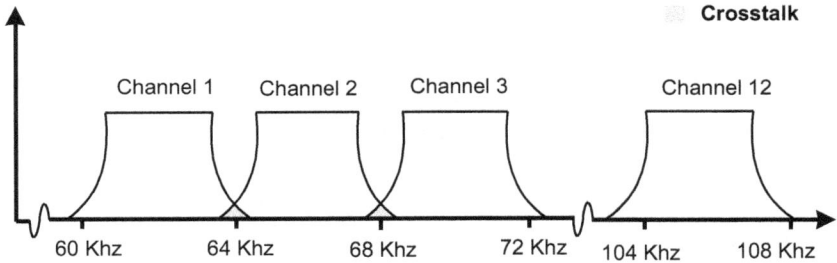

Fig. 5.6 *Frequency Division Multiplexing*

5.3.2 Time Division Multiplexing

As discussed in Chapter 3, the circuit switching technique primarily relies on Time Division Multiplexing (TDM) to carry user information from one node to another. TDM is a technique in which individual input channels are time-multiplexed onto a single physical channel. The multiplexed channel is divided into frames and each frame is further sub-divided into slots. Each input channel gets a fixed time slot in each frame. Chapter 3 also discussed a variant of TDM, known as *Statistical Time Division Multiplexing (STDM)*. which is used as a means of avoiding wastage due to unused time slots in pure TDM. Refer to Chapter 3 for more details on TDM and its use in circuit switching.

5.3.3 Wavelength Division Multiplexing

Wavelength Division Multiplexing (WDM) is a multiplexing technique employed for fiber-optic cables. The basic idea behind WDM is to partition the bandwidth of a fiber into multiple channels, wherein each channel carries a light signal of different wavelength. Thus, WDM for fibre is analogous to FDM as used in microwave systems.

Since multiple WDM channels are carried in parallel in the same fiber cable by using different wavelengths, transmission rates get multiplied by the number of channels carried. Thus, a much higher rate is obtained by using WDM, as compared to the transmission rates of any non-WDM system. For example, by using a Quad-WDM (four-channel) system, the transmission rate can be quadrupled.

A WDM-based network is classified into two categories, namely *single-hop* network and *multi-hop* network. In a single-hop network, each node is directly connected to every other node of the network. Once data is transmitted as a light signal, it is converted back into electrical signals only at the destination. Single-hop networks pose two important challenges. First, the transmitters and/or receivers should quickly tune to a particular channel for data transmission. Second, the arbitration and allocation of channels should be done through efficient protocols.

Multi-hop networks avoid some of the problems posed by single-hop networks. The problem is avoided by allowing each node to access only a set of wavelength channels. If a node (say A) cannot communicate directly with another node (say B), then it uses an intermediate node (C) whose receiver is tuned to A's transmitter's frequency and whose transmitter is tuned to B's receiver frequency. It is also possible that more than one hops are used for communication between two nodes. This is the reason why such networks are called multi-hop networks.

5.3.4 Code Division Multiplexing

As discussed in the previous sections, the TDM scheme divides the total available bandwidth between multiple users on a time basis (time-sharing). Similarly, the FDM scheme divides the total available bandwidth by allocating different frequency channels to different users. However, unlike TDM and FDM, the Code Division Multiple Access (CDMA) scheme, offers the entire frequency spectrum to each user, for the entire duration of time, as per the user's requirement. This leads to the question of how multiple access is provided to multiple users simultaneously, if neither time nor frequency is used as a parameter for multiplexing.

CDMA introduces a concept of spreading codes, which are applied as part of a secondary modulation of user signals. These codes are used to transform the user signals into a spread-spectrum-coded version of the original signal, before transmission. The receiver carries out the reverse process to recover the original signal from the coded signal. Different users are allocated different spreading codes, and the signals of multiple users are thus differentiated on the basis of the spreading codes. Spreading codes can therefore be considered as the third possible parameter on which sharing of resources can be achieved, besides time slots and frequency channels. CDMA systems use this spreading code to provide multiple access to users. Section 5.5 describes the concepts related to CDMA in more detail, when discussing the use of CDMA in third generation mobile networks called UMTS.

5.4 EXAMPLE: FDMA/TDMA IN GSM NETWORKS

The Global System for Mobile Communication (popularly known as GSM) is amongst the most popular second-generation mobile communication systems. GSM uses a scheme that derives from a combination of the FDM and the TDM schemes discussed in sections 5.3.1 and 5.3.2, respectively. This combination of FDM and TDM is also commonly referred to as the hybrid FDMA/TDMA scheme. However, before we can discuss the hybrid FDMA/TDMA scheme, it is important to understand the concept of cells in a GSM network.

In the GSM network, the only wireless link is the link between the mobile handset and the radio tower. A radio tower in GSM is called the Base Transeiver Station (BTS). Since the coverage area for each BTS can only be limited (signals fade when transmitted over longer distances), a large number of BTS are required to provide service in a big geographical area. These BTS are, in turn, connected to each other via a network of switches, all of which are connected via guided transmission media. The geographical area serviced by one BTS in GSM is called a cell (hence the name cellular communication!).

Figure 5.7 depicts a GSM network comprising three cells, serviced by three different BTS.

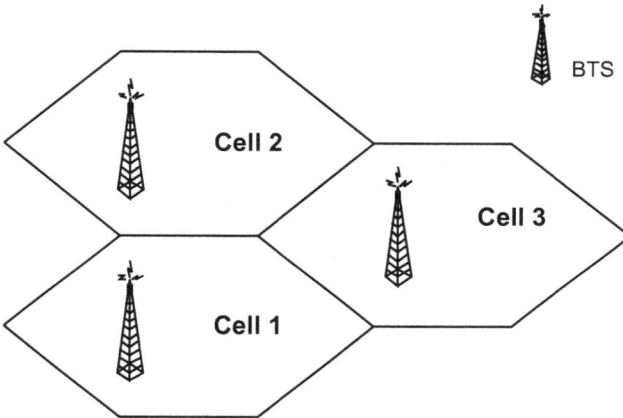

Fig. 5.7   *Cells in a GSM network*

In the hybrid FDMA/TDMA scheme followed in GSM networks, the entire frequency spectrum available for cellular communication is first divided into multiple smaller frequency bands. These frequency bands are distributed for use in cells in a manner such that no two neighbouring cells use the same frequency band. Two cells that are not neighbouring cells may, however, re-use the same frequency band. In this sense, FDM is used between neighbouring cells to differentiate between the communication links, and to thus prevent interference in communication occuring simultaneously in neighbouring cells.

However, within each cell region, there can be multiple subscribers who need to be provided with a wireless communication channel. To achieve this, the frequency band assigned to each cell is time-shared between multiple users in the cell. This principle of time-sharing is the same as that followed in a pure TDM system. Hence, in GSM, a time slot on a particular frequency band identifies the communication channel assigned to a GSM subscriber. In other words, each GSM subscriber who requires a communication channel is allocated a time slot within a particular frequency band of the cell (as in TDM), where the frequency band of the cell is itself a part of a larger frequency spectrum (as in FDM). Hence, the multiplexing scheme in GSM derives from both the FDM and the TDM multiplexing schemes, and is thus called the hybrid FDMA/TDMA scheme.

5.5 EXAMPLE: CDMA IN UMTS NETWORKS

The Universal Mobile Telecommunications System (popularly known as UMTS) is an evolution of the GSM system, and is a third-generation mobile communication system. While GSM networks are based on the hybrid FDMA/TDMA principle, UMTS networks are designed to use Code Division Multiple Access (CDMA) as the scheme for multiple access. As briefly discussed in section 5.3.4, CDMA uses codes as a means of performing the multiplexing and de-multiplexing of multiple communication channels. This is unlike GSM, which uses frequency and time as parameters on the basis of which communication channels are multiplexed and de-multiplexed. In CDMA systems, each communication channel uses the same frequency spectrum, for the entire lifetime of the communication channel. Codes are used to distinguish between the different channels. Two types of codes are used in CDMA-based UMTS networks, namely, *spreading codes* and *scrambling codes*. The following sections briefly explain the concepts of spreading and scrambling, and the associated codes used for the processes of spreading and scrambling.

5.5.1 Spreading and De-spreading

Spreading is a technique used to transform the user's original signal into a signal form that is spread over a larger bandwidth than what is required for the original signal. Codes used for this transformation of the signal are called spreading codes (also known as *channelization codes*). This section explains the process of spreading and de-spreading.

The process of spreading involves transformation of the user signal by multiplication (XOR) with bits in a spreading code. Each user is assigned a unique spreading code. The bits in the spreading code are termed as chips, while the bits within the user signal are called signals. Figure 5.8 depicts the process of spreading. The transformation involves bit-wise XOR of the

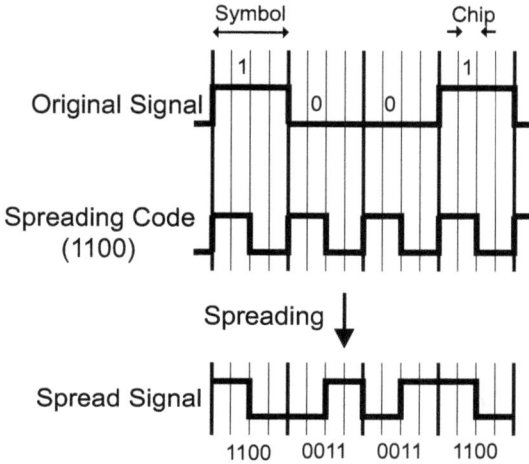

Fig. 5.8 *Spreading (with a spread factor of 4)*

user signal with the bits in the spreading code to form the spread signal.

The process of spreading can also be seen as using the chips in the spreading code to chop the user signal into smaller parts. The spread signal is actually spread over a larger bandwidth as a result of the spreading process. The ratio between the bandwidth required to transmit the spread signal to the original bandwidth requirements is called the spreading factor. Values of the spreading factor within the UTRAN lie between 4 and 512. In Fig. 5.8, a spreading factor of 4 is depicted.

De-spreading is the reverse process of spreading. It involves recovery of the original signal at the receiving end from the spread signal. The same sequence of the spreading code is used in the de-spreading process. The process involves bit-wise XOR of the spread signal with the spreading code to recover the original signal. Figure 5.9 depicts the process of de-spreading.

Having gone through the description of spreading and de-spreading, at first thought, the entire process seems to be a waste of bandwidth. The available bandwidth on the air interface is a scarce resource, and must be efficiently utilized amongst multiple users. Why would CDMA then use spreading to transform the original signal into a signal that would require more bandwidth than the original signal? There are many reasons why spreading is so important for CDMA. However,

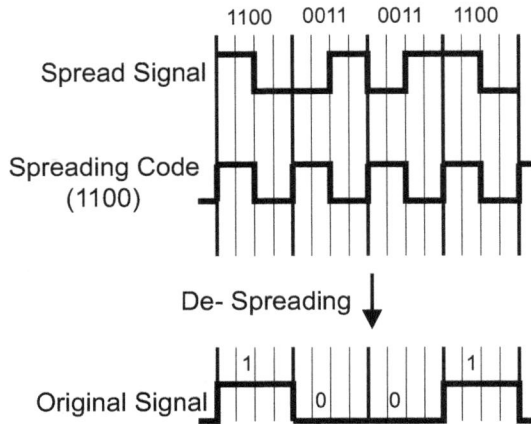

1100   0011   0011   1100

Spread Signal

Spreading Code
(1100)

De- Spreading ↓

Original Signal

Fig. 5.9   *De-spreading*

before delving into the benefits obtained by spreading, it is important to understand two key concepts: *auto-correlation* and *cross correlation*.

5.5.2 Auto-correlation and Cross Correlation

The propagation of a radio signal over the radio interface (between the mobile handset and the radio tower) is generally characterized by multiple reflections and diffractions. The reflections and diffractions are the result of obstacles within the path from the mobile handset to the radio tower (or vice versa). These obstacles could be tall buildings, hills, etc, which reflect/diffract the signal, resulting in a concept known as *Multipath propagation*. As a result of multi-path propagation, the same signal is received by the receiver (mobile handset or radio tower) more than once, at different time intervals, and with different power levels. Thus, in such a scenario, the receiver has to have the intelligence of receiving the same signal multiple times, and then using these signals to obtain the original signal. For this purpose, a rake receiver is used at the receiving end.

The rake receiver consists of multiple rake fingers (just like a garden rake has multiple fingers), with each of them receiving a multi-path signal. The receiver itself acts as a correlator, correlating the signal received by each rake finger. It is here that the concept of auto-correlation finds its significance. Auto-correlation measures the amount of correlation between the received signal and a delayed version of the same signal received later in time. The higher the auto-correlation, the easier it is to receive a multi-path signal. Since a signal is received through multiple paths in the form of various components, these (components) may result in interference at the receiver's end. A spreading code that provides good auto-correlation properties of the spread signal can resist this interference.

Cross correlation, on the other hand, measures the correlation between a signal spread using a particular spreading code, to the same signal spread with some other pseudo-random code. Spreading codes should have a low cross correlation with other spreading codes, which results in lower interference between signals received by a receiver. Thus, a good spreading code should have high auto-correlation properties, but lower cross correlation with other spreading codes. However, it is generally not possible to have both high auto-correlation and lower cross correlation at the same time. A trade-off between the two properties is normally required.

5.5.3 Benefits of Spreading

The use of spreading techniques provides many benefits. Firstly, the use of spreading (codes) makes it possible to provide a multi-access environment. Spreading codes are unique to each user. As a result of this, each user's signal is transformed differently before transmission on the air interface. This allows multiple users to use the same frequency channel at the same time, simultaneously, by transmitting the signal after transformation with different spreading codes. This is what forms the basis of CDMA code-based distinction between multiple simultaneous users.

Spreading codes have low cross correlation among them. Hence, when the spread signals of different users are received by the receiver, which have been spread by different spreading codes, then the signals can be easily separated from each other. Since the receiver knows the spreading code used by the sender, the original signal can be recovered by using this spreading code. Any other noise received along with the spread signal cannot be recovered by using this spreading code, and is hence separated from the original signal as noise.

Figure 5.10 depicts the concept of recovery of the original signal by separating it from the interference/noise. An interfering signal present in the same band typically appears as a higher power, narrow band signal [see Fig. 5.10(a)]. At the receiving end, the de-spreading process recovers the original signal (de-spread signal) and spreads out the interference instead [see Fig. 5.10(b)].

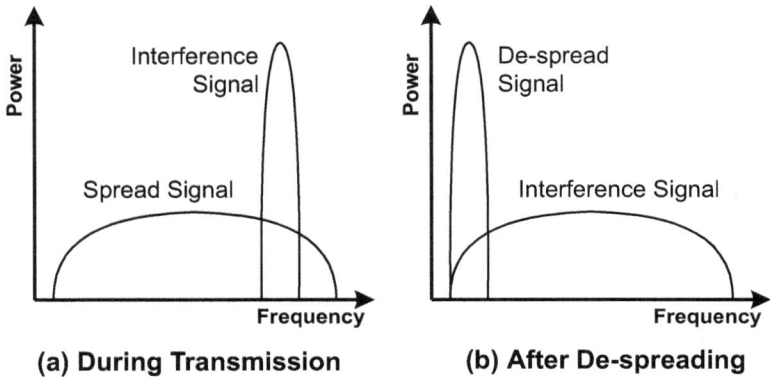

(a) **During Transmission**   (b) **After De-spreading**

Fig. 5.10   *Recovery of the spread signal*

Secondly, the spreading of the signal over a larger bandwidth results in a frequency re-use factor of one. This means that the same frequency can be re-used in adjacent cells, resulting in a high spectral efficiency. On the other hand, the FDMA/TDMA hybrid scheme of GSM results in a frequency re-use factor of at least four, implying that the same frequency can at best be re-used every fourth cell.

Thirdly, the higher auto-correlation between the spreading codes helps tackle the problem of multi-path interference. Using a rake receiver can constructively correlate signals received from multiple paths. Resolving multi-path interference was not so simple in GSM systems, which did not use the concept of spreading.

Lastly, the spreading of user signals before transmission results in the improved security of the transmitted signal. Since the spreading code used to spread the signal is known only by the sender and the receiver, it is not possible for any receiver in between to capture this signal, other than the intended receiver. However, this only provides security from non-resourceful hackers. A resourceful hacker can always capture a user signal and use brute-force mechanism to derive the original signal using all spreading codes, especially since the number of spreading codes is fixed.

Besides hacking, another significant property of spread signals is that unlike the original signal, spread signals cannot be jammed. This property stems from the fact that while jamming a particular frequency could have jammed the original signal, the spread signal is spread over a much larger frequency band, requiring a jamming of the entire frequency band. This is more difficult than jamming the transmission at a particular frequency. Hence, spreading also finds a great deal of importance in military communications.

5.5.4 Scrambling

The previous section discussed the benefits of using spreading before transmission of the signals over the radio interface. However, spreading alone is not sufficient to provide an efficient solution to the transmission problem. The following issues still remain to be addressed:

- The spreading codes are orthogonal in nature. This means that two spreading codes would have a negligible cross correlation between them, provided they are synchronized in time. However, time synchronization between signals cannot be guaranteed in the uplink direction, where multiple mobile handsets can be communicating asynchronously with the radio tower. In such a case, the signals from multiple mobile handsets may interfere with each other, resulting in difficulty in de-spreading and separating the original signals from different mobile handsets. This problem is generally not observed in the downlink direction, where one radio tower is co-ordinating transmission to multiple mobile handsets. In the latter case, the radio tower can ensure that the timing synchronization is maintained.
- Secondly, as a result of reflections and diffractions, the multi-path components of signals can be received at the receiver. The orthogonality of these multi-path components cannot be guaranteed, since these can have distorted orthogonality as compared to the original signal. Again, this results in higher cross correlation between

signals received at the radio tower, and hence, difficulties in separating the different user signals.

- Lastly, CDMA-based systems normally have a frequency re-use factor of one. This means that adjacent cells can be using the same frequency. This leads to a problem in transmission on the downlink direction. The problem stems from the fact that the number of spreading codes is finite, and two different mobile handsets in adjacent cells can be allocated the same spreading codes. In such a case, a mobile handset on the border of two cells (where it can receive transmission from both radio towers) cannot figure out if the transmission is for it, or for some other mobile handset in the adjacent cell, which uses the same spreading code.

The process of *scrambling* is used to solve these problems, which provides a means to distinguish between signals from multiple mobile stations in the uplink, and helps to reduce inter-radio tower interference in the downlink. Scrambling is done after the process of spreading at the transmitting end, using pseudo-random codes called scrambling codes. This is as depicted in Fig. 5.11.

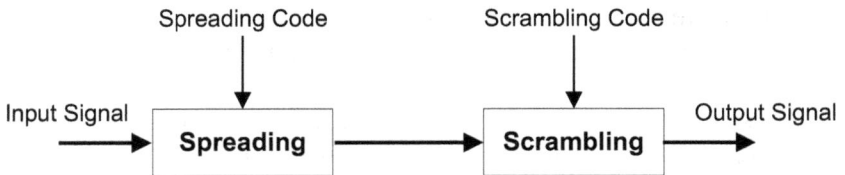

Fig. 5.11 *Spreading and scrambling*

In the uplink direction, each mobile handset is assigned a unique scrambling code. Similarly, in the downlink direction, each radio tower is assigned a unique scrambling code. Like spreading codes, scrambling codes also have high auto-correlation properties. The process of Scrambling a signal using scrambling codes is similar to the process of spreading as depicted in Fig. 5.8. Thus, scrambling is used in CDMA-based systems in addition to spreading, to solve the problems imposed due to lack of synchronization between mobile handsets, multi-path reception of signals, and re-use of spreading codes in adjacent cells.

5.6 CONCLUSION

This chapter looked at two important functions of a communication network, namely, information transmission and providing multiple access. Concepts related to information transmission included a discussion on the nature of the information to be transmitted (analog or digital) and the conversion of information from one form to another (analog to digital and vice versa). Also, various transmission media used for information transfer were discussed. The chapter introduced the concept of bandwidth and capacity, and described two principles, the Nyquist Principle and the Shannon Theorem, which define a relationship between the bandwidth of a transmission media, and its capacity.

In the second part of the chapter, concepts related to multiple access were discussed. One of the prime responsibilities of a communication network is to allow multiple users to simultaneously use the network resources. Mechanisms to provide access were discussed, which include FDM, TDM, WDM and CDM. Two examples, for GSM and UMTS mobile networks, were discussed to explain the concept of multiple access, and in particular, to look at the practical use of FDM, TDM and CDM principles.

REVIEW QUESTIONS

Q 1. Describe some of the commonly used mechanisms for conversion of signals from digital form to analog form.

Q 2. Even though increasing phases in the M-ary PSK scheme gives a higher data rate, what are the practical problems in increasing the phases beyond a limit?

Q 3. Digital information is to be transmitted over a transmission media as binary signals. The bandwidth of the transmission media is 4 KHz. What is the maximum data rate that can be achieved by using the Nyquist Principle?

Q 4. Consider the same problem as in the question above. What is the maximum data rate that can be achieved as per the Shannon Theorem, if the SNR value for the transmission media is 100 dB?

Q 5. Why is GSM said to follow the hybrid FDMA/TDMA scheme?

Q 6. UMTS networks (described in section 5.5) define two modes of operation, namely Time Division Duplex (TDD) and Frequency Division Duplex (FDD). In the FDD mode of operation, separate 5 MHz carrier frequencies are used in the uplink (mobile handset to radio tower) and the downlink (radio tower to mobile handset) direction. In the TDD mode of operation, only one 5 MHz carrier is time-shared between the uplink and the downlink traffic. It is sometimes said that the FDD mode of operation can be considered as a hybrid scheme that uses the CDM and FDM schemes. On the other hand, the TDD mode of operation is said to use a hybrid scheme based on CDM, FDM and TDM. Justify the statement.

Q 7. UMTS networks use the principles of CDMA to provide multiple access. Two different types of codes are defined, namely, spreading codes and scrambling codes. Discuss the reason for the need for two different types of codes.

FURTHER READING

[Gen A.Tanenbaum] is an excellent reference for more information on concepts related to network data transmission and multiple access. [Gen W.Stallings] can also be referred to for gaining a better understanding of concepts related to data transmission. The book covers concepts related to digital and analog transmission in sufficient detail. For the GSM case study presented in the chapter, [Wireless M.Mouly] and [3GPP 05.02] can be referred to for more details. [Wireless S.Kasera], [Wireless A.Viterbi] and [Wireless H.Holma] are some of the good reference books that can be referred to for more details on the UMTS case study presented in the chapter. Also, for the UMTS case study, [3GPP TS 25.214], [3GPP TS 25.401] and [3GPP TS 25.922] are some of the 3GPP specifications that can be referred to.

DATA LINK CONTROL

6.1 INTRODUCTION

The scope of the physical layer is limited to the transmission of signals over a physical media (e.g. coaxial cable). However, in order to ensure reliable and efficient communication between the nodes connected to the physical media, a layer needs to exist in the communicating nodes, which can supplement the physical layer with the following functions:

- To remove communication errors that occurred during transmission over the physical media—in other words, this function ensures that the data received at the other end of the transmission link is error-free.
- To adapt transmission as per the finite data rate of the communication link.
- To adapt the transmission rate at the transmitting end according to the processing capacity of the receiving end.

The data link layer is thus used above the physical layer (at layer 2) to provide the above-mentioned functions for data link control. This chapter provides a detailed description of the data link layer and its associated data link control functions. However, before a discussion on the data link layer, the concept of data link and its associated line configurations is first discussed in the following section.

6.2 DATA LINK LINE CONFIGURATIONS

A data link can be defined as a communication link (or physical media) connecting two or more network nodes directly. The line configuration of a data link (or communication link) is defined in terms of two characteristics, namely its *topology* and its *duplexity*. Each of these concepts is introduced in the following sections.

6.2.1 Topology

The topology of a data link refers to the physical arrangement of nodes on the link. A link connecting only two nodes is called a point-to-point link [refer to Fig. 6.1(a)]. In Fig. 6.1(a), two point-to-point links are depicted, which are connecting Node 1 to Node 2 and Node 3, respectively. A second kind of topology is one in which a single communication link connects multiple nodes. This is referred to as a multi-point link topology [refer to Fig. 6.1(b)]. In this topology, a shared single link is used to connect more than two nodes in the network. Ethernet, one of the most common data link layer technologies, uses the multi-point topology.

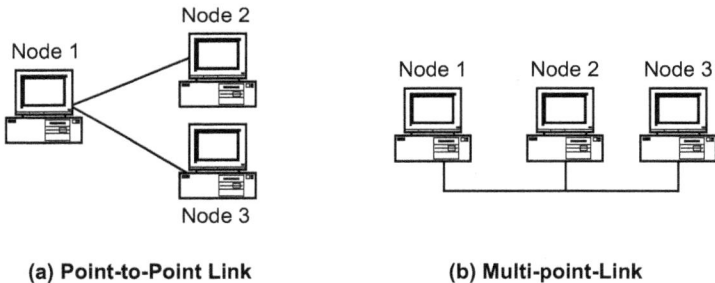

(a) Point-to-Point Link   (b) Multi-point-Link

Fig. 6.1   *Data link topologies*

6.2.2 Duplexity

The duplexity of a data link refers to the direction and timing of the signal flow on the link. In *simplex* transmission, the signal flow is always in one direction. For example, an input device such as a card reader can be attached to a host machine via a simplex data link, such that the card reader device can only transmit to the host, but never receive information from the host. Hence the flow of signals is only in one direction on the link, from the card reader to the host.

A *half-duplex* link, on the other hand, can transmit and receive in both directions. However, at any node, the transmission and reception functions cannot both take place simultaneously. In other words, only one node in the link can transmit, while the other(s) can only receive information. They cannot transmit while one receiver is already transmitting on the link. Signals on the link can thus flow in both the directions, but the flow in both directions cannot be simultaneous. This mode is also referred to as *two-way alternate*, suggestive of the fact that two stations on a half duplex link must alternate in transmitting.

A third category of data links with respect to duplexity includes *full duplex* links. In full duplex links, two nodes connected by a point-to-point link can simultaneously send and receive data from each other. Thus, this mode is also referred to as a *two-way simultaneous* mode. When signals are transmitted as digital signals on the physical media, full duplex usually requires two separate transmission paths (e.g. two twisted pair cables), one for each direction of transmission.

6.3 DATA LINK LAYER FUNCTIONS

The data link layer implements the following functions:

- Grouping of bits on the physical layer into frames.
- Detecting and correcting transmission errors.
- Regulating the flow of frames (and hence the data rate) on the data link such that fast senders do not swamp receivers.
- Carrying out functions for data link management

Each of these functions performed by the data link layer is described in the following sections.

6.3.1 Framing

At the physical layer, information is transmitted as a raw bit stream. Due to transmission errors in the physical path, the bits received at the other end of the physical media may have different values than what was transmitted. One of the functions of the data link layer is therefore to detect, and if possible, correct the errors in the bit stream received from the physical layer. The upper layers, thus, do not need to worry about bit-error detection, since this function is performed at the data link layer.

In order to achieve this function, the data link layer computes a checksum over each chunk of information that it transmits over the physical layer. The chunk of information over which the checksum is computed is called a *Frame*. While the error control algorithms for computing a checksum on frames is covered in the following section (section 6.3.2), this section discusses the process of framing, which converts the raw bit stream into discrete frames at the data link layer.

Multiple approaches are possible for the framing of information. Some of these approaches are:

- **Insertion of time gaps:** This approach is similar to the approach used to differentiate between words in a sentence by inserting spaces. However, the problem with this approach is that it doesn't take into consideration the variations in the transmission link delays. Due to variations in the propagation delays on the transmission link, one cannot rely on the time gaps between consecutive bits to delineate the boundary of frames.

- **Use of character count:** In this approach, a header is attached to the beginning of each frame. A field in this header stores the length of the frame as a count of the number of characters in the frame. At the destination, the data link layer sees the character count, and it knows how many characters follow the header. Thus, frame boundaries can be identified by using the header and the character count. The problem with this approach is that in case the count of any single frame gets corrupted, the receiver would lose synchronization with the transmitter and would not be able to delineate frame boundaries (refer to Fig. 6.2).

- **Character stuffing approach:** This approach caters to the problem of loss of synchronization between the transmitter and receiver, when the character count method is being used for framing, after a corrupted frame is received. Re-synchronization between the sender and the receiver can be achieved by having a special sequence of ASCII characters at the start and end of the frame, which can explicitly identify the frame boundary. In the character stuffing approach, a special sequence of ASCII characters (DLE STX) is used at the start of the frame and

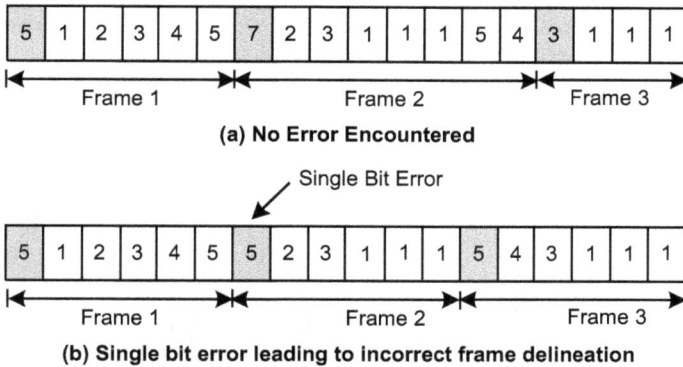

(a) No Error Encountered

(b) Single bit error leading to incorrect frame delineation

Fig. 6.2 *Character count method*

another sequence of ASCII characters (DLE ETX) is used at its end. DLE is an acronym for Data Link Escape, STX for Start of Text, and ETX for End of Text. To re-synchronize, the receiver has to look for DLE STX or DLE ETX characters. Even in case of an error in the frame, the receiver can search for these characters to achieve synchronization once again.

The only problem with the approach described above is that it is possible that the DLE characters appear in the actual data to be transmitted. If this happens, it would interfere with framing, since the receiver will incorrectly interpret the DLE characters in the actual data as the end of the frame, or the start of a new frame. One way to solve this problem is to have the sender's data link layer insert an additional ASCII DLE character just before each occurrence of the DLE character in the actual data. This extra DLE character can be used by the data link layer at the receiving end to identify the DLE character as part of the data. The receiving end data link layer can then remove the extra DLE character inserted by the sending end, before the data is given to the network layer. Thus a DLE STX or DLE ETX used for framing can be distinguished from a DLE STX/ETX in the data itself, by looking for the presence (or absence) of an additional DLE character. In simple terms, DLE characters in the data are always doubled, i.e. an additional DLE will precede each DLE character in actual data. Because of this additional stuffing of DLE characters within the actual data, this technique is called character stuffing (refer to Fig. 6.3).

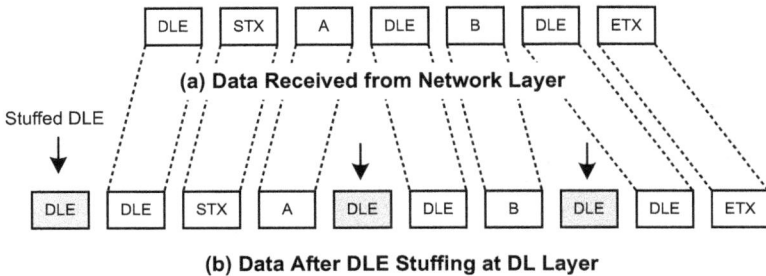

(a) Data Received from Network Layer

Stuffed DLE

(b) Data After DLE Stuffing at DL Layer

Fig. 6.3   *Character stuffing*

A major disadvantage of using this framing method is that it is closely tied to 8-bit characters or the ASCII character code, and cannot be used with any other scheme for the encoding of data.

- **Bit stuffing approach:** The bit stuffing approach can be used for framing any kind of data, i.e. it need not be based on ASCII character coding. Each frame begins and ends with a special bit pattern, namely 01111110 (a zero, followed by a sequence of six ones, followed by another zero bit). This is somewhat similar to the DLE characters used in the character stuffing approach for frame delineation. In case the actual data stream also contains this special bit pattern (01111110), bit stuffing is used to break this pattern. To do this, the sender's data link layer parses the incoming data stream and looks for any sequence of five consecutive 1's in the bit stream. If it encounters any such sequence of five consecutive 1's, it automatically stuffs an additional 0-bit into the outgoing bit stream. This pre-empts the presence of the special bit pattern (01111110) in the actual data stream.

  At the receiving end, whenever the receiver encounters a sequence of five consecutive 1 bits, followed by a 0 bit, it automatically deletes the 0 bit. Just as in the character stuffing approach, bit stuffing and stripping is completely transparent to the network layer at both the ends (the transmitting and receiving ends), and is performed at the data link layer (refer to Fig. 6.4).

0 1 1 0 1 1 1 1 1 1 1 1 1 1 1 1 1 1 1 1 1 0 0 1 0

(a) Data Received from Network Layer

Stuffed Bits

0 1 1 0 1 1 1 1 ⬚0⬚ 1 1 1 1 1 ⬚0⬚ 1 1 1 1 1 ⬚0⬚ 1 0 0 1 0

(b) Data after Bit Stuffing at DL Layer

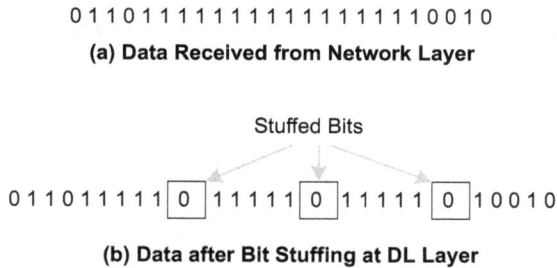

Fig. 6.4 *Bit stuffing*

6.3.2 Error Control

Errors are commonly observed when transmitting over communication links. It is possible that the receiving node incorrectly receives a bit in a frame as zero, when it was originally transmitted as a one, and vice versa. Such bit errors are introduced by signal attenuation and electromagnetic noise in the communication link. A major source of noise for data transmission is impulse noise. These pulses or spikes on the line typically have a duration of 10 msec. To human ears, such noises would sound like little chickens. To a 9600 bps line, they would sound like the death knell for 96 bits. Arcing of relays and other electromechanical wheezes at switching offices causes most of these spikes or pulses. Another major source of errors arises from the fact that the amplitude, propagation speed and phase of signals are all frequency-dependent. The telephone system, for example, Fourier-analyzes all signals, distorts each frequency component separately and then re-combines them at the end. This causes distortions and errors in the system. Also, cross talks can exist between two wires that are physically adjacent, and this can be a third source of errors.

In order to overcome these transmission errors, protocols are introduced in the data link layer, which detect such errors and correspondingly take appropriate action (e.g. discard the frame, send negative acknowledgement to sender of frame, correct the frame if possible, etc.). Thus error control includes one or more of the following steps:

- Detection of corrupted frames
- Correcting the errors in the corrupted frame, in case sufficient information is present in the received frame for error correction

- Sending negative acknowledgement to the sender, in case the received error frame cannot be recovered, which may trigger re-transmission at the sender's end
- Managing timers and sequence numbers so as to ensure that each frame is ultimately passed to the network layer at the destination exactly once.

At the data link layer, two kinds of strategies exist for dealing with errors. One strategy is to include enough redundant information along with each block of data that is sent. This redundant information is used to enable the receiver to deduce the required correction in the received information, in case an error occurs. Thus, in other words, this strategy supports both error detection and correction. The second strategy, on the other hand, is to only include enough redundant information to allow the receiver to detect any error in the received information, but not enough to correct the error. Thus, by using this strategy, the receiver can detect errors, and optionally, request the sender for re-transmission of the error frame. It cannot, by itself, correct the error in the received information.

The former strategy is based on the use of error-correcting codes, while the latter is based on the use of error-detecting codes. Further details of error detection and correction, and the corresponding codes used, are covered as part of section 6.9.

6.3.3 Flow Control

Flow control is a means used to synchronize between the sender's sending capacity and the receiver's receiving capacity. In order to understand the importance of flow control, consider the case of an automobile plant assembly line. Two basic issues govern the design of this assembly line. First, each unit in the assembly line must get sufficient time to finish its job. Second, if a faulty unit begins to perform below capacity, either replacements must be made available for the faulty unit, or the production rate must be suitably reduced to take into account the reduction in the processing rate of this faulty unit. Everything will work smoothly as long as either of these two criteria is met. However, if none of the criteria is met, the faulty unit will become a bottleneck in the entire assembly line.

Incomplete cars, waiting to be serviced by the faulty unit will overload the warehouse. The net effect will be a chain reaction moving backwards from the faulty unit, resulting in the reduction of the production capacity of all the preceding units.

The above example aptly highlights the need for proper flow control in the assembly line of an automobile plant. The same theory of flow control is applicable in the case of communication networks as well. In either scenario, the over-riding concern is to ensure that no entity in the line of action is stressed beyond its servicing capability.

In communication networks, flow control is a mechanism used to control the flow of data between the sender and the receiver so that the receiver's buffers do not overflow. Flow control is necessary when a fast processor (sender) is communicating with a slow processor (receiver). In this case, the buffers at the receiver will slowly begin to build up. Since the storage capacity at the receiver is also limited, flow control is required to ensure that the buffers at the receiver are not exhausted, which would lead to loss of information at the receiver.

Flow control mechanisms are broadly classified into two categories: *window-based flow control* mechanisms and *rate-based flow control* mechanisms. In a window-based flow control mechanism, the focus is on tackling the memory-related bottlenecks. In rate-based flow control mechanisms, on the other hand, the focus is on tackling the link and processing speed bottlenecks instead. Further details on the two categories of flow control mechanisms are provided in section 6.8.

6.3.4 Link Management

Another important function of the data link layer is to manage the administration of the data link, and to discipline the use of the data link. For a half duplex link, this means ensuring that only one station transmits at any point of time. On either a half or a full duplex link, this means that a station should transmit only when it knows that the intended receiver is prepared to receive. For connectionless services, this administration is minimal. However, for connection-oriented services, it is far more complex.

Consider the case of a multi-point data link where several different machines share a common communication channel (e.g. shared bus). Typically, communication on this common channel is made feasible by having one of the machines on the common channel perform the role of a master node (also known as primary node) and all the other machines work as slave nodes (also known as secondary nodes). The primary node first sends a short frame, called a poll frame, to the first secondary node on the communication channel, asking if it has any data to send. If the secondary node has data to transmit, it sends the data; otherwise the primary node polls the next secondary node, and so on.

The master-slave protocol described above is only one kind of protocol used for link management. Multiple other varieties of protocols exist for link management. For example on LANs where there is no concept of master-slave nodes, protocols like CSMA/CD and token ring are used. The specifics of these link management protocols are covered later as part of section 6.7.

6.4 SERVICES OFFERED TO NETWORK LAYER

As per the OSI reference model, a lower layer in the reference model is responsible for providing certain well-defined services to its higher layers. Accordingly, the data link layer provides its service to the network layer for the transfer of information between any two nodes connected via a data/communication link. The data link layer can offer the following different categories of services to the network layer:

- **Unacknowledged connectionless service:** This service allows the sending node to send independent frames towards the receiving node, without requiring the receiving node to acknowledge them. If a frame is lost due to noise in the communication link, no attempt is made by the data link layer to recover the lost frame by using re-transmission means. Thus, the responsibility of handling the loss of frame is left entirely with the network layer.

  This class of service is appropriate for two kinds of situations. The first scenario where this class of service can be used is where the transmission links are assumed to have a very low error rate and frame loss rate. In such a scenario, it is assumed that most frames will reach the destination correctly, without errors. Thus, the data link layer is freed of the overheads of frame level bookkeeping. Instead, the upper layers handle recovery from any rare loss of frames. The other scenario where this class of service is used is for transmission of real-time traffic, such as speech. For real-time traffic, it is assumed that re-transmission of lost frames is not an ideal solution, since re-transmission causes delays, which are generally not tolerable for real-time traffic. Thus, the upper layers, depending on the kind of real-time traffic, handle any loss of frames. The data link layer is freed from any kind of re-transmission of lost frames.

- **Acknowledged connectionless service:** This service differs from the unacknowledged connectionless service in the sense that the receiving node acknowledges the independent frames sent by the sending node. In this way the sender knows whether or not a frame has reached the correct destination. If the frame has not reached the destination within a specified time interval, the sending node can re-transmit it.

- **Connection-oriented service:** In this service, each frame sent over the communication link is tagged with a sequence number. The sequence number is used by the data link layer to ensure that each frame sent by the sender is indeed received at the other end of the data link. Furthermore, the use of sequence numbers allows the data link layer to ensure that each frame is delivered to the network layer only once (no duplication of frames, as a result of re-transmissions). Also, sequence numbers allow the data link layer to re-sequence out of sequence frames, before delivery to the network layer, to ensure that all frames are received at the network layer in the order/sequence in which they were transmitted.

The salient features of this type of service (in-sequence, exactly once delivery of frames) closely resemble the features of a circuit-switched service. In some sense, it can be construed that the sending and receiving data link layers establish a logical connection before any data is transferred between them. Thus, there are three different phases in the use of this service. In the first phase, the logical connection is established by having both sides initialize their variables and counters (sequence numbers, etc.) needed to keep track of the frames which have been received and which have not. In the second phase, the frames are actually transmitted. In the third phase, the connection is released, freeing up the buffers, clearing variables and freeing up other resources.

6.5 DLC PROTOCOL LAYERING

The IEEE 802 standard defines the data link layer to consist of two sub-layers. This sub-layering of the data link layer is more evident in the case of Local Area Networks. The sub-layers of the data link layer are:

- Logical Link Control (LLC) sub-layer
- Media Access Control (MAC) sub-layer

The functions, features, protocol and services of the Logical Link Control layer are defined in the [IEEE 802.2]. Media Access Control layer protocols are defined in different IEEE standards. Some of the popular protocols used in the MAC sub-layer include CSMA/CD protocol defined in [IEEE 802.3]. CSMA/CD is the protocol used in Ethernet. Another important MAC sub-layer protocol is the token passing protocol, used in token ring networks, and defined in [IEEE 802.5].

Figure 6.5 depicts the sub-layering of the data link layer, using the protocol defined by the IEEE 802.3 standard as the protocol for the MAC sub-layer. As depicted in Fig. 6.5, the LLC sub-layer is the upper sub-layer of the data link layer.

The LLC sub-layer performs the following functions:

- Managing the communication on the data link
- Link addressing
- Defining Service Access Points (SAPs)
- Sequencing.

The LLC provides a mechanism for the upper layers (network layer and above) to interface with any type of MAC sub-layer. Similar concepts related to protocol layering apply to the sub-layers of Data Link Layer as well, i.e. the data field of the MAC sub-layer frame contains the LLC sub-layer PDU.

Network Layer	Unix IP (SAP: 80)	NetBios (SAP: F0)	Novell IPX (SAP: E0)
Data Link Layer	IEEE 802.2 Logical Link Control Layer		
	IEEE 802.3 CSMA/CD Media Access Control Layer		
Physical Layer	802.3 -10Base5	802.3a -10Base2	802.3i -10BaseT

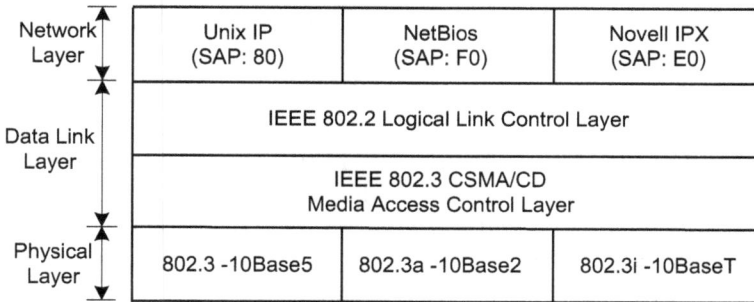

Fig. 6.5 *Data link control sub-layers*

The MAC sub-layer is required in local area networks (LANs), where most often, the physical communication link is a shared medium. This is typical of LANs which have the multi-point link topology, discussed in section 6.2.1. Such networks typically use broadcast mechanisms to transmit data over the shared communication link. However, the key issue in such networks is to control the access to the communication link. Since there can be multiple nodes on the network which wish to simultaneously transmit over the shared communication link, some sort of protocol is required to share the link between all potential senders. The MAC sub-layer performs this important task of resolving the contention on the available shared channel and providing access to all nodes to use the channel. The protocols used for this purpose are sometimes called *Media Access Protocols* or *Random Access Protocols*.

With the above description, it becomes evident that the MAC sub-layer is required only for networks having multi-point link topologies. This sub-layer is generally not required for point-to-point link topologies, since in these topologies, there is no shared medium on which the access resolution is required. Thus, the MAC sub-layer is generally relevant in the case of LANs, but not so much for WANs, which generally use point-to-point topologies.

6.6 LOGICAL LINK CONTROL (LLC)

Logical Link Control (LLC), as defined in IEEE 802.2 standards, forms the upper sub-layer of the data link layer. The main function of the LLC sub-layer is to present a uniform interface to the upper layers of the protocol stack, usually the network layer.

The LLC sub-layer defines the concept of a *Service Access Point (SAP)*. A SAP is a kind of port used by the network layer protocol to access services provided by the data link layer. The SAP also acts as an identifier for the user of the data link layer. In case of multi-protocol network layers, each network layer protocol will have its own SAP with the data link layer (see Fig. 6.5). For example, Unix's TCP/IP, Novell's SPX/IPX and IBM's Netbios protocol use different SAPs to access the data link layer services.

Figure 6.6 depicts the PDU format for the LLC sub-layer. As depicted in the figure, the LLC sub-layer appends an 8-bit (Destination Service Access Point (DSAP)) and Source Service Access Point (SSAP) label to the network layer PDU. While DSAP represents the network layer protocol at the receiving node, SSAP identifies the network layer protocol at the transmitting node.

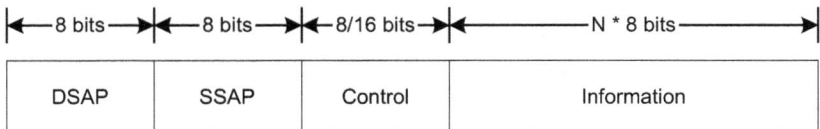

←—8 bits—→	←—8 bits—→	←8/16 bits→	←——————N * 8 bits——————→
DSAP	SSAP	Control	Information

Fig. 6.6 *LLC PDU format*

The control field consists of one or two octets that are used to define command and response operation codes, and for containing sequence numbers, when required. The contents of the control field are described in greater detail in section 6.6.3. The information field in the LLC sub-layer PDU consists of an integral number (including zero) of octets, which are received from the upper layers. The following sections discuss the types of LLC operations and classes, and the types of LLC PDUs in greater detail.

6.6.1 Types of LLC Operations

As discussed in section 6.4, the data link layer provides three different forms of service to the network layer. Each of these forms of service is classified by using different operation types, as described below:

- **Type 1 operation:** LLC Type 1 operation provides the unacknowledged connectionless mode service to the network layer, which was discussed in section 6.4. In the unacknowledged connectionless mode service, the data transfer can be point-to-point, multi-cast, or broadcast.

- **Type 2 operation:** LLC Type 2 operation provides the connection-oriented mode service to the network layer. The connection-oriented mode service was also discussed in section 6.4. This set of service provides the means for establishing, re-setting, and terminating data link layer connections. LLC Type 2 operation supports only point-to-point connections, and does not have the inherent support for multi-cast or broadcast. The following are some of the services offered by the LLC Type 2 operation:

    - The connection establishment service provides the means by which a network entity can request, or be notified of, the establishment of data link layer connections.

    - The connection-oriented data transfer service provides the means by which a network entity can send or receive information over a data link layer connection. This service also provides data link layer sequencing, flow control and error recovery.

    - The connection re-set service provides the means by which established connections can be returned to the initial state. When returning to the initial state, all variables and states maintained for the connection (e.g. sequence number, etc.) are also re-set to the initial value.

    - The connection termination service provides the means by which a network entity can request, or be notified of, the termination of data link layer connections.

    - The connection flow control service provides the means to control the flow of data associated with a specific data link layer connection, across the network layer/data link layer interface.

- **Type 3 operation:** LLC Type 3 operation provides the acknowledged connectionless mode service. In this mode of operation, the LLC sub-layer implements an acknowledgement scheme in which the receiving LLC sub-layer acknowledges the receipt of each data frame to the sending LLC sub-layer. Like the LLC Type 2 operation, the LLC Type 3 operation is also only valid for point-to-point links.

6.6.2 Classes of LLC

On the basis of the operations that an LLC sub-layer supports, the following four types of LLC classes are defined:

- **Class I LLC**: A Class I LLC supports only LLC Type 1 operations.
- **Class II LLC:** A Class II LLC supports both LLC Type 1 and Type 2 operations, but not Type 3 operations.
- **Class III LLC**: A Class III LLC supports both LLC Type 1 and Type 3 operations, but not Type 2 operations.
- **Class IV LLC**: A Class IV LLC supports all types of LLC operations (i.e. LLC Type 1, Type 2 and Type 3 operations).

From the very definition of the four types of LLC classes, it is evident that all LLC sub-layers on a LAN shall have at least one type of operation in common, which is the Type 1 operation. However, unlike Class I LLC sub-layers, Class II, Class III and Class IV LLC sub-layers are capable of going back and forth between using Type 1 operation or Type 2 operation (for LLC Class II and IV only) or Type 3 operation (for LLC Class III and IV only) on a PDU-to-PDU basis, if necessary.

6.6.3 LLC Sub-layer Protocol Data Units

At the LLC sub-layer, three kinds of LLC PDUs are defined. All these PDUs have the same common PDU format that was depicted in Fig. 6.6. However, different PDUs perform different functions, and differ slightly with respect to the PDU content. The three kinds of LLC PDUs are:

- **Un-numbered (U-Format) PDU:** U-format PDUs define the use of an 8-bit control field. This PDU format is used mainly for a Type 1 (connectionless) operation. The PDUs are not numbered, and they are sent with the hope that they will successfully reach their destination. U-Format PDUs can be used to transfer LLC sub-layer commands and responses or to transfer user level information data.

- **Information transfer (I-Format) PDU:** An I-format PDU defines the use of a 16-bit control field. It is used for the purpose of transfer of user information for Type 2 (connection-oriented) operations. Hence, the control field of an I-frame PDU also carries the sequence number for the frame. For Type 2 operations, only I-format PDUs are allowed to carry user information, and not U-format PDUs.
- **Supervisory (S-Format) PDU:** An S-format PDU also defines the use of a 16-bit control field. S-format PDUs are used for supervisory functions at the LLC sub-layer, which includes the handshake function for connection establishment in case of Type 2 operations. The control field in the S-Format PDUs is used for acknowledging the I-Format PDUs, requesting for re-transmission of PDU, and for requesting a temporary suspension of transmission (for example, in case of a buffer-full state at the receiving end).

The control field in the LLC sub-layer PDU is used to decide the format (U, I or S-format) of the LLC PDU.

6.7 MEDIA ACCESS CONTROL (MAC)

The MAC sub-layer is the lower sub-layer of the data link layer, and it sits below the LLC sub-layer. As the name suggests, the main function of the MAC sub-layer is to provide media access control. In order to understand this concept further, consider the case of a broadcast network in which all connected nodes are capable of transmitting data frames. It is likely that two (or more) nodes transmit a data frame at the same time, thus leading to a collision. When this happens, the receiving nodes only receive garbled data, which results from the collision of two or more original frames. Typically when there is a collision, the signals of the colliding frames become inextricably tangled together, and all colliding frames are lost. Thus the bandwidth of the broadcast channel can be considered wasted during the collision interval. In this scenario, it is evident that if multiple nodes wish to transmit frames frequently and simultaneously, many of these transmissions will result in collisions, leading to much loss of link bandwidth.

In order to overcome this problem, it is necessary to co-ordinate and regulate the data transmission of active nodes. This co-ordination of data transmission between multiple active nodes is the responsibility of multiple access protocol, which is implemented as part of the MAC sub-layer in each transmitting node. Different mechanisms exist to co-ordinate data transmission over a shared channel, and correspondingly, different media access protocols are possible. Some of these protocols are explained in the following sections.

6.7.1 Static Channel Allocation Protocols

One way to co-ordinate between multiple active senders is to perform a partitioning of the shared channel in the time or frequency or code domain, and then assign these partitions to different active senders. The static channel allocation protocols like time division multiple access, frequency division multiple access and code division multiple access were discussed in detail in Chapter 5.

6.7.2 Random Access Protocols

Random access protocols are based on the principle that each transmitting node on the shared channel also acts as a receiver of the transmitted information. In case of a collision due to multiple transmitting nodes, each transmitting node will receive garbled data, and will hence be able to detect the collision. Once the collision is detected, each node involved in the collision chooses a random value, which is the waiting period, at the end of which the node tries re-transmission. In other words, each node involved in the collision introduces a random delay before re-transmitting the frame. Since the individual nodes choose the random delays independently, it is possible that one of the nodes will pick a delay value that is sufficiently less than the delays of the other colliding nodes and will therefore be able to sneak its frame into the channel without collision.

Many popular random access protocols exist for use in local area networks. All of these use the same inherent principles for transmission (and re-transmission), as discussed in this section. However, subtle differences in these different approaches leads to differences in the throughput achieved by each protocol. The following sections discuss some of these protocols in greater detail.

6.7.2.1 Aloha Protocol

The simplest known protocol for random access is the Aloha protocol, which works on the same principles as discussed in section 6.7.2. Its operation is carried out in phases as described below:

- **Phase 1—Transmission phase:** In this phase, a transmitting node immediately transmits a fresh frame (i.e. a frame for which no earlier transmission attempts were made) entirely onto the broadcast channel, without any delay. If no collisions are detected, no re-transmissions are required for the frame, and the procedure ends with Phase 1 itself. However, if the transmitted frame experiences a collision with one or more other transmissions, Phase 2 of the two-phase operation is performed.

- **Phase 2—Re-transmission phase:** In this phase, the transmitting node tries re-transmission of the collided frame in a probabilistic manner. The transmitting node either immediately re-transmits the collided frame with a probability of p or waits for a frame transmission time interval (with probability $1-p$), before re-transmission. Here, the frame transmission time interval is defined as the time it takes to transmit one frame of information, if it were assumed that all frames are of equal size. At the end of the frame transmission time interval, the node performs the same calculation, i.e. either it transmits the frame with probability p, or waits for another frame time with probability $1-p$. The value of p differs for different nodes, thus giving them different priorities for re-transmission. However, there is still a possibility that re-transmitted frames collide with frames that are transmitted by other nodes as part of Phase 1 transmission, or with nodes which simultaneously perform re-transmission. If this happens, all nodes with colliding frames (re-)enter phase 2 operations.

If we assume that the frame transmission time interval is one unit of time, and if that a node 'i' begins the frame transmission at time t_o, then a collision may take place if any other nodes also transmits a frame within the time interval $[t_o\text{-}1, t_o\text{+}1]$ (see Fig. 6.7).

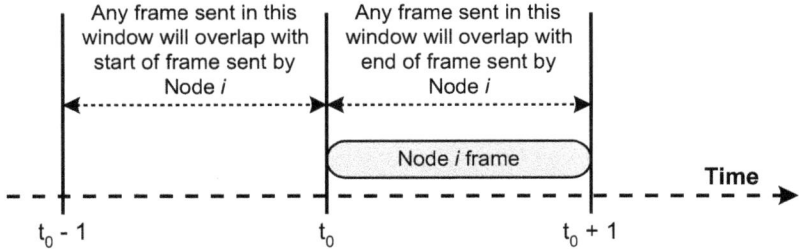

Fig. 6.7 *Interfering transmission in pure Aloha*

With this background, we can calculate the overall probability of a successful re-transmission in phase 2 of Aloha as follows. Assume that in phase 1, N nodes suffered a collision. Thus, all N nodes will enter phase 2 of Aloha, and perform frame re-transmission. Also, to simplify calculations, let us assume that no other node performs frame transmission (as part of phase 1) during this time interval. Now, at any given point in time, the probability that a particular node transmits a frame is p. Suppose this frame transmission begins at time t_o. In order to ensure that this frame is successfully transmitted, no other nodes should begin their transmission in the time interval $[t_o\text{-}1, t_o\text{+}1]$. The probability that all other nodes do not begin transmission in this time interval is $(1\text{-}p)^{2(N\text{-}1)}$. Thus the probability that a successful re-transmission occurs for a particular node is $p(1\text{-}p)^{2(N\text{-}1)}$.

6.7.2.2 Slotted Aloha

The Aloha protocol discussed in section 6.7.2.1 is also sometimes called pure Aloha. An improved variant of the pure Aloha protocol is the *Slotted Aloha* protocol. As shown in Fig. 6.7, in pure Aloha, nodes can begin transmission at any point of time, and in order for to ensure that a frame is successfully transmitted, no other nodes should begin their transmission in the time interval $[t_o\text{-}1, t_o\text{+}1]$. Slotted Aloha improves upon this aspect of pure Aloha, by introducing the following constraints:

- Time is divided into slots, where-in each slot is equal to the time it takes to transmit one frame of information (frame transmission time interval).
- Nodes are only allowed to begin frame transmission at the start of a time slot. If information is available at a node only in the middle of a time slot, the node needs to wait for the start of the next time slot before it can transmit the frame.
- All nodes are synchronized with each other so that each node knows the time slot boundaries.
- If two or more frames collide in a slot, then all the nodes detect the collision event before the end of the time slot.

Thus, unlike pure Aloha, where-in for a successful frame transmission, no other node must transmit frame in time interval $[t_0\text{-}1, t_0+1]$, in slotted Aloha, this time interval is reduced to half. In slotted Aloha, the only requirement for a successful frame transmission is that no other node should begin frame transmission in the same time slot (e.g. in $[t_0, t_0+1]$) As a result of this change, the probability of a successful frame transmission is doubled, and becomes $p*(1\text{-}p)^{(N\text{-}1)}$. Thus, slotted Aloha provides almost double the throughput achieved with a pure Aloha protocol.

6.7.2.3 CSMA/CD

In both slotted and pure Aloha protocols, a node's decision to transmit is made independently of the activity of the other nodes attached to the shared channel. In other words, a node neither pays attention to whether another node is already transmitting when it begins to transmit, nor does it stop its transmission if another node begins transmission in the overlapping time interval. While this makes the Aloha family of protocols simpler to implement, it also leads to lower overall throughputs that are achieved over the shared channel. Another family of protocols, which are quite different from the Aloha protocols, is the Carrier Sense Multiple Access (CSMA) and Carrier Sense Multiple Access with Collision Detection (CSMA/CD) protocols. The CSMA family of protocols is based on the following principles:

- **Listen before transmission:** In networking terms, this is also called *carrier sensing*, i.e. a node listens to the channel before it begins transmission. If another node is already transmitting a frame onto the channel, the node waits for a random amount of time before it begins to sense the channel again for transmission. Only if the channel is sensed to be idle, a node begins its frame transmission. Both CSMA and CSMA/CD implement carrier sensing.
- **Stop transmission in case of collision:** In networking terms, this is also called *collision detection*, i.e. a transmitting node also listens to the channel while it is transmitting. If it detects that another node is also transmitting simultaneously, it stops its transmission. A re-transmission protocol is used to determine when to attempt re-transmission of the frame. Collision detection is implemented in the CSMA/CD protocol.

While carrier sensing enables a node to determine when a shared channel is free for transmission, there might still be a collision once the node begins transmission after carrier sensing. Although rare, this scenario may occur when two or more nodes simultaneously sense the carrier, and find it to be free. They can then all start transmission together, leading to a collision. Hence, CSMA/CD also implements mechanisms for collision detection, apart from carrier sensing.

It is simple to calculate the performance of a CSMA/CD network where only one node attempts carrier sensing and subsequent transmission at any point in time (i.e. no two nodes simultaneously sense the carrier, and begin transmission). In this case, near 100 per cent utilization of the shared channel can be achieved, providing almost 10 Mbps of throughput on a 10 Mbps LAN. However, when two or more nodes perform simultaneous carrier sensing, and begin transmission after finding the shared channel free, the throughput of the shared channel decreases as a result of collisions. The fall in utilization and throughput occurs due to wasted bandwidths as a result of collisions and back-off delays. In practical deployments of CSMA/CD, a busy 10 Mbps Ethernet network (which is a shared channel) typically achieves 2–4 Mbps of throughput. The mathematical calculations of the throughput are, however, beyond the scope of this text, and are left for the interested readers to perform.

As with any other random access protocol, with CSMA/CD too, as the number of nodes on the network increases and the utilization of the network increases, an overload condition may occur. In this case, the throughput achieved on the network can be reduced considerably, and much of the capacity will be wasted as a result of collisions and idle back-offs. This is one reason why many engineers use a threshold of 40 per cent utilization to determine if a LAN is overloaded. A LAN with a higher utilization will observe a high collision rate, and possibly an extremely variable transmission time (due to back-offs after collisions). In such scenarios, separating the LAN into two or more segments by using bridges or switches is likely to provide significant benefits.

A drawback of using random access protocols on a shared medium is that the sharing is not necessarily fair. In case each node on the network has small amounts of data to send, the network will exhibit a fair sharing of network bandwidth. However, if one node on the network begins to transmit an excessive number of frames, it may hog the network resources. In Ethernet (which uses CSMA/CD), this effect is known as *Ethernet Capture*. To overcome the problem of Ethernet capture, higher speed transmission links are used (e.g. 100 Mbps). Also, the use of full duplex cabling eliminates the effect altogether. A different family of protocols (from the random access protocols), which provide more fairness in bandwidth sharing, is the family of *Taking-Turns Protocols*. This is discussed in the following section.

6.7.3 Taking-Turns Protocols

As discussed in the previous section, the main drawback of random access protocols is their inability to share the channel resource fairly between all nodes on the network. Taking-Turns protocols provide more fairness in the sharing of network resources as compared to random access protocols. The two desirable properties of a protocol belonging to this family of protocols are:

- When only one node on the network is active (i.e. it has data to transmit), the active node should be able to achieve a throughput of R bps, if required, where R is the maximum bandwidth of the transmission media.

- However, in case M nodes on the network are active, then each active node should be able to realize a throughput of nearly R/M bps.

Most of the random access protocols (Aloha and CSMA protocols) fulfil the first criteria, but not the second. As a result, random access protocols do not provide fairness in the sharing of network resources. This is one important feature that makes all the difference between random access protocols and taking-turns protocols.

Many different protocols belonging to the taking-turns family of protocols are known to exist. The first example is that of a *Polling Protocol*. A polling protocol requires that one of the nodes on the network be designated as a master node. The master node polls each of the other nodes on the network in a round robin fashion, to determine if they wish to transmit data. If any node has data to transmit, it responds to the poll request, and begins data transmission. Once a node has finished transmission, the master node polls the next node for transmission, and this protocol continues endlessly. While the polling protocol eliminates collisions that are common to random access protocols, it also has the following drawbacks:

- **Polling delay:** The protocol introduces polling delay, which is defined as the amount of time required for polling a node and for the node to respond to the poll. As a result of this delay, even if only one node is active, the throughput achieved by the node will be less than the total bandwidth of the shared channel.
- **Single point of failure:** In the polling protocol, if the master node fails, the entire channel becomes inoperative. Thus, the network is known to have a single point of failure.

Another variety of the taking-turns protocol is the *token-passing protocol*. The token-passing protocol eliminates the requirement for having a master node on the network. Instead, a small special purpose frame, known as a token, is circulated among all the nodes in the network, in some pre-defined order. When a node receives a token, it holds onto the token only if it has some frames to transmit; otherwise it immediately forwards the token to the next node. Only a node in possession of a token is allowed to transmit frames on the network. Thus, the token-passing protocol efficiently and fairly shares the network bandwidth, without requiring a centralized controlling node.

While the token-passing protocol looks fairly simple to implement, it has its own share of complications. Consider, for example, what would happen if a node having the token were to fail accidentally. In such a scenario, some recovery procedure would have to be implemented to determine this condition, regenerate a token and get it back into circulation. Nevertheless, despite these complications, the token-passing protocol aptly fulfils the requirements of a taking-turns protocol.

6.8 FLOW CONTROL PROTOCOLS

As discussed in section 6.3.3, flow control is a means to synchronize between the sender's sending capacity and the receiver's receiving capacity. Flow control mechanisms are broadly classified into two categories: *window-based flow control* mechanisms and *rate-based flow control* mechanisms. Each of these two categories of flow control mechanisms is discussed in detail in the following sections.

6.8.1 Window-based Flow Control

Window-based flow control mechanisms limit the maximum amount of data that a sender can send at a given point of time, thereby preventing memory buffer overflows at the receiver. In a typical window-based flow control implementation, the data source maintains a transmission window (see Fig. 6.8). The size of this transmission window is kept equal to (or lesser than) the maximum available buffer space at the receiver. At any point of time, the source can only send as many packets to the receiver, as the size of the transmission window. Thereafter, the sender of the data awaits the acknowledgement message from the receiver for each of these transmitted packets. With each acknowledgement received, the sender is allowed to transmit one more packet of data to the receiver. This protocol ensures that the maximum number of packets that the receiver has to buffer will never be greater than the transmission window-size of the sender.

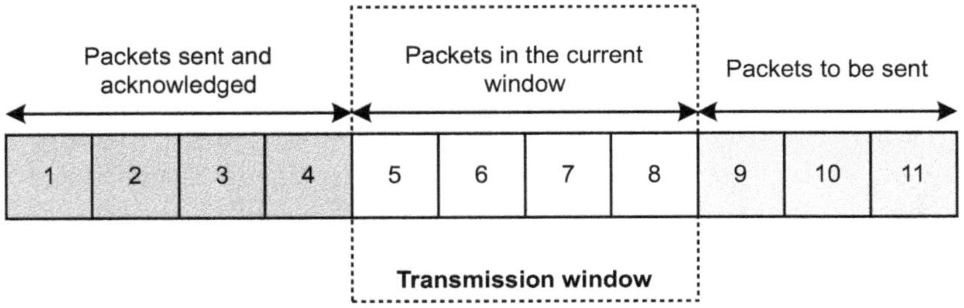

Fig. 6.8 *Flow control using window of size 4*

In Fig. 6.8, packets 1 to 4, which are towards the left of the transmission window, are those that have been acknowledged by the receiver. In other words, these are packets that have been successfully received/processed at the receiver. Within the transmission window, there are two types of packets. The first type are those packets that have been sent towards the receiver, but are waiting for an acknowledgement from the receiver. The second type of messages are those that can be sent to the receiver without waiting for any previous acknowledgements, but are yet to be transmitted. Towards the right of the transmission window are packets that can only be sent after acknowledgement of one or more packets of the current transmission window. Thus, packets to the right of the transmission window are kept buffered at the sender. If the receiver is short of buffer space, it can withhold the transmission of acknowledgements, thus checking the flow of packets from the sender.

Consider what would happen if an acknowledgement is received for packet 5 in the transmission window. In this case, packet 5 will move out of the transmission window, towards its left, since the receiver has already acknowledged it. Also, since the sender can now send one more packet to the receiver, packet 9 will enter the transmission window. This operation can be seen as the transmission window sliding to the right by one packet. Hence, this protocol is also called the *Sliding-Window Protocol*.

In window-based schemes, the window size only acts as a means to prevent the overflow of the buffer space available at the receiver. The sender, however, is free to send the packets within its transmission window at any rate. Since the rate at which the sender can send is not controlled, there can be sudden bursts of data transfer. However, the amount of data that can be sent in a burst is still limited by the window size. In other words, irrespective of the rate at which the data is transferred, a window-based flow control ensures that there is no buffer overflow at the receiver as a result of incoming packets. This explains why window-based flow control schemes tackle memory bottlenecks but not link speed bottlenecks.

The simplest form of the window-based flow control mechanism is called the *stop-and-wait* scheme. In this scheme, the source waits for an acknowledgement from the receiver after the transmission of each packet [see Fig. 6.9(a)]. The acknowledgement from the receiver acts as a permission for sending the next packet. In other words, the stop-and-wait scheme is a special case of window-based flow control, where-in the size of the transmission window is kept fixed as one packet. This scheme has the obvious disadvantage that successive packets can be sent only after a delay of one round trip propagation time, thereby reducing the data rates considerably.

The stop-and-wait scheme can be improved by making only minor modifications. Instead of allowing just one packet, the source is allowed to send multiple packets, wherein the number of packets that the sender can send, without receiving an acknowledgement, is referred to as the window size. The window size is kept static for the entire duration of the transmission, just like in the stop-and-wait scheme. Such a scheme is called a *static window based flow control* scheme [see Fig. 6.9(b)]. As is evident from the figure, the static window-based flow control method reduces the wait time at the source, as compared to the stop-and-wait method.

The choice of window size has a significant effect on the performance of the static window-based flow control scheme. If a small value is chosen, the source frequently waits for acknowledgements. In the extreme, case when the window size is 1, the source waits for an acknowledgement of every packet. On the other hand, if the window size is extremely large, the receiver must have ample buffers to buffer the arriving packets. In the absence of adequate buffers, there will be packet loss.

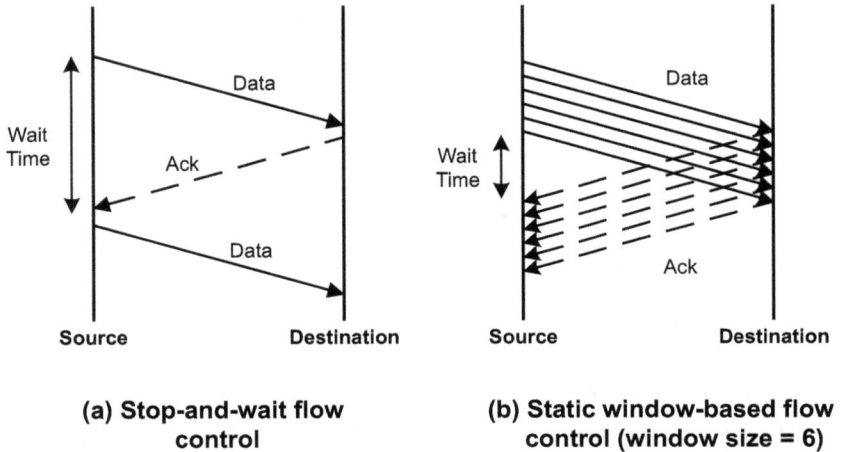

(a) Stop-and-wait flow control

(b) Static window-based flow control (window size = 6)

Fig. 6.9 *Flow control mechanisms*

One of the ways to ensure that the receiver has adequate buffer space and that there is no packet loss, is to use a *credit-based flow control* scheme (see Fig. 6.10). In credit-based flow control, the receiver continuously sends a credit value to the sender. The credit value indicates the amount of data that the sender can send at that point of time. Once the sender has sent the data corresponding to the credit value, it has to wait till it receives more credits from the receiver. The credit-based scheme can also be viewed as a *dynamic window-based flow control* mechanism, wherein the size of the window at any given time depends upon the credits available to the sender.

6.8.2 Rate-based Flow Control

In rate-based flow control mechanisms, instead of limiting the amount of data sent by a sender, the rate at which data is sent is controlled. In this flow control technique, the sender starts with some initial data transmission rate. This initial rate is fixed between the sender and the receiver by using some handshake protocol. Depending upon the resource availability, the receiver may later, request the sender to perform any of the following actions:

- Increase the data transmission rate
- Decrease the data transmission rate
- Continue with the current data transmission rate.

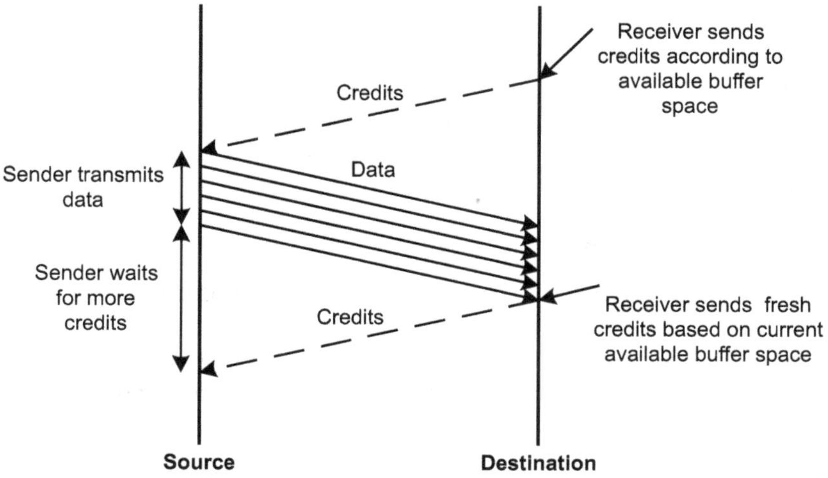

Fig. 6.10   *Credit-based flow control*

The sender then alters its data rate according to the indication received from the receiver.

Rate-based flow control has one major disadvantage. The use of this scheme can lead to a major buffer space problem at the receiver in case an indication requesting the source to reduce transmission rate is lost. Since no acknowledgements or credits are exchanged (as in window-based flow control), the sender will continue to send packets at its current transmission rate. In order to prevent such situations from arising, the receiver is made to periodically transmit data rate information, even if it does not want to alter the sender's data rate. In case the sender does not receive a rate indication for a sufficiently long interval of time, it interprets the receiver to be overloaded and accordingly reduces its data transmission rate.

6.9 ERROR DETECTION AND CORRECTION MECHANISMS

As discussed in section 6.3.2, errors are commonly observed during transmission over communication links. In order to overcome these transmission errors, protocols are introduced in the data link layer, which detect such errors and take appropriate action. Error detection and correction techniques allow the receiver to detect and optionally correct bit errors.

Three different techniques for detecting errors are discussed in the following sections. These techniques use *parity checks*, *check summing methods* and *cyclic redundancy checks* for error detection and correction.

6.9.1 Use of Parity Bits

The use of parity bits is one of the most commonly used mechanisms for error detection. In the simplest of parity schemes, one extra bit of information is appended with each data packet. Two different kinds of one-bit parity schemes are defined, namely, *one-bit even parity* and *one-bit odd parity*. A one-bit even parity scheme is depicted in Fig. 6.11. As depicted in the figure, the data to be transmitted consists of d bits of information. In a one-bit even parity scheme, the sender appends one additional bit of information, called the *parity bit*. The value of the parity bit is chosen such that the total number of bits with value 1 in the resulting $d+1$ bits is even. Similarly, in the case of one-bit odd parity scheme, the parity bit value is chosen such that there is an odd number of 1's in the resulting $d+1$ bits of information.

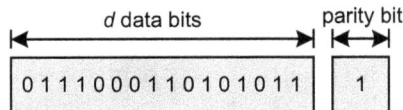

Fig. 6.11 *One-bit even parity*

With a single-bit parity scheme, error detection is a trivial operation at the receiver. In order to detect an error, the receiver is only required to count the number of 1s in the received $d+1$ bits. If an odd number of 1-value bits are found, and a one-bit even parity scheme is being used, the receiver knows that an odd number of bit errors have occurred. Similarly, the receiver can detect an odd number of bit errors with a one-bit odd parity scheme. One-bit parity schemes cannot, however, detect an even number of errors. Thus, in case the received information has an even number of bits in error, the error is not detected at the receiver. Also, as is evident from the description of a one-bit parity scheme, error correction is not possible by using this scheme, since it is not possible to detect which bit is in error.

Generally, it is observed that errors occur in bursts. In case of burst errors, the probability of undetected errors in case of a one-bit parity scheme is approximately half. Clearly, such a scheme for error detection is not acceptable, and a more reliable error-detection scheme is required. Also, along with error detection, it is also desirable to pinpoint the bit in error, so that error correction is feasible. Figure 6.12 depicts a two-dimensional parity scheme, which is an extension of the one-bit parity scheme. In a two-dimensional parity scheme, the d bits in the data frame are divided into i rows of j columns each. A one-bit parity value is then computed for each row and for each column. Next, a single-bit parity is computed for the row and column parity bits themselves. The resulting $i+j+1$ parity bits comprise the data frame's error detection bits.

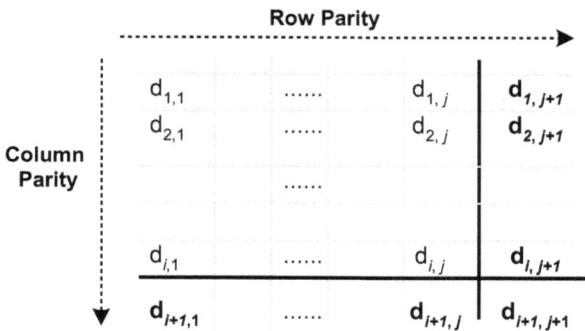

Row Parity

		$d_{1,1}$	......	$d_{1,j}$	$d_{1,j+1}$
		$d_{2,1}$	......	$d_{2,j}$	$d_{2,j+1}$
Column Parity			......		
		$d_{i,1}$	......	$d_{i,j}$	$d_{i,j+1}$
		$d_{i+1,1}$	......	$d_{i+1,j}$	$d_{i+1,j+1}$

Fig. 6.12 *Two-dimensional parity scheme*

With a two-dimensional parity scheme, consider what would happen in the case of a single bit error (see Fig. 6.13). With a single bit error, the parity of both the column and row containing the error bit will incur a mismatch. This will help in not only detecting the bit in error, but also in correcting the error bit. In general, a two-dimensional parity scheme will be able to perform the following operations:

• Detect an odd number of bit errors in the information.

```
1 0 1 0 1  | 1                    1 0 1 0 1  | 1
                   Bit Error
1 1 1 1 0  | 0                    1 0 1 1 0  | 0   Parity Error
           ───transmission⟩
0 1 1 1 0  | 1                    0 1 1 1 0  | 1

0 0 1 0 1  | 0                    0 0 1 0 1  | 0
                                  Parity Error
```

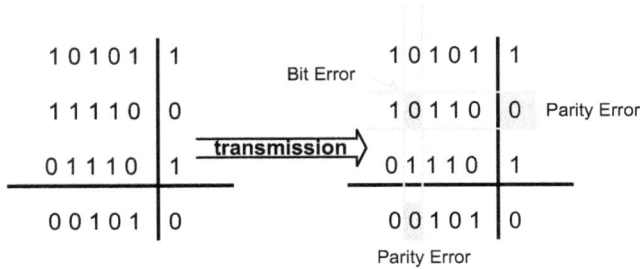

Fig. 6.13 *Error correction with two-dimensional parity scheme*

- Detect and correct a single bit error in the received information.
- Detect and correct all multi-bit errors, as long as no row or column of the information has more than one bit in error. Thus, an even number of bit errors can also be detected and corrected, subject to the mentioned constraint.

Clearly, a two-dimensional parity scheme is more reliable and robust than a one-bit single-dimensional parity scheme. A two-dimensional parity scheme not only detects a higher number of errors, but it also enables error correction. In networking terminology, the ability of the receiver to both detect and correct errors is called *Forward Error Correction (FEC)*.

FEC techniques are commonly used in audio storage and playback devices such as Compact Discs. FEC techniques are valuable within a transmission network, because they can decrease the number of packet re-transmissions required. More importantly, they permit immediate correction of the error at the receiving end. This avoids the round-trip propagation delay introduced as a result of re-transmission of the packet. This is a very important advantage for most real-time networking applications.

6.9.2 Checksum Techniques

With checksum techniques, the d bits of information are divided into a sequence of k-bit integers. A simple checksum technique is to add these k-bit integers and use the bits in the resulting sum as the error detection bits. The checksum solution used in the Internet is based on this approach. In the solution used in the Internet, bytes of data are treated as 16-bit integers and then added. The 1's complement of this sum then forms the Internet checksum, which is carried in the packet header. Figure 6.14 depicts the use of a checksum technique for error detection.

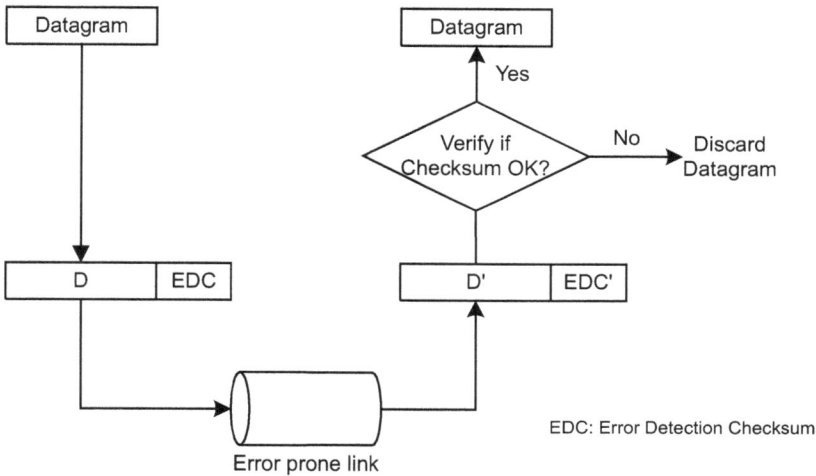

Fig. 6.14 *Error detection using checksum techniques*

Checksum methods are often used in the transport layer protocols. Both TCP and UDP protocols define the use of a checksum method. The benefit of using a checksum technique is that it introduces little packet overhead. However, when compared with some of the other error detections schemes, checksum techniques provide relatively weaker protection against errors. A more robust scheme for error detection is the use of CRC, which is commonly used at the data link layer. Thus, practical implementations of networking protocols define the use of a checksum technique at the transport layer, along with a more reliable CRC method at the data link layer. Another reason for choosing such an implementation technique is that checksum is easier to implement in software, as compared to CRC, which is easier to implement in hardware.

6.9.3 Cyclic Redundancy Checks (CRC)

Cyclic Redundancy Check (CRC) is one of the more powerful methods for error detection. CRC is implemented by grouping the bytes of data into a block and then calculating a CRC code on this block. The calculated CRC code is then appended to the data frame before transmission. In networking terms, CRC codes are also sometimes known as polynomial codes. This name for CRC codes stems from the mechanism used for the calculation of CRC codes. For CRC code calculation, the information to be transmitted is viewed as a polynomial, where the coefficients of the polynomial are the 0 and 1 values in the information bit stream. The operations performed on the bit stream for the calculation of the CRC code can thus be interpreted as polynomial arithmetic.

Consider for example the bit stream 100101. The bit stream can then be viewed as the following polynomial in x:

$$1*x^5 + 0*x^4 + 0*x^3 + 1*x^2 + 0*x^1 + 1*x^0 = x^5 + x^2 + 1$$

The CRC code of a bit stream is calculated by performing a modulo-2 division of the bit stream with a *generator polynomial*. The CRC code is then the remainder obtained after performing the division. Consider, for example, a data stream consisting of d-bits of information, which is to be transmitted over the network. The sender and the receiver of information mutually agree on an $r+1$ bit pattern, known as the generator polynomial. The d-bit of information is then divided with the $r+1$ bit generator polynomial, to obtain an r-bit remainder. These r-bits of information are appended to the d-bit data stream, resulting in a $d+r$ bit pattern. The $d+r$ bit pattern is now exactly divisible by the generator polynomial, using modulo-2 arithmetic. The sender then transmits this $d+r$ bit pattern to the receiver. Figure 6.15 depicts the process of calculation of the CRC code.

At the receiving end, the process of error detection is trivial. The receiver divides the received $d+r$ bits with the mutually agreed upon generator polynomial. If the remainder of the division operation is non-zero, the receiver knows that an error has occurred; otherwise the data is accepted as error-free.

All CRC calculations are done using modulo-2 arithmetic without any carry-over operation in addition or borrow operation in subtraction. This means that addition and subtraction are identical operations, equivalent to the bit wise exclusive-or (XOR) operation. Multiplication and division are

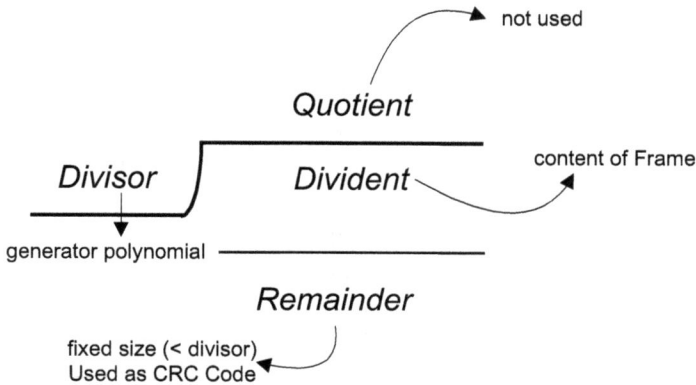

Fig. 6.15 *CRC code calculation*

the same as in base-2 arithmetic, except that the required addition or subtraction is done without carry-over and borrow operations. Figure 6.16 depicts an example for CRC calculation.

Three well-known generator polynomials are most commonly used in communication networks. These are:

Example:
Data to be transmitted = 101110
Generator polynomial =1001 (i.e. r = 3)

Steps:
• Append r number of 0's to the data to make the dividend d+r bits (101110000).
• Divide the dividend with the generator polynomial to obtain 011 as the remainder (division operation is left for the reader to perform).
• The resulting data to be transmitted is thus 101110011. This is completely divisible by 1001.
• On the receiver end, the received data is divided by the generator polynomial. If the remainder is non-zero, this means an error has occurred.

Fig. 6.16 *CRC example*

- CRC-16 $= x^{16} + x^{15} + x^2 + 1$ (used in HDLC, refer to section 6.10)
- CRC-CCITT $= x^{16} + x^{12} + x^5 + 1$
- CRC-32 $= x^{32} + x^{26} + x^{23} + x^{22} + x^{16} + x^{12} + x^{11} + x^{10} + x^8 + x^7 + x^5 + x^4 + x^2 + x + 1$ (used in Ethernet)

Each of the CRC standards can detect burst errors of fewer than $r+1$ bits. This means that all bit consecutive errors of r bits or fewer will be detected. Also each of the CRC standards can detect any odd number of bit errors. As an example, the CRC-16 is able to detect all single-bit errors, all double-bit errors, all odd numbers of errors and all errors with a size of less than 16 bits in length. In addition 99.9984 per cent of the other possible error patterns can also be detected with CRC-16.

6.10 EXAMPLE: HDLC

Amongst the oldest layer 2 protocols is the data link layer protocol that was used in IBM's System Network Architecture (SNA). The protocol was designed by IBM in 1975, and was known as the Synchronous Data Link Control (SDLC) protocol. After developing SDLC, IBM submitted it to ANSI and ISO for acceptance in the US and the international standards, respectively. ANSI modified SDLC and renamed it as the *Advanced Data Communication Control Procedure (ADCCP)*. Similarly, ISO also modified SDLC, and this modified version of the protocol came to be known as the *High-level Data Link Control Procedure (HDLC)*.

The HDLC protocol was later adopted by the CCITT and was modified to evolve the *Link Access Procedure (LAP)* protocol as part of the X.25 standards. LAP was modified further to evolve *Link Access Procedure-Balanced (LAP-B)* to make it more compatible with later versions of HDLC. Today, a lot of data link layer protocols exist, that have evolved from the HDLC. HDLC can, in that sense, be known as the mother of all data link layer protocols. Besides LAPB, the flavour of HDLC is also evident in some of the other protocols, namely:

- Link Access Procedure for Modems (LAPM) provides HDLC services for V.42 modems.

- Link Access Procedure for D-Channel (LAPD) is used on the ISDN D-channel for packet delivery.
- Logical Link Control (LLC) was developed by the IEEE 802.2, and is used to provide HDLC style services on a LAN (refer to section 6.6).
- Frame relay uses a protocol that is similar to LAPD.
- The Point-to-Point Protocol (PPP) is a derivation of HDLC and encapsulates Protocol Data Units for transport across point-to-point links.

All of the above protocols are based primarily on similar principles, namely, bit-oriented protocols, using bit stuffing for data transparency. In this section, concepts related to the HDLC protocol are discussed in further detail.

The HDLC protocol is considered as an umbrella protocol, under which many different wide area protocols exist. The HDLC protocol was developed by ITU-T (formerly known as CCITT) in 1979. HDLC defines three different types of stations, namely:

- **Primary station:** A primary station, if it exists, acts as a control node for the communication link. The primary station controls all data link layer operations, and issues commands to other secondary stations connected via the link layer. It has the ability to hold separate sessions with different secondary stations on the link.
- **Secondary station:** In case an HDLC configuration has a primary station, there must also exist corresponding secondary stations. Secondary stations are under the control of the primary station. They have no responsibility for controlling the communication link. Secondary stations only become active when requested by the primary station to do so.
- **Combined station:** A combined station is a communication node that includes the functions of both a primary station and a secondary station. In other words, all combined stations are able to send and receive commands and responses independently, without requiring any permission from other stations on the link.

A communication channel in HDLC that is used by a station can be configured in one of the following three different ways:

- **Unbalanced:** This configuration allows one primary station to communicate with a number of secondary stations over either of half duplex, full duplex, switched, unswitched, point-to-point or multi-point paths.
- **Balanced:** This configuration is used when two combined stations communicate over a point-to-point link, which is either full/half duplex, or switch/unswitched.
- **Symmetrical:** The symmetrical configuration is not commonly used today. This configuration assumes two independent point-to-point links, each working in the unbalanced mode. In other words, on one link, one station acts as the primary station, and the other as the secondary station. The roles are reversed on the other link. Thus, each station is logically considered as two stations.

When participating in communication sessions, stations are defined to exist in one of the following three different modes:

- **Normal Response Mode (NRM):** In this mode, secondary stations require permission from the primary station before transmitting data. This mode is mainly used on multi-point links.
- **Asynchronous Response Mode (ARM):** In this mode, secondary stations can send data without receiving permission from the primary station. Although defined by the HDLC protocol, practically, this mode is hardly ever used.
- **Asynchronous Balanced Mode (ABM):** In this mode, either station participating in communication can initiate a transmission without requiring permission from the other station. This mode of operation assumes the presence of combined stations. This is the most commonly used mode on point-to-point links. For example, LLC (IEEE 802.2) uses the Asynchronous Balanced Mode of HDLC.

HDLC protocol defined three different forms of frame formats, as depicted in Fig. 6.17. In each of the frame formats, the HDLC frame is delineated with a begin and end flag that has a value of 0x7E. The control field in the HDLC frame defines the three different frame formats, which are:

- **Information frames:** Information frames are used for data transfer between stations. The Send Sequence

Flag (0x7E)	Address (8bits * n)	Control (8bits * n)	Data	FCS (16/32 bits)	Flag (0x7E)

Information Frame	0	Send Sequence Number (3 bits)	Poll/ Final (1 bit)	Receive Sequence No. (3 bits)

Supervisory Frame	1	0	Supervisory code (2 bits)	Poll/ Final (1 bit)	Receive Sequence No. (3 bits)

Un-numbered Frame	1	0	Unnumbered bits (2 bits)	Poll/ Final (1 bit)	Unnumbered Bits (3 bits)

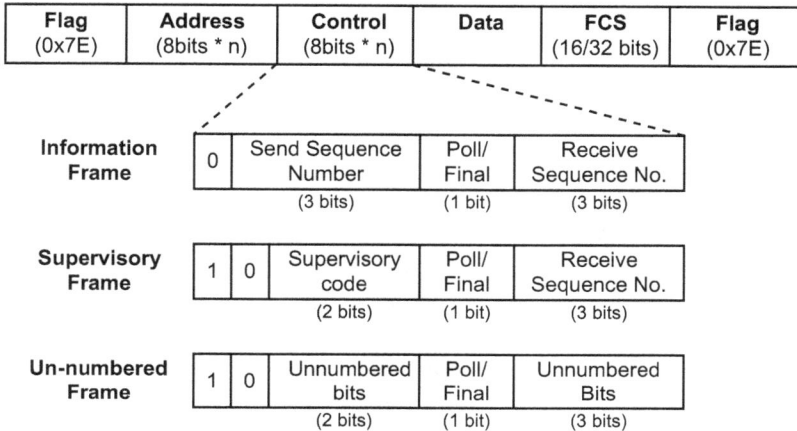

Fig. 6.17 *HDLC frame format*

Number and the Receive Sequence Number maintain the frame sequence numbers in either direction of transmission. The poll/final bit is set to value poll when the frame is used by the primary station to obtain a response from a secondary station. It is set to a value final when used by the secondary station to indicate a response or the end of transmission.

- **Supervisory frames:** Supervisory frames are used for multiple purposes, which include acknowledging received frames to the sender, requesting the sender for frame re-transmission or requesting the sender for suspension of transmission. Within the supervisory frame, the supervisory code indicates the type of the supervisory frame.

- **Un-numbered frames:** Unnumbered frames are used for link initialization and link disconnection. The un-numbered bits indicate the type of un-numbered frame being used.

The HDLC protocol implements most of the data link layer procedures that were discussed as part of this chapter. The rest of the data link layer protocols, which are derived from the HDLC protocol, borrow heavily from the HDLC protocol with respect to the frame formats and the other concepts related to communicating modes and channels. In this respect, the HDLC protocol can be considered as the ancestor of all later developed data link layer protocols.

6.11 CONCLUSION

The focus of this chapter was layer 2 of the OSI reference model, which is better known as the data link layer. The main functions of the data link layer include error detection and correction over noisy, error-prone communication links, and flow control to prevent a fast sender from over-burdening a slow receiver. Techniques for flow control include the sliding window-based protocols, which were discussed in this chapter. Similarly, multiple techniques exist for error detection and correction, which include use of parity bits, checksum techniques and CRC codes.

For networks, which support inherent broadcast capabilities, e.g. Ethernet-based LANs, the data link layer also performs the important function of media access control. As part of media access control, it is required to somehow co-ordinate between the transmissions of multiple active nodes, so as to overcome collisions on the shared communication channel. This co-ordination job is the responsibility of the multiple access protocol, which is implemented in the Media Access Control (MAC) sub-layer of the data link layer. The chapter discussed some common MAC layer protocols, which include the Aloha-family of protocols and the CSMA family of protocols. Also, the taking-turns protocols were discussed, which include some well-known protocols like the token ring protocol.

Towards the end of the chapter, concepts related to the HDLC protocol were discussed, which is amongst the oldest known protocols for the data link layer.

REVIEW QUESTIONS

Q 1. Describe the services provided by the data link layer to the network layer.

Q 2. The media access protocol used in Ethernet is based on the CSMA/CD protocol. Calculate the minimum length of an Ethernet frame, which is required for successful operation of the CSMA/CD protocol. Assume that the Ethernet speed is 10Mbps, the Ethernet cables are copper cables on which the propagation speed is 77c m/s (c = speed of light = 3 * $10 \wedge 8$ m/s) and the repeaters are put after every 600km. [Ans: 64bytes]

Q 3. A group of N stations share a 56kbps pure Aloha channel. Each station transmits 1000-bits per frame on an average of once every 100 sec, irrespective of whether the previous frame has been successfully transmitted or not. Calculate the maximum value of N for operation of the channel at peak-load. [Hint: Assume that the peak load in pure Aloha is only 18 per cent of the channel capacity.] [Ans: 1008]

Q 4. Describe one most important aspect that makes the Taking-Turns protocols differ from the Random Access protocols.

Q 5. Describe the different classes of the LLC protocol.

Q 6. Describe the concept of a sliding window protocol. Sliding window protocols can help to achieve both flow control and error control. Explain this statement with the help of an example.

Q 7. Describe the techniques used for error detection and correction. Why is checksum used more often in the transport layer, while CRC is more popular in the data link layer?

Q 8. Using the CRC technique, describe the contents of the frame transmitted by the data link layer, if the data from the higher layer is D = 1101011011, and the generator polynomial is G = 10011.

FURTHER READING

[Gen A.Tanenbaum] provides a good reference for concepts related to flow control, error control and media access control. [DLC U.Black] provides a good survey of various data link control protocols, and is also an excellent reference on the topic. [IEEE 802.2] describes the LLC part of the Data Link Control layer, while specifications [IEEE 802.3] and [IEEE 802.5] describe the most commonly used MAC protocols, namely, CSMA/CD and Token Ring respectively. [ITU-T Q.921] can be referred to for details on the LAPD protocol. These protocols were discussed as part of the HDLC case study presented in the chapter.

BRIDGING

7.1 INTRODUCTION

Bridging is a technique used to extend or segment a Local Area Network (LAN) for easier network management and access control. Smaller network segments can be bridged together to form a bigger LAN. Similarly, a big physical local area network can be segmented into smaller LAN segments, each connected to each other by using a bridge. In either case, the network administrator can define access control rules between the multiple smaller segments of the same LAN on the basis of the access requirements.

Bridging is performed at layer 2 of the OSI reference model, unlike 'Routing' (discussed in Chapter 11), which works at layer 3. Primarily, the function of routing is to connect multiple LANs to each other via backbone networks. However, sometimes, bridging is also used to connect two distinct LANs, often implementing different layer 2 protocols. This is normally the case when the two (or more) LANs bridged together fall under the same administrative domain, and the network administrator decides to avoid the overheads of routing at layer 3 by connecting the two (or more) LANs using bridging at layer 2.

As an example of bridging, consider the case of an office complex spread across multiple floors of a single building. Each floor of the office building could be housing different departments of the same organization. In this scenario, the local area network of the organization can be organized as a collection of different LAN segments, wherein each LAN segment covers one floor of the office building. These LAN segments can be bridged together by using devices known as bridges, to form the complete LAN for the organization (see Fig. 7.1). Access rules can be defined between the LAN segments to define the nature of communication and the rules governing the communication between different segments. Further, the layer 2 protocol followed in each of the LAN segments can be distinct (Ethernet, token ring, etc). The following sections describe two of the most commonly used techniques for bridging, namely, *transparent bridging* and *source route bridging*.

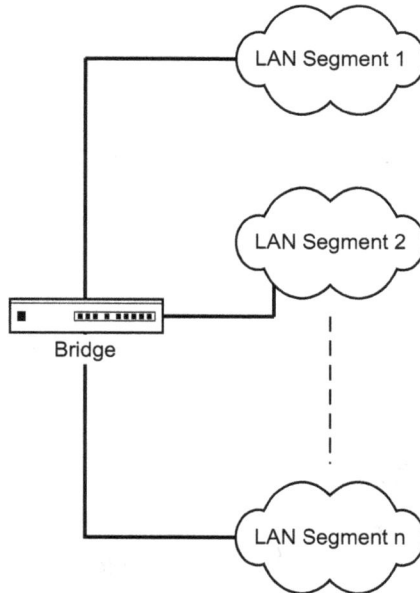

Fig. 7.1 *Local Area Network using bridging*

7.2 TRANSPARENT BRIDGING

Transparent bridging is a concept in which hosts on the LAN are not aware of the presence of bridges connecting LAN segments. Bridges connecting LAN segments are therefore, assumed to require no additional information or support from end-hosts to perform their functions. In other words, the segmentation of LAN and its interconnection using bridges is transparent to end-hosts. Transparent bridging is amongst the most often used technique for bridging, and is used by Ethernet for connecting LAN segments.

In order to understand how transparent bridging works, first try to understand the functions performed by a solitary bridge operating in transparent mode. This is discussed in the following section. However, in case of a LAN where more than one bridge is used to connect LAN segments, and wherein multiple (more than one) paths exist between any two LAN segments, a more complex approach is required. Multi-bridge operation in transparent bridging and its associated issues are discussed in section 7.2.2.

7.2.1 Functions of a Bridge in Transparent Bridged LANs

Figure 7.2 depicts the use of a solitary bridge, operating in transparent mode, to connect three LAN segments. Each LAN segment is depicted to contain two end-hosts each, with their MAC addresses as shown in the figure. The following text describes the functions performed by a bridge operating in transparent bridging mode with reference to this figure.

In transparent bridging, a bridge connecting two or more LAN segments has to broadly perform two main functions,

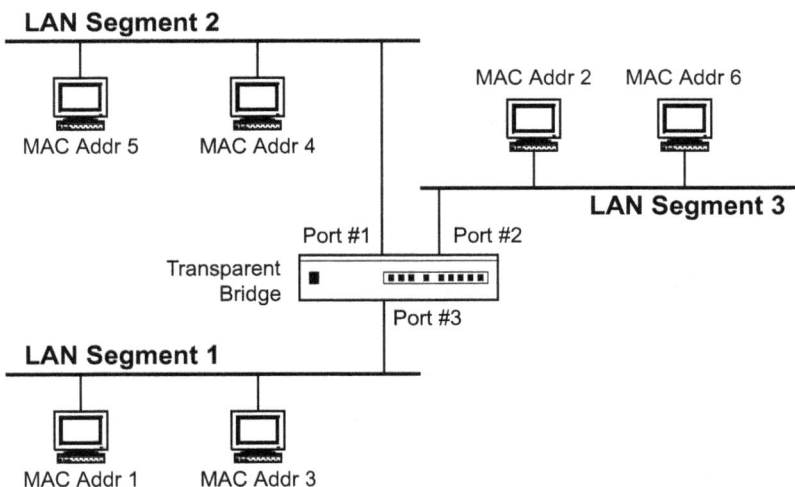

LAN Segment 2

MAC Addr 5   MAC Addr 4

MAC Addr 2   MAC Addr 6

LAN Segment 3

Port #1   Port #2

Transparent Bridge

Port #3

LAN Segment 1

MAC Addr 1   MAC Addr 3

Fig. 7.2 *Operation of a solitary bridge in transparent bridged LANs*

namely maintenance of a destination lookup table and forwarding of layer 2 packets based on the look-up table. While these functions are described in a particular order below, note that both these functions are performed simultaneously by the bridges, and in no specific order.

7.2.1.1 Management of Destination Look-up Table

In order to perform packet forwarding between LAN segments, the transparent bridge depicted in Fig. 7.2 maintains a destination look-up table having a mapping between the MAC addresses of the end-hosts, and the LAN segment in which these end-hosts reside. Since the physical port identifier on the bridge is sufficient to identify a LAN segment, the destination look-up table simply maintains a mapping between the MAC addresses of the end-hosts and the physical ports on the bridge via which these end-hosts are reachable. In transparent bridging, the bridge operates in transparent mode and hence, it needs to form the destination look-up table on its own, with no active support from the end-hosts. In this sense, transparent bridges are *self-learning bridges*. To form a destination look-up table, the bridges follow one of the following two approaches:

- **Destination lookup table updation during packet forwarding:** In this approach, bridges update their destination look-up tables during the process of packet forwarding. While only the destination MAC address in the layer 2 packet header is used for packet forwarding, bridges also read the source MAC address from the layer 2 packet header. The source MAC address, along with the port on which the packet was received, is used to update the destination look-up table for this source MAC address. This mapping is later used whenever the bridge receives a packet destined for the source MAC address under discussion. For example, in the configuration in Fig. 7.2, if the host with MAC address 5 sends a packet to MAC address 2, then the bridge learns from the source address that port 1 can be used to reach MAC address 5. Thus, if a packet for MAC address 5 is later received by the bridge, it knows it has to forward it on port 1.

- **Destination look-up table updation by using promiscous mode:** In this approach, bridges listen to all packets being exchanged on each LAN segment, irrespective of whether the packet is forwarded by the bridge or not. Thus, even packets exchanged within the same LAN segment, which do not even traverse the bridge, are read by the bridge for updation of the destination look-up table.

Note that the two approaches mentioned above are not very different from each other. It is only a matter of the approach choosen by the network administrator for the updation of the destination look-up table.

Besides addition of mapping records in the destination look-up table, it is also imperative that bridges prune invalid mapping records. Mapping records become invalid in case end-hosts detach from the LAN segment, or move from one LAN segment to another. Mapping records in the destination look-up table are hence assigned a validity lifetime, on expiry of which the record is pruned. However, if during the lifetime of the record, the mapping is refreshed (as a result of receipt of a layer 2 packet with the same source MAC address as the address in the record entry), the validity lifetime is extended.

7.2.1.2 Layer 2 Packet Forwarding

On the basis of the information available in the destination look-up table, the transparent bridge forwards packets between LAN segments. Note that because the bridge learns the mapping between the MAC addresses and the LAN segments over a period of time, at any instant of time, the destination look-up table need not have a mapping for all the end-hosts in the LAN. This is also true because of the fact that entries in the destination look-up table are also pruned as a result of aging. Hence, the packet-forwarding algorithm implemented in the bridges has to take into account this specific case of MAC addresses for which no mapping exists. Whenever a bridge receives a packet on any of its incoming ports, it undertakes the following operations for packet forwarding:

1. Performs a look-up of the destination look-up table for the destination MAC address in the received layer 2 packet. If no entry is found in the table for the destination MAC address, then the packet is flooded over all outgoing ports

of the bridge (except the port on which the packet was received). If however, a match is found in the destination look-up table, then the subsequent steps are performed.

2. Verifies if the outgoing port in the look-up table is the same as the port on which the packet was originally received. If so, the bridge performs no action, and simply discards the packet. This is done because the intervention of the bridge is not required within hosts belonging to the same LAN segment. If however, the outgoing port is distinct from the incoming port, then the packet is sent to the outgoing port mentioned in the look-up table.

7.2.2 Transparent Bridging in Multi-Bridge LAN

Section 7.2.1 discussed the functions of a solitary bridge in transparent bridging. However, most often, LAN segments are connected together to each other via more than one single bridge. While there could be multiple reasons for using more than one bridge to connect LAN segments, one of the prime reasons for this configuration is to provide redundancy in the network by removing the single point of failures.

Since the communication between LAN segments depends on the proper functioning of the bridge connecting them, a malfunctioning bridge, or a failure of the link connecting a LAN segment to the bridge can jeopardize the communication within the LAN. Hence, to provide redundancy against such failures, more than one bridge (and hence, more than one path) is normally available between any two LAN segments in a LAN. An example of such a redundant set-up is depicted in Fig. 7.3.

As depicted Fig. 7.3, three LAN segments of a LAN are connected to each other via three bridges. The configuration of the LAN is such that there is no single point of failure. Consider, for example, the case when bridge 1 fails. In this case, LAN segment 1 and LAN segment 3 are still connected to each other, via bridges 2 and 3. Hence, a single point of failure will not lead to a disrupted communication in the LAN depicted in Fig. 7.3.

Even though multi-bridge operation is preferred by network administrators to provide immunity to single point failures, it introduces some complexities for transparent bridges. As is evident from Fig. 7.3, redundancy in the LAN comes at the cost

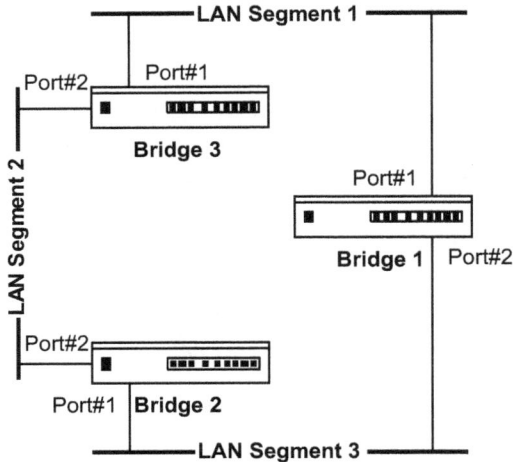

Fig. 7.3 *Multi-bridge redundant configuration for transparent bridging*

of having loop(s) or cyclic connection(s) between LAN segments. As a pitfall of such cyclic connections, packets can continue to remain in the network forever, looping around the cyclic connection.

In order to overcome this limitation, a *Spanning Tree Approach* is followed in multi-bridge LANs that use transparent bridging. The basis of this approach is to de-activate certain links in the LAN that cause cyclic connections to form, thus forming a tree-type structure. The spanning tree approach is discussed in detail in section 7.4, which discusses bridging in Ethernet-based networks.

7.3 SOURCE ROUTE BRIDGING

Source route bridging is a technique of bridging that is in sharp contrast to the concepts of transparent bridging. It is used in token ring networks, and is documented in [IEEE 802.5]. Unlike transparent bridging, source route bridging places the responsibility of determining the path from the source to the destination on the originating end-host. Source route bridges simply forward the packet towards the destination, on the basis of the path that is provided by the originating end-host as part of the layer-2 packet header.

In order to undertake source route bridging, the end-hosts within the LAN have to perform the following functions:

- **Determination of the LAN segment of the destination host:** The originating end-host has to first determine whether the destination host is on the same LAN segment, or on another. If the destination is on the same LAN segment, then the packet can be sent directly, without any path information. However, if the destination is on a LAN segment other than the LAN segment of the source, then path information (via connecting bridges) has to be provided by the source in the layer-2 packet header. In order to determine the location of the destination host, the originating end-host first sends a test frame/packet on its own LAN segment. If the destination host is on the same LAN segment, it sends back the packet to the source, with an indication that the destination node has seen the packet. If however, the originating end-host does not receive such an indication, it deciphers that the destination end-host is on another LAN segment.

- **Determination of the path to the destination host:** In case the originating end-host discovers that the destination is on another LAN segment, it sends an explorer frame on its LAN segment, with the MAC address of the destination end-host. Each bridge on the LAN that receives the explorer frame forwards the same on all its outgoing interfaces. Also, before forwarding the frame, each bridge adds its address in the layer-2 packet header. This serves two purposes:

    - Firstly, this helps the bridge to determine if the same explorer frame reaches it for the second or subsequent time. If the address of the bridge is already present in the layer-2 packet header, then the bridge does not broadcast the frame on its outgoing interfaces, and simply discards it.
    - Secondly, this helps in maintaining the route that the explorer frame took from the source to the destination. When the destination host receives the explorer frame, it knows the route the frame took from the source to the destination, by looking at the ordered list of bridge addresses in the layer-2

packet header. It then sends this path information back to the source host.

- **Transmission of the layer-2 packet with the source-determined path:** In a LAN, multiple paths may exist between the source and the destination. For example, in Fig. 7.3, a node in LAN segment 1 can reach another node in LAN segment 3 via two paths: a direct path via bridge 1, or via a two-hop path over bridges 3 and 2. The path-determination approach described above is such that the destination end-host will send a response to the source node for each explorer frame it receives. Hence, the source node will become aware of all the possible paths to the destination. Using this information, the source node can choose one of the paths, and add this path in the header of the actual layer-2 data packet. This packet can then be transmitted in the LAN segment of the source, and intermediate bridges will then forward it towards the destination using the path in the packet header. The specifications do not define how a source should choose a path to the destination, from a list of possible paths. However, in general, the source could choose one of the following paths from a list of possible paths:

  - That described by the first response to the explorer frame received from the destination
  - One with the minimum number of hops.
  - One with the maximum allowed frame size—this criterion is important in LANs where the maximum frame size differs on different LAN segments.

Any combination of the above-mentioned criteria could also be used by the source to determine the path to the destination.

7.4 EXAMPLE: BRIDGING IN ETHERNET LANS

Ethernet LANs use transparent bridging for connecting multiple Ethernet LAN segments to each other. transparent bridging followed in Ethernet LANs is documented in [IEEE 802.1] standard and was discussed in section 7.2.

As mentioned in section 7.2.2, a spanning tree approach is used in Ethernet LANs that have multiple bridges connecting LAN segments in a redundant fashion. The spanning tree approach aims towards forming a tree of Ethernet links and bridges, such that the tree spans the entire bridged LAN. The spanning tree approach works as per the following steps:

1. One bridge in the Ethernet LAN is first chosen as the root of the spanning tree. The choice of bridge as the root of the spanning tree can be either network administrator-controlled, or automatic based on some other criterion (e.g. the bridge with the smallest MAC address). Initially, the spanning tree thus contains only the root bridge.
2. For each LAN segment in the LAN, a set of links and bridges via which the LAN segment can connect to the spanning tree are identified. As a result of redundant links in the network, more than one path can exist to connect to the spanning tree. Normally, the path chosen is the one with the least cost to connect to the tree.
3. Once a path (via a set of links and bridges) has been chosen, all other links to other bridges on this LAN segment are de-activated.

Figure 7.4 depicts the spanning tree formed for the LAN configuration depicted in Fig. 7.3.

In simple terms, the way the spanning tree approach works is to identify and isolate bridge-interfaces that cause loops to form in the bridged-LAN. While the spanning tree approach enjoys the advantage that it makes the LAN topology loop-free, and prevents LAN bandwidth from being wasted due to looping packets, it also has certain disadvantages. Transparent bridging, along with the spanning tree approach, as followed in Ethernet, is not efficient in LANs that have fluid topologies (i.e. topologies that can change frequently). Since bridges in transparent bridging are self-learning bridges, some time is required for the bridges to re-configure their destination look-up tables in case of changes in the topology. Further, frequent changes in the LAN topology would mean that the spanning tree is to be formed more frequently, thus wasting LAN bandwidth.

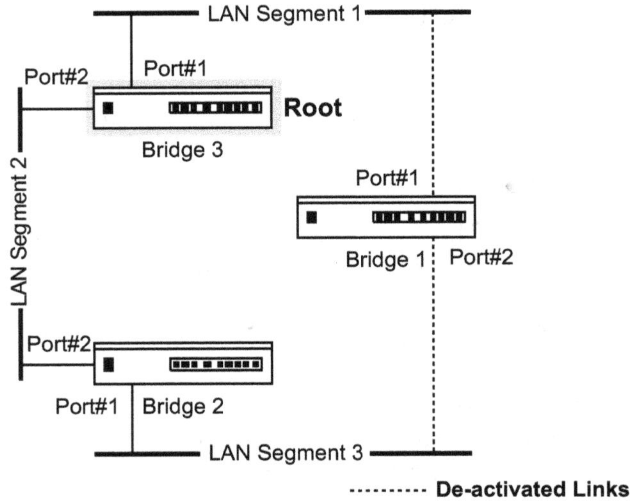

Fig. 7.4   *Spanning tree in multi-bridge transparent bridged LANs*

7.5 CONCLUSION

This chapter discussed the concepts of bridging in LANs, which is used to segment a LAN into LAN segments. Bridging works at layer 2 of the OSI reference model, unlike routing, which operates at layer 3. Two of the commonly used approaches for bridging were discussed, namely, transparent bridging and source route bridging. The functions of bridges and end-hosts in each of the bridging method were also discussed. Ethernet, one of the most popular LAN technology, uses transparent bridging for connecting LAN segments. Along with transparent bridging, Ethernet also uses the spanning tree approach to keep the LAN topology free from any forwarding loops, which can lead to wastage of LAN bandwidth.

REVIEW QUESTIONS

Q 1. What are the prime differences between the approach used for transparent bridging and source route bridging? Discuss in relation to the functions performed by bridges and end-hosts in either approach.

Q 2. In transparent bridging, what possible actions can be taken by a bridge when it gets a packet? Explain this in relation to: a) Update of destination look-up table, and b) Packet forwarding.

Q 3. With the help of an example, describe a scenario in which packets can keep looping around the LAN, in case transparent bridging is used without spanning tree approach.

FURTHER READING

[Bri R.Perlman] provides an excellent reference on concepts related to bridging. Also, [Gen W.Stallings] has a dedicated chapter describing the bridging concept, and is also a good reference for the topic. [IEEE 802.1D] describes the protocol architecture for MAC bridges, while [IEEE 802.5] describes the concept of source routing in detail. For more details related to transparent bridging, the reader is referred to [IEEE 802.1D], which describes bridging in Ethernet LANs and the concept of spanning trees in bridging.

SWITCHING

8.1 INTRODUCTION

A *switch* can be viewed as a network element that transfers packets from input ports to the appropriate output ports. Here, a port refers to a point-of-attachment in the switch. Accordingly, *switching* is the process of transfer of packets from input ports to output ports. This transfer is also referred to as *internal routing* of packets. Switching or internal routing forms the core functionality of the switch. A switch with an equal number of input and output ports is called a symmetric switch. It is, however, not a pre-requisite for a switch to have an equal number of input and output ports. Nonetheless, for the sake of simplicity, in this chapter, it is assumed that switches are symmetric in nature. It is further assumed that the port speed of all input ports and output ports is the same (and equal to V), and that it takes one time slot to switch a packet from an input port to an output port. In order to simplify the presentation, an input port is referred to as an inlet and an output port as an outlet. Also, the discussion assumes that a switch is used for virtual circuit switching.

The phenomenal growth in the speeds of optical transmission systems has caused severe pressure on switch designers to develop switches that match the transmission capability of optical fiber. By current estimates, in terms of speed, switches will lag the transmission systems for many decades to come. It is said that without high-speed switches, it is like building a ten-lane highway interspersed by a single-lane tunnel. Against this backdrop, it is imperative to design switches that are capable of forwarding packets speedily, are highly scalable and easy to implement. This chapter discusses issues related to high-speed switching.

The two core switching issues are *internal routing* and *buffering*. Both these issues have been adequately covered. Another important issue that has become very significant today is multi-casting capability. Techniques to incorporate multi-casting are also discussed. Besides discussing different switching issues, the three basic switching architectures, viz. *shared-memory architecture*, *shared-medium architecture* and space-division architecture are explained.

8.2 COMPONENTS OF A TYPICAL SWITCH

A typical switch has three basic components: *input modules, switching fabric* and *output modules*, as shown in Fig. 8.1. These components are detailed below.

- **Input module (IM):** The input module accepts packets arriving at the input ports. The input module then extracts the virtual circuit number (i.e. the logical address or label) from the packet header of the arrived packets. The extracted label is used to index a virtual circuit table and to determine the outgoing label and the output port of the arrived packet. This function may also be performed by the output module. Other functions of the input module include de-framing and unscrambling, synchronization, classification of the packet into various categories (like

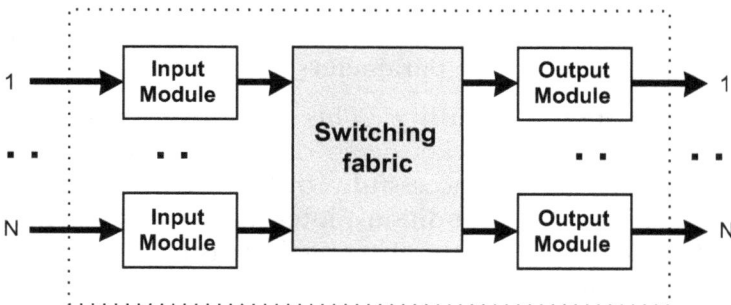

Fig. 8.1 *A typical switch*

signalling packets, management packets, etc.), error checking and internal tagging (i.e. associating route information with packets so that they can be forwarded to their destined output). Other functions of the input module are dependent on the transmission media and the underlying technology.

- **Switching fabric:** The switching fabric provides the means to route packets from input ports to output ports. The switching fabric consists of transmission links and switching elements. A transmission link is a passive entity (without any decision-making intelligence), which merely carries packets. A switching element, on the other hand, performs functions like internal routing. The switching fabric is also referred to as the *interconnection structure*.

- **Output module:** The functionality of the output module is tightly linked to the transmission media and the underlying technology. Depending upon the overall design of the switch, it can perform functions like label swapping, error control and data filtering.

8.3 PERFORMANCE MEASURES IN SWITCH DESIGN

Before discussing various switching issues and switching architectures, it is important to define parameters based on which different approaches are appraised. These parameters (referred to as *performance measures*) can be classified either as QoS-related parameters or non-QoS parameters. QoS-related parameters include *packet loss probability*, *switching delay* and *throughput*. Non-QoS parameters include *design complexity*, *scalability* and *cost*. These parameters are defined below.

- **Packet loss probability:** This parameter refers to the probability of a packet being dropped within the switch before it can be successfully transferred to its destined outlet. Overload conditions, internal design and shortage of buffer space are the three main reasons for packet loss in a switch. The desirable values of packet loss probability range between 10^{-8} and 10^{-12}.

- **Delay:** This refers to the time required to switch a packet, right from the time it enters the input module to the time it leaves the output module. The delay includes the time required to process a packet at every intermediate switching element. Thus, the greater is the number of switching elements, the greater is the switching delay. The delay parameter also includes the waiting time of a packet in the switch buffers.

- **Throughput:** There are two ways in which people define the throughput of the switch. According to one school of thought, throughput is equal to the port speed of an input link. According to others, throughput is the sum total of link rates of all input links. However, both the measures may not reflect the maximum data rates achievable under different traffic conditions. A better measure is the ratio of output traffic rate and input ratio rate. This ratio takes into account the cases when the internal design of a switch does not permit 100 per cent throughput under unity load.

- **Design complexity:** Switch designers are generally wary of making complicated switch designs. A complex switch design invariably translates into either higher costs or limited switching capability. For this reason, a simple design with slightly lesser capability is preferred over complex designs.

- **Scalability:** Scalability, in the present context, refers to the ease with which a switch can be expanded. There are two aspects associated with scalable switch design. First, a scalable switch must be capable of supporting additional inlets without any deterioration in other performance measures. Second, it must also be capable of supporting any increase in the port speed of its inlets.

- **Cost:** Needless to say, cost is a significant factor in deciding the success or failure of a switch design. Quite often, performance-efficient switch designs prove to be a commercially unviable option, thereby relegating the design to a place in research papers only. Thus, the cost aspects related to the commercial production of switches must always be borne in mind while proposing new switch designs.

8.4 SWITCHING ISSUES

Most of the issues in switch design pertain to two important aspects: *buffering* and *internal routing*. The first aspect relates to the placement of buffers in the switch, while the second relates to the design of the switching fabric (specifically speaking, the design of the interconnection structure). These two are, however, not independent concepts, because fixing one of them imposes constraints on the other. In fact, there are two ways of designing a switch. Either one can choose a particular buffering technique and then decide from one of the interconnection structures possible. Alternatively, one can choose a particular interconnection structure and then decide upon the buffering technique that suits it best. In this section, the first approach is adopted.

First, the need for buffering is explained and then various buffering techniques and associated interconnection structures are discussed. A few other switching issues in this section are also discussed. In the following sections, three different interconnection structures are explained and the applicability of different buffering techniques is discussed.

8.4.1 Conflicts and Contentions

One of the fundamental challenges in designing a high-speed switch is tackling the problem of *conflicts*. Conflict is a phenomenon occurring during the switching process, whereby the switching process is hindered. Conflicts are of two types: *external conflicts* and *internal conflicts*. In order to understand external conflict, consider the following case: 'N' packets simultaneously arrive at 'N' different inlets, all destined for the same outlet (see Fig. 8.2). Here, all N packets cannot be switched onto the same outlet in a single time slot. In cases like this, there is a conflict (or contention) between different inlets for an outlet. This phenomenon is termed as an external conflict. The most common way to handle external conflict is to buffer the excess packets. In the scenario presented above, $N-1$ packets have to be buffered, while one packet is allowed to go through.

In contrast to external conflicts, internal conflicts are the result of the way the interconnection structure is designed.

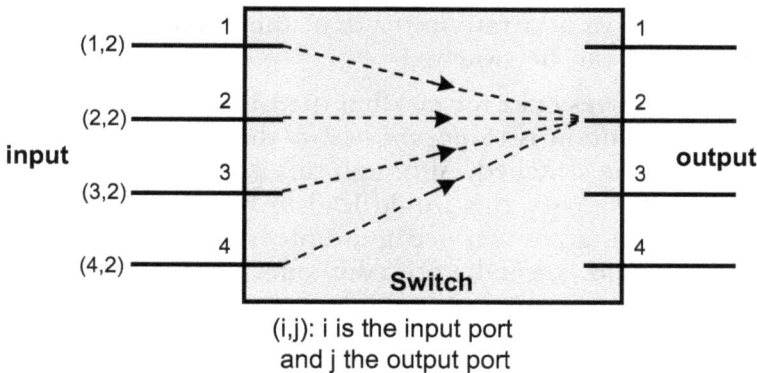

(i,j): i is the input port
and j the output port

Fig. 8.2 *External conflict*

Internal conflict occurs when a switch is incapable of forwarding arrived packets even when an external conflict has not occurred; i.e. a conflict occurs even when 'N' arrived packets are destined for 'N' different outlets. Generally, any internal resource of a switch that is shared by packets coming from multiple paths can lead to internal conflicts. For example, in a multi-stage switching architecture, the internal switching elements are potential candidates for internal conflicts. The problem of internal conflicts for multi-stage switches is analysed in section 8.8 of this chapter.

8.4.2 Blocking and Buffering

Depending upon whether there are contentions/conflicts in the switching process, a switch can be classified as *blocking* or *non-blocking*. In a non-blocking switch design, as long as an outlet is free, packets coming from any inlet can be switched onto the free outlet. In contrast, a blocking switch design leads to conflicts, either external or internal. The external conflicts cause external blocking and internal conflicts cause internal blocking.

There are mainly three conditions under which blocking (internal and external) occurs. These are:

1. Packets from multiple inlets arrive simultaneously, all destined for the same outlet.
2. There are more than one contenders for a shared resource like memory, system bus and internal switching elements.

3. Packets arrive at a rate faster than the maximum rate at which they can be switched.

Many alternatives exist for tackling the problem of blocking. The simplest solution is to *discard packets* that are in excess of that which can be switched. Although this is a simple solution, in case of traffic bursts, this would lead to huge packet losses, thereby disrupting the service guarantees of a connection. Thus, this method is usually not adopted.

The next alternative is to have *dedicated switched paths* within the switch. This method is used in a majority of telecommunication switching systems. This option, however, would preclude the use of statistical multiplexing. This is because in statistical multiplexing, instead of allocating dedicated switched paths, resources are shared among multiple connections and a path is found only when packets arrive. Thus, dedicated internal switched paths cannot be used in technologies that employ statistical multiplexing (e.g. ATM).

The third option to avoid blocking is to use a *back-pressure* mechanism. In this technique, new inputs are allowed only if it can be ensured that this will not lead to packet loss. For applying back-pressure, feedback signals are sent to the previous stage for throttling the input rate. Instead of solving the problem, back-pressure merely shifts the problem to a preceding stage. This shifting process may aggravate the problem to such an extent that packets may have to be queued up at the input of the switch. As one shall observe shortly, input buffering is not the most preferred form of buffering. Thus, though switches using the back-pressure mechanism and giving optimal throughput have been designed, this technique is not very popular.

Finally, the best option is to *buffer* excess packets that cannot be switched at one go. These buffered packets are then served, as and when the resources are available. Indeed, buffering or queueing is a key issue in all switch designs. The most important design considerations are where to buffer and what buffer size to maintain. The first consideration (i.e. *where to buffer*) influences the throughput and the scalability of the switch. There are basically three approaches to the placement of buffers, viz. *input buffering*, *output buffering* and *central buffering*. All three approaches are detailed in the following sub-sections.

   The second consideration (i.e. *what buffer size to maintain*) also influences the throughput of the switch. Along with throughput, the buffer size also affects two important performance measures, viz. *loss* and *delay*. Choosing the optimal value of buffer size is an important design decision. This decision is essentially a trade-off between loss and throughput on one side, and delay and cost on the other. A large value of buffer size means that even during heavy loads, there is very little loss, thereby giving a better throughput. However, a large value also means that the mean waiting time of a packet in the buffers increases, thereby increasing the average delay of the packet. Moreover, keeping large buffers also increases the cost of the switch. The inverse logic applies to a small buffer size.

8.4.2.1 Input Buffering

In switch designs employing *input buffering*, the blocking problem is solved by buffering the packets at each inlet of the switch. The packets are buffered until the control logic allows the packets to cross the switching fabric. In this scheme, each inlet maintains its separate buffers. Figure 8.3 shows a schematic diagram for input buffering.

   The performance analysis of input buffering is tightly coupled with the *scheduling strategy* used to serve the buffered

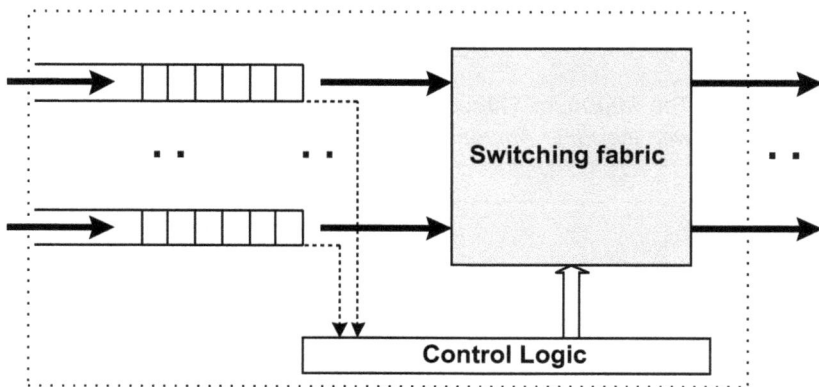

Fig. 8.3 *Input buffering*

packets. Some of the packet-scheduling approaches are discussed below.

One of the simplest packet scheduling strategies for input buffering is the *First-in-first-out (FIFO)* approach. In this approach, the first packet to enter the queue is also the first one to leave. This makes the design of the input queues very simple. However, the FIFO approach suffers from a serious problem called the *'head-of-line' (HOL) blocking*. Consider a case in which the first packet (say X) in one of the input queues is waiting for its destined outlet to be free. Now, packets queued up behind packet X cannot be switched even if their respective outlets are free. In such cases, it is said that the first packet is causing a head-of-line blocking. Due to this head-of-line blocking problem, a significant portion of the switch capacity is wasted. Thus, the maximum throughput is in the range of 55-70 per cent of the switching capacity only. Table 8.1 shows the maximum throughput achievable by using the FIFO approach for different number of input ports, (N). For as less as two inputs, the saturation throughput is only 75 per cent. This highlights the impact that HOL blocking has on the throughput of an input queued switch. Figure 8.4 shows the mean waiting time of packets as a function of the offered load. Offered load is the percentage of time that an input port receives packets for forwarding. It is observed from Fig. 8.4 that as the offered load approaches the 60 per cent mark, the waiting time approaches infinity. Thus, the throughput does not exceed the 60 per cent mark due to the HOL problem mentioned above.

Table 8.1 *The Maximum Throughput Achievable Using Input Buffering with the FIFO Approach [(Swi M. G. Hluchyj © 1988 IEEE)]*

N	Saturation throughput
1	1.0000
2	0.7500
3	0.6825
4	0.6533
5	0.6399
6	0.6302
7	0.6234
8	0.6184
∞	0.5858

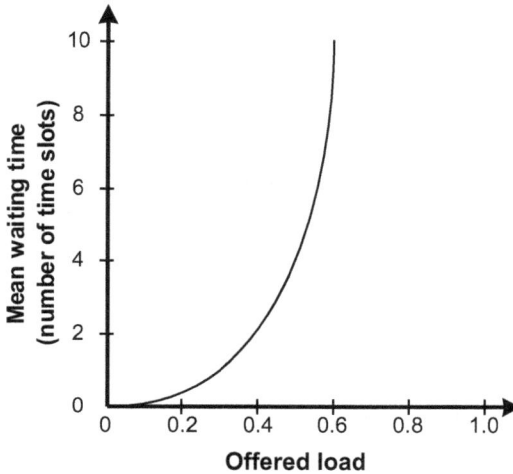

Source: [(Swi M.G. Hluchyj) © 1988 IEEE]

Fig. 8.4 *Mean waiting time for input buffering with the FIFO approach, for N = ∝*

Another issue associated with the FIFO approach needs to be considered here (in fact, this issue applies for the next scheduling approach as well). In case two or more packets at the head of different queues contend for the same outlet, one packet has to be selected out of the many contending packets. This selection can be done on a fixed priority basis; i.e. each queue has a static priority associated with it and the output contention is solved by choosing the packet of the highest priority. The other option is to assign priorities to queues on a dynamic basis, thereby precluding the possibility of starvation packets of low-priority queues. These two approaches constitute a trade-off between design complexity and fairness.

Another approach used for scheduling is the *multiple* queueing approach. In this approach, instead of maintaining a single queue, multiple queues are maintained at each inlet. In fact, to completely solve head-of-line blocking, the number of queues maintained at each inlet is equal to the number of outlets. By keeping multiple queues, significant improvements are observed in the throughput of the switch with near-optimal capacity utilization. This improvement, however, comes with a problem. The buffer requirements immediately shoot up by an order of magnitude. For a 256 × 256 switch, this would mean

65,536 queues and a total of 65536*B buffers, where 'B' is the number of buffer elements in each queue.

The third strategy used is the *Maximum matching approach* that tends to maximize the number of matches between inputs and outputs (i.e. finding the largest set of packets arriving at input ports that do not have any contention for the output ports), thereby yielding the best throughput achievable. There are, however, a couple of problems associated with this approach. First, it leads to starvation under certain traffic conditions. For example, if there is a packet lying in buffer 1 destined for outlet 2, and packets continuously arrive at buffer, 1 and 2, destined for outlets 1 and 2, respectively. Under this scenario, to maximize throughput, while the newly arrived packets will be allowed to pass through, the buffered packet will wait indefinitely. This will lead to starvation. Second, the computation involved in determining the maximum match can be very time-consuming at times, rendering the approach unsuitable for high-speed switching. Thus, the preferred approach is to go for heuristic, which provides a near-optimal solution (i.e. maximal solution), but not the exact solution. *Parallel Iterative Matching (PIM)*, *iterative round-robin matching with slip (SLIP)* and *Least Recently Used (LRU)* are examples of maximal matching approach. Interested readers are referred to [Swi Nick McKeown] for details.

8.4.2.2 Output Buffering

In the *output buffering* approach, the blocking problem is solved by buffering the packets at each outlet of the switch. Note that the placement of buffers is always defined in relation to the switching fabric. Thus, 'buffering at the output' implies that the buffered packets have crossed the switching fabric and are waiting for their destined outlets to be free, that is, the decision of 'where to switch' has already been made, and only then have the packets been buffered. Figure 8.5 shows a schematic diagram for output buffering.

Output buffering is preferred over input buffering because of its better performance in terms of throughput and delay, for the reasons explained below. However, this technique mandates that the switches be capable of internally switching multiple packets in a single time slot. This pre-condition is necessary to

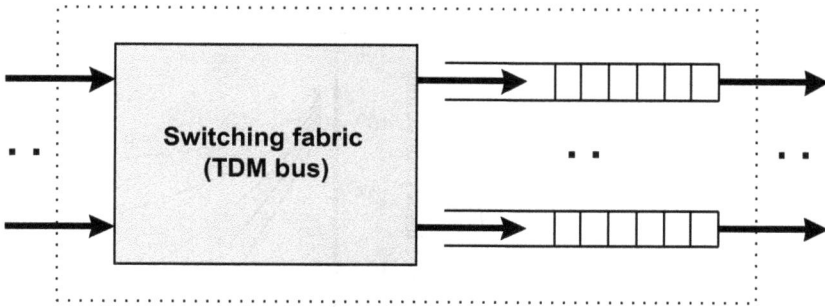

Fig. 8.5 *Output buffering*

avoid input buffering. In order to understand why this condition exists, reconsider the case when 'N' packets arrive at 'N' different inlets, all destined for the same outlet. If only one cell is switched out in a single time slot and the switch is not employing input buffering, where and how are the remaining N-1 packets be buffered? If packets are to be buffered at the output, the N-1 packets must have to cross the switching fabric. Now, to do this, the switching fabric must operate at a speed that is N times the port speed. In other words, the switch must be capable of switching internally multiple packets in a single time slot. This is the reason why a *speed-up factor* is required in switches employing the output buffering technique.

The packets in output queues are scheduled, typically using a FIFO approach. Since there is no HOL problem similar to the one that exists in input buffering, the throughput in the output-buffered switch is near-optimal. Figure 8.6(a) shows the mean waiting time of packets as a function of the offered load. The figure establishes the fact that with output buffering, a near 100 per cent throughput is achievable. At first look, the graph depicted in Figure 8.6(a) does not seem to be very clear. This is because by increasing the buffer size, there is an associated increase in the mean waiting time of packets. For a buffer of infinite size, the mean waiting time approaches infinity. But if one compares this graph with that for input buffering, one will observe that in the latter case, the mean waiting time reaches infinity near the 60 per cent mark.

Thus, the maximum throughput in an input-buffered switch cannot exceed 60 per cent. In contrast, the maximum throughput in an output-buffered switch can reach nearly 100

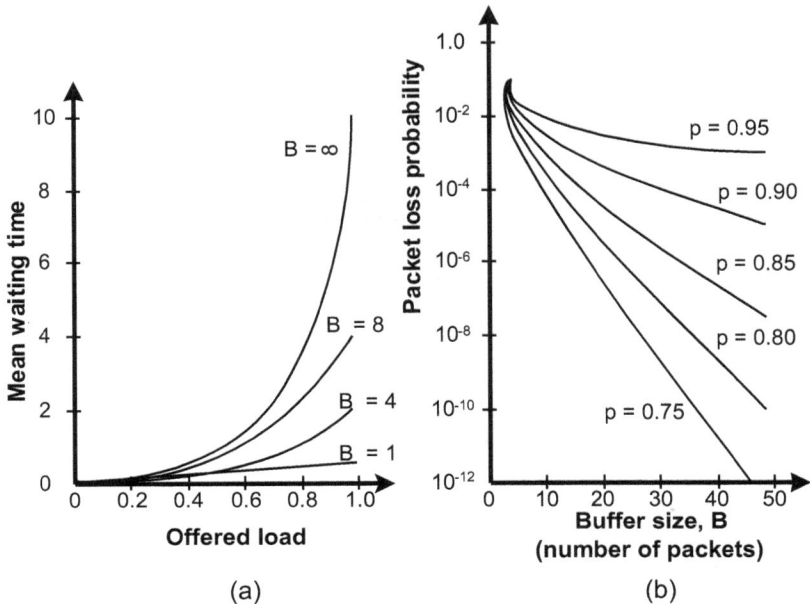

Source: [(Swi M.G. Hluchyj) © 1988 IEEE]

Fig. 8.6   *Performance of output buffering (a) Mean waiting time as a function of the offered load p, for N = ∞. (b) Packet loss probability as a function of buffer size B, for N = ∞*

per cent. As far as the mean waiting time is concerned, the increase in this value is attributed to the increase in buffer size. The rationale for keeping a larger buffer can be appreciated by looking at Fig. 8.6(b). A larger buffer results in a lower packet loss. For achieving a given packet loss, (say, 10^{-10}) at a given load (say, $p = 0.90$), a very large buffer is needed. In other words, the switch must have ample buffers to guarantee a packet loss of 10^{-10} under heavy loads ($p = 0.90$). The two graphs explain why the buffer size is a compromise between throughput and packet loss, on one side, and delay (mean waiting time), on the other.

   Although output buffering is performance-efficient, the speed-up factor leads to severe scalability problems. In order to solve this problem, instead of having a single buffer, multiple buffers are kept at each outlet, one for each inlet. After doing this, it is no larger necessary for the memory to operate at 'N' times the port speed. This case, however, is not very different from the multiple input buffering approach discussed earlier.

Thus, output buffering also leads to scalability problems because of excessive buffer requirements.

8.4.2.3 Shared Buffering

One of the drawbacks of both the input buffering and output buffering techniques is that buffers in both these approaches have to be tailored to handle the most severe of traffic bursts. Thus, the size of each buffer must be above a given threshold, wherein the threshold value depends upon the loss probability desired from the switching fabric. This leads to poor memory utilization of the buffers because the average traffic behaviour is much less severe. The low memory utilization because of keeping dedicated buffers for each output has prompted the switch designers to look for low-cost alternatives. The *shared buffering approach* is an excellent option in this regard. In this approach, a large pool of memory is shared between all inlets and outlets. This leads to substantial cost reduction because of the sharing of memory buffers. Figure 8.7 shows a schematic diagram for shared buffering.

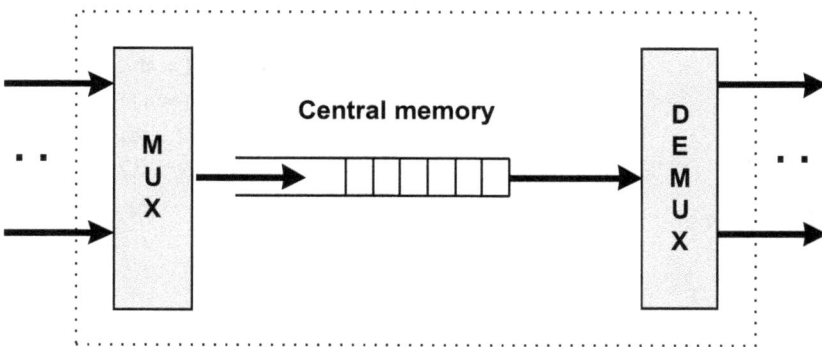

Fig. 8.7 *Shared buffering*

Although it is cost-effective, the shared-memory approach leads to some design complications. First, the memory management required for a shared buffering system is much more complicated as compared to the memory management requirements for other buffering approaches. This is because though a single memory is used for maintaining all queues, distinct logical queues for each outlet must still be maintained.

Two alternatives exist for maintaining logical queues. The first option is to have a fully shared memory, in which case the whole memory is available for the queue of a single outlet. The other option is to maintain an upper limit for the queue of each outlet (partitioning). The two approaches constitute a trade-off between lower packet loss and greater fairness.

Second, the memory of the shared buffer must be equal to at least the total bandwidth of all the inlets, and moreover, it must operate at N times the port speed. All these factors, especially the memory management complexity, make the shared-memory switch difficult to design.

Keeping the design complications aside, the performance of shared buffering is almost as good as that of output buffering. If all the parameters are kept the same, the packet loss probability in the former is much better than in the latter. Figure 8.8(a) shows the mean waiting time of packets as a function of the offered load. Fig. 8.6 again establishes the fact

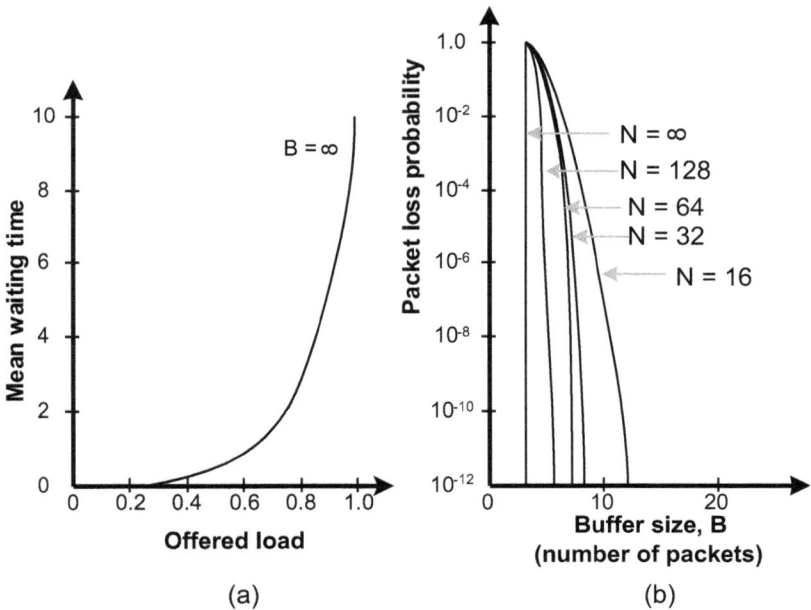

Source: [(Swi M.G. Hluchyj) © 1988 IEEE]

Fig. 8.8   *Performance of shared buffering (a) Mean waiting time as a function of the offered load, for N = ∞ (b) Packet loss probability as a function of buffer size B, for p = 0.90*

that with infinite buffering, a near 100 per cent throughput is achievable. If one compares Fig. 8.6(b) with Fig. 8.8(b) for $p = 0.9$, then shared buffering fares much better. In order to achieve a given packet loss, the number of buffers required in shared buffering is much less than that required in output buffering. This result is not surprising because, in the case of the former, buffers are not dedicated for each outlet as is the case in the latter. Thus, even if one of the buffer queues is sufficiently loaded, the spare capacity of other queues can be used to buffer packets. Note that Fig. 8.8(b) applies for a complete sharing approach. If buffer partitions are used, then obviously, the result will tend towards that of output buffering. In the extreme case of shared buffering, when a distinct partition is maintained for each output queue, the result will converge with that of output buffering.

8.4.3 Memory Look-up

Memory in switches is used for two purposes—to *store connection-related information* (like outgoing labels and connection priority), and to *buffer packets*. Thus, for each packet, memory is accessed for retrieving connection-related information and in some cases, for buffering. Thus, memory management and efficient memory look-up have important implications for switch design.

One of the techniques that is quite commonly used to speed up the access time for information retrieval is to employ Content Addressable Memory (CAM). For application areas that require some form of searching or translation, CAMs prove to be an attractive alternative to software solutions like searching and hashing. The trade-off is between the speed of the CAM (albeit at a higher cost) and the simplicity, reliability and the low-cost of software options. Earlier, because of the prohibitive costs of CAMs, people used to opt for the low-cost software option. However, with a fall in prices, there has been a phenomenal growth in the sales of CAMs. Now CAMs can be seen in switches, bridges and routers. Some of the application areas of CAMs are as follows:

- MAC address look-up in layer 2 bridges/switches
- VPI/VCI translation in ATM switches

- DLCI translation in frame relay switches
- Other address translation and protocol conversion in gateways
- Longest prefix matching for network layer protocols like IP4
- ARP cache in network servers

CAM is significantly different from the way data is stored and accessed in Random Access Memory (RAM). For reading data from a RAM, the specific memory address is supplied. Again, for writing something to the memory, the memory location must be supplied. In contrast, reading data from CAM does not require the memory address of any location. The read operation is similar to the way one recalls the names of the people whom one has met. The picture of the person is compared with a number of visual images stored in the brain. When a match is found, the information associated with the image is fetched and one recalls the name, age, and other details of the person. Similarly, for reading something from the CAM, some information is placed in a Special Purpose Register (SPR). The information here can be a virtual circuit identifier, an MAC address or any other piece of data. This information is then compared with the information stored in all the entries of CAM. Note that information stored in the SPR is compared with only a part of every CAM entry (that is, one compares a picture with a picture only, not with the name). Once a match is found, a particular flag is set to inform the user that the given information was located in the CAM. It is also possible that more than one match is found. In that case, some priority is accorded to sort the matching entries. The entry with the highest priority is then returned to the user.

8.4.4 Multi-casting

Traditional communication techniques normally involve the exchange of data between two entities only. Today, however, new applications like audio/video conferencing and collaborative computing demand simultaneous data exchange between more than two entities. This has led to new communication paradigms, one of which is *multi-casting*. Multi-casting is an attractive alternative to multiple, point-to-point connections, because it helps in reducing the network traffic significantly.

Multi-casting is essentially a one-to-many transmission technique, in which instead of sending a copy of data to each recipient, data is sent to a group of receivers using a multi-cast address. For multi-casting data, intermediate network elements (like switches) interpret the multi-cast addresses and make multiple copies of data only when required.

Depending upon the way a switch is designed, multi-casting can be inherently natural or can involve additional design complexity. As a general rule, it is much easier to multi-cast in those switch designs that broadcast packets, as compared to other switch designs. The manner in which multi-casting is handled is explained when different switching architectures are discussed.

8.5 SWITCHING ARCHITECTURES

Different writers tend to classify switching architectures differently. Most of them, however, categorize all switching architectures under three broad categories, viz. *shared-memory architecture*, *shared-medium architecture* and *space-division architecture*. The following sections discuss each of the three architectures. In each section, first the *basic aspects* of the architecture are covered, followed by a discussion on issues pertaining to *buffering*, *multi-casting* and *scalability*.

8.6 SHARED-MEMORY ARCHITECTURE

A shared-memory switch uses a common pool of memory to buffer packets for all inlets and outlets. The shared memory is typically a dual port memory. A dual port memory is advantageous because it permits simultaneous read and write operations, thereby reducing the time required to perform a given set of operations by half. However, it occupies more space than a single-port memory. Figure 8.9 shows a schematic diagram of the shared-memory architecture. Examples of switches employing the shared-memory approach include CNET's Prelude switch and Hitachi's shared buffer switch.

The operation of a shared-memory switch is relatively straightforward. Packets from all the inlets are multiplexed into

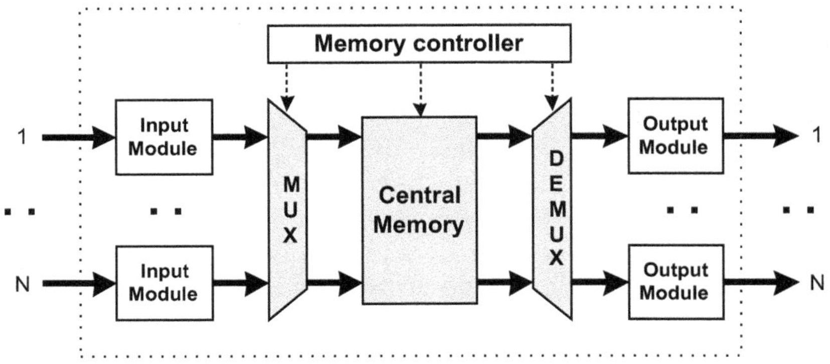

Fig. 8.9 *Shared-memory architecture*

a single stream and written to the dual port memory. At the output end, the packets are read out sequentially and de-multiplexed on the basis of their outgoing port identifier. It is the task of the memory controller to decide upon the order in which the packets are written and read out. Some aspects of shared-memory architecture are detailed below.

8.6.1 Buffering

As is obvious from its name, a shared-memory switch uses the shared buffering approach. Thus, a shared-memory switch has all the advantages and disadvantages associated with shared buffering. The most important advantage is the significant cost reduction because of lower memory requirements. The most important disadvantage is that complicated memory management is required.

8.6.2 Multi-casting

Shared-memory switches are not very well suited for broadcasting/multi-casting. This is because of the centralized approach adopted. Still, there exist techniques to incorporate multi-casting in a shared-memory switch. This, however, requires additional logic in the switching fabric. Two of these techniques are:

- **Copy networks:** One of the simple (but inefficient) methods to implement multi-casting, without complicating the internal design, is to employ copy networks. A copy network makes multiple copies of a multi-cast packet before the packet enters the switch. After the copies are made, each packet is treated as a unicast packet. Note that a copy is made for each outlet for which the original multi-cast packet was destined, with each copy having its distinct label. Thus, though copy networks simplify the internal design of the switch, complicated logic is now required at the beginning of the switch. Moreover, copy networks result in unnecessary wastage of buffers and the internal capacity of the switch. After all, multi-casting is useful only if the packets are duplicated as late in the switching process as possible.

- **Address buffering:** The drawbacks of a copy network can be removed by making multiple copies of the packet addresses, rather than copies of the packets themselves. In this approach, when a multi-cast packet arrives, a pointer entry is copied in each of the logical queues of the outlet for which the multi-cast packet is destined. Since, instead of the whole packet, only a small address is copied, considerable memory can be saved. This copying, however, requires that the shared memory must now operate at N^2 times the port speed. This is because in the worst case, if N multi-cast packets arrive, each destined for all the outlets, the central memory must do N^2 write operations in a single time slot. An $O(N^2)$ speed-up factor can cause severe scalability problems. One of the ways to handle this problem is to have a multi-bank memory (with each output queue having its own memory). By keeping multi-bank memories, the speed-up factor decreases by an order of N. This, however, increases the total cost of the system.

8.6.3 Scalability

The main problem in scaling shared-memory switches is the memory look-up time. For completely non-blocking behaviour, both the read and write operations must each take place at N times the port speed. Thus, for an $N \times N$ switch having links operating at a speed V, and bus width w, the memory speeds required for single-port and dual-ported memory are $(2NV/w)$ Hz and (NV/w) Hz, respectively. Implementing multi-casting further aggravates the problem. Overall, shared-memory switches do not scale very well.

8.7 SHARED-MEDIUM ARCHITECTURE

In a shared-medium architecture, a broadcast medium is used to switch packets to their destined outlets. The broadcasting is done typically through time division multiplexing (using either a bus or a ring interconnect). Fig. 8.10 shows a schematic diagram of the shared-medium architecture. FORE System's AS-X200 and NEC's ATOM are examples of switches using the shared-medium approach.

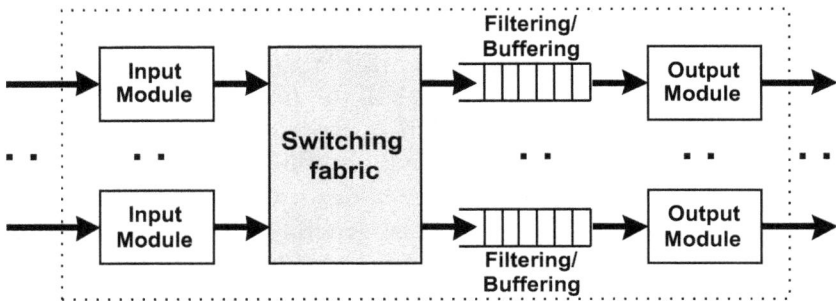

Fig. 8.10 *Shared-medium architecture*

The operation of a shared medium switch can be explained as follows:

As a packet arrives, the input module determines the outgoing port of the packet, using the packet's virtual circuit number. The outgoing port number, along with other information like the multi-cast identifier, packet priority, etc. is attached to the packet in the form of a routing tag. This tag is used for subsequent processing of the packet. The packet is then broadcast to each outlet, using the broadcast medium. At the output side, the address filters decode the outgoing port address of each packet. If the packet is destined for that outlet, the packet is copied to the output buffer. Otherwise, the packet is discarded. The buffered packets are served as and when the outgoing transmission link is free.

Some aspects of shared-medium architecture are discussed below.

8.7.1 Buffering

In order to avoid external blocking, output buffering is used in shared-medium architecture. Thus, this switch architecture is optimal in terms of throughput. It also follows from the type of queuing that the shared medium (i.e. the bus or the ring) must operate at N times the port speed. Each filtering and buffering module also requires this speed-up factor of N.

8.7.2 Multi-casting

As mentioned earlier, multi-casting is inherently natural for switching fabrics (like shared-medium) that use the broadcast mechanism for transferring packets. In order to broadcast a packet to all outlets (or to a sub-set of it), the only requirement is that each address filter must be capable of recognizing a multi-cast address. This can be easily done through a bit-map vector, where-in a set bit in one of the fields implies that the outlet corresponding to that bit must accept the packet.

8.7.3 Scalability

The speed-up factor required for the address filters, memory buffers and the broadcast medium hamper the scalability of a shared medium switch. For an $N \times N$ switch having links operating at a speed V, the bus must operate at (NV/w) hz, where w is the width of the bus. This means that a 16-port switch with OC-3 (155 Mbps) links, the operating speed of the bus must be 80 Mhz and 40 Mhz for 32 bits and 64 bits, respectively. As the number of ports is increased to 256, the new speed requirements turn out to be 1.28 Ghz and 0.64 Ghz for the corresponding bus width. This multiplicative increase in speed requirements has rendered the shared-medium approach unsuitable for large switching systems. Nonetheless, switches up to few Gb/s, employing the shared-medium approach, are not uncommon.

8.8 SPACE-DIVISION ARCHITECTURE

Due to the speed-up factor involved in both the shared-medium approach (the bus/ring must operate at N time the port speed) and shared-memory approach (read/write operations must be at N times the port speed), neither of the two architectures scale very well. The main reason for this is the use of a resource (like TDM bus or central memory) that must be capable of handling all the packets from all the inlets. Viewing at the problem differently, one realizes that neither of the two architectures exploits the availability of parallel switching paths.

Space-division architecture helps in solving problems associated with both these architectures. In this sub-section, two categories of space-division switches, viz. crossbar switches and banyan-based switches, are discussed.

8.8.1 Crossbar Switching

Crossbar switches have been used since ages. They are still pervasively used in telecommunication networks. The design of a crossbar switch is relatively straightforward. It consists of an $N \times N$ matrix of *crosspoints*, each of which can either be enabled or be disabled. When a crosspoint is enabled, a dedicated path is set up between an inlet and outlet. For the proper functioning of the switch, only one crosspoint corresponding to an output link can be enabled. The converse, however, is not mandatory, because an input can be simultaneously connected to more than one output link (provided no other input is connected to the given set of outputs). Figure 8.11 provides a schematic diagram of a crossbar switch.

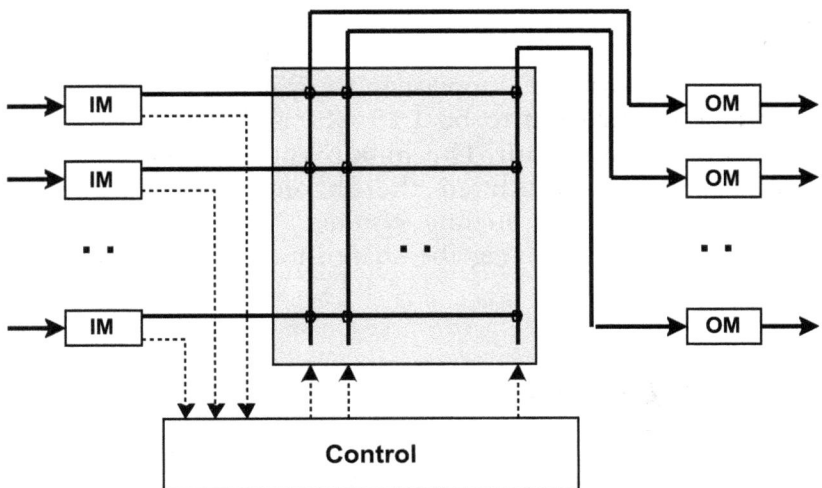

Fig. 8.11 *Crossbar switch*

Crosspoints can be implemented by using transmission gates, which can exist in two states. In a *cross state*, there is no connection between horizontal and vertical lines, i.e. the vertical input is connected to the vertical output and the horizontal input is connected to the horizontal output. When idle, a crosspoint remains in this state. In *bar state*, the vertical input is connected to the horizontal output and horizontal input is connected to the vertical output. Now, it can be verified that for connecting input line i to output line j, it is sufficient to keep bar switches (i, k), $k = 1, 2, \ldots (j - 1)$ and (m, j), $m = (i + 1), \ldots, N$ in the cross state and the bar switch (i, j) in the bar state.

8.8.1.1 Buffering

In a crossbar switch, buffering can be done in several ways. The simplest option is to have FIFO queues at each inlet (i.e. input buffering). This scheme is depicted in Fig. 8.12(a). Since this leads to HOL blocking, variants of input buffering can be used instead. The relative merits and demerits of various forms of input buffering have been discussed earlier. For input buffering, besides tackling the HOL blocking, there must also exist means to arbitrate HOL packets in different queues destined for the same outlet (i.e. to solve the external blocking problem). The solution to this problem leads to two different approaches for the design of the switching fabric. One approach is to have a centralized controller. The task of the

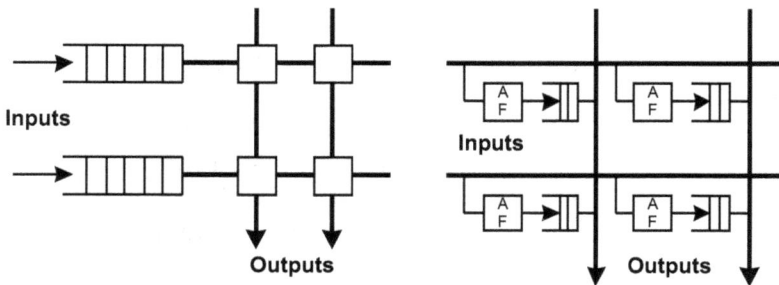

(a) Crossbar switch with input buffering.

(b) Crossbar switch with crosspoint buffering

Source: [(Swi H. Ahmadi) © 1989 IEEE]

Fig. 8.12   *Buffering in crossbar switches*

centralized controller is to find a matrix configuration that maximizes the number of packets being switched in a single time slot. The requirement is that not only must the solution be optimal, the computation must also be done in real-time. In order to achieve this aim and to solve the problem of external blocking, the controller should have global knowledge of the packets present in the input queues and the destined outlets of each packet. This requirement complicates the design of the controller, thereby increasing the cost of the switch.

Another approach is to opt for a distributed controller. In this approach, each output maintains its separate arbiter. The arbiter inspects packets coming from all the inputs and chooses one of them by some fair algorithm. Control signals are then sent to all but one inlet, requesting to block the transmission of packets. For details, readers are referred to [Swi Fouad A. Tobagi].

The next buffering option for the crossbar switch is to use output buffering. Output buffering mandates that the crossbar must operate at speeds greater than the port speed. This is required to prevent either input buffering or packet loss (due to packet discard). The actual speed-up factor required depends upon the traffic characteristic. For bursty traffic with peak rates that are small as compared to port speed, a speed-up factor of 2 is sufficient. As the ratio of the peak rate to the port speed is increased, the speed-up factor required also increases.

Another option is to use, what is termed as, *crosspoint buffering* (see Fig. 8.12(b)). This option is attractive because it alleviates problems associated with both input buffering (HOL blocking) and output buffering (the speed-up factor). In this approach, packets are buffered at each crosspoint. Crosspoint buffering, however, leads to severe scalability problems. This is because an $N \times N$ switch already has N^2 crosspoints. With crosspoint buffering, the total number of buffers required goes up to N^2B. This enormous memory requirement and associated design complexity renders the crosspoint buffering approach unsuitable except for very small systems.

8.8.1.2 Multi-casting

Multi-casting in crossbar switches is much more difficult than it may seem to be in the first place. The following will make this point more clear. Multi-casting seems easy because each inlet is connected to all the outlets, and hence, theoretically, a packet can easily be broadcast to all the outlets. However, consider what happens if input buffering (with FIFO) is used and one or more destined outlets of the multi-cast packet is not available. If the multi-cast packet is held till all the destined outlets are free, the packet can get excessively delayed (which will disrupt its delay and jitter bounds). Moreover, the FIFO approach in input buffering will lead to the HOL blocking problem. If the multi-cast packet is forwarded to one or more of the free outlets (without waiting for all the outlets to be free), the control logic must keep track of the outlets to which the packet has already been sent, and the outlets to which the packet is due.

Keeping track of all this is not a simple task. If output buffering is used instead of input buffering, then in the worst case, the crossbar must operate at N times the port speed. This worst case scenario results in the crossbar switch losing all the advantage it had over the shared-medium approach.

8.8.2 Banyan-based Switching

One of the major drawbacks of crossbar switching is that the number of crosspoints required for an $N \times N$ switch is N^2. This drawback prompted designers to look for alternatives that would require fewer numbers of crosspoints than N^2. *Banyan-based multi-stage switches* are one such alternative. These switches derive their name because of their tree topology. It is customary to refer to a multi-stage Banyan switch as a *Banyan network*. Banyan networks are characterized by the following attributes:

- **Multi-stage network:** All Banyan networks are composed of $b \times b$ crosspoint switches. The total number of stages in a Banyan network is $\log_b N$ with N/b crosspoint switches in each stage. This gives a total number of $b^2*(N/b)*\log_b N$ crosspoints. For a 32-input/32-output Banyan network composed of 2×2 crosspoint switches, this comes to 320 crosspoints. This is almost one-third of the crosspoints required for a single-stage crossbar switch ($32 \times 32 = 1024$).
- **Self-routing:** In a Banyan network, there exists only one path between an input and output. This path is represented by a k-bit output address $(a_1, a_2, \ldots, a_k)$, where

k is the number of stages in the Banyan network. In each intermediate stage j, the jth bit of the output port address is used to make a forwarding decision. For example, consider a Banyan network composed of binary crosspoint switches. In this network, if $a_j = 0$, the packet is forwarded to the upper output of the crosspoint switch (at the jth stage), and if $a_j = 1$, the packet is forwarded to the lower output. Since, once the port address of a packet is decided, the packet is automatically routed to the specified output, Banyan networks are said to exhibit *self-routing behaviour*. Fig. 8.13 shows how a packet with output address 010 is routed to the destined output port. Note that the leftmost bit is used in the first stage, the second bit from the left in the second stage, and so on.

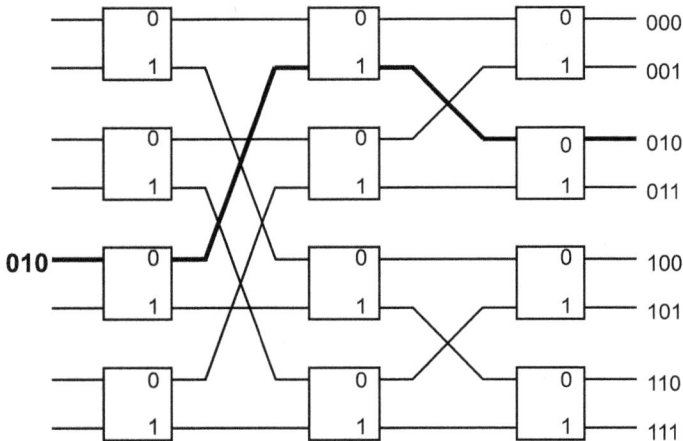

Fig. 8.13 *Self-routing in a Banyan network*

The above characteristics impose the following limitations on a Banyan network:

- **Internal and external blocking:** One of the most severe limitations of Banyan networks is *internal* and *external blocking*. Internal blocking occurs when the inputs of an intermediate crosspoint switch contend for the same output link (see Fig. 8.14(a)). This problem can be solved by a number of means like buffering conflicting packets, internal speed-up and sorting. External blocking, on the

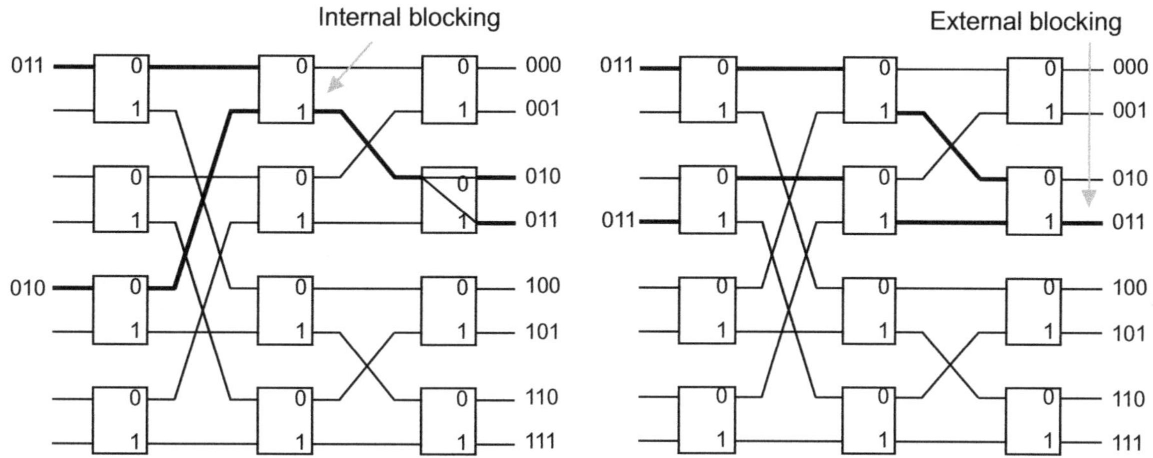

Source: [(Swi H. Ahmadi) © 1989 IEEE]

Fig. 8.14 Internal and external blocking in Banyan networks

other hand, occurs when two or more packets are destined for the same output port in a single time slot (see Figure 8.14(b)). Buffering is used to resolve the problem of external blocking.

- **Non-realizable permutations:** The reduced requirement of crosspoints in a Banyan networks has an assistant performance limitation. It is that not all permutations are now realizable. Recall that in an $N \times N$ crossbar matrix, without external blocking, there can be N simultaneous paths for data transfer and a total of $N!$ possible permutations. In contrast, a banyan network allows only $2^{kN/2}$ permutations, where $k = \log_2 N$, number of stages in the switch. This factor can be calculated by knowing that each crosspoint switch can be in one of the two states and that there are a total of $(N/2)*\log_2 N$ crosspoint switches. In essence, this factor means that there can be cases in which there can be a non-realizable permutation without external blocking.

8.8.2.1 Blocking Solutions in Banyan Networks

In Banyan networks, the problem of blocking can be solved by using one of the following options: *buffering, internal speed-up*, and *sorting*. In this section, buffering and sorting techniques are discussed.

- **Buffering:** Both internal and external blocking problems can be solved by buffering conflicting packets. In Banyan networks, three types of buffering are possible—*input buffering, output buffering* and *internal buffering*. Internal buffering is again classified into input buffering, output buffering and shared-memory buffering. The general issues pertaining to each of these techniques have already been discussed. Nonetheless, a few comments on internal buffering are in order. With internal buffering, one of the two conflicting packets in a 2×2 switch is allowed to pass through, while the second one is buffered. In the next time slot, for the buffered packet to move forward, either the buffer of the next stage should be empty or the buffered packet of the next stage should move forward. In case neither these conditions holds true, some mechanism is required to inform the packet of the preceding stage to remain in its buffer. This is achieved by using upstream control signals. Indeed, internal buffering has its own share of problems. As the number of intermediate stages increases, internal buffering disrupts the delay and jitter guarantees. Moreover, determining the optimal buffer size is not easy, because this tends to be a function of traffic behaviour. For traffic distributed uniformly across all the links, a fewer number of buffer elements are required than when traffic is concentrated across a few links.

- **Sorting:** Banyan networks have an interesting property in that when the packets at inputs are sorted according to their output port addresses, there is no internal blocking. Going back to Fig. 8.14, one will observe that a packet with higher output address (011) is above the packet with a lower output address (010). For the same inputs, Fig. 8.15 shows how the internal blocking is removed by introducing a *Batcher-Bitonic sort network* before the Banyan network. By introducing a sorter before the Banyan network, the network becomes internally non-blocking. The problem of external blocking, however, still remains.

8.8.2.2 Multi-casting

Due to the connection-oriented approach of Banyan networks, it is difficult to incorporate multi-casting capability. One way to incorporate multi-casting in Banyan networks is to introduce a copy network before the Banyan network. In this model, the switch consists of a copy network, followed by a distribution network and a routing network. The copy network makes multiple copies of multi-cast packets. Then, the distribution network, which essentially is a buffered Banyan network, distributes packets evenly across all inputs of the routing network. Between the copy network and distribution network are the Broadcast and Group Translators (BGTs). The purpose of a BGT is to translate the destination address of the copies of packets to the appropriate destination address. The routing network forwards the packets to the appropriate output port.

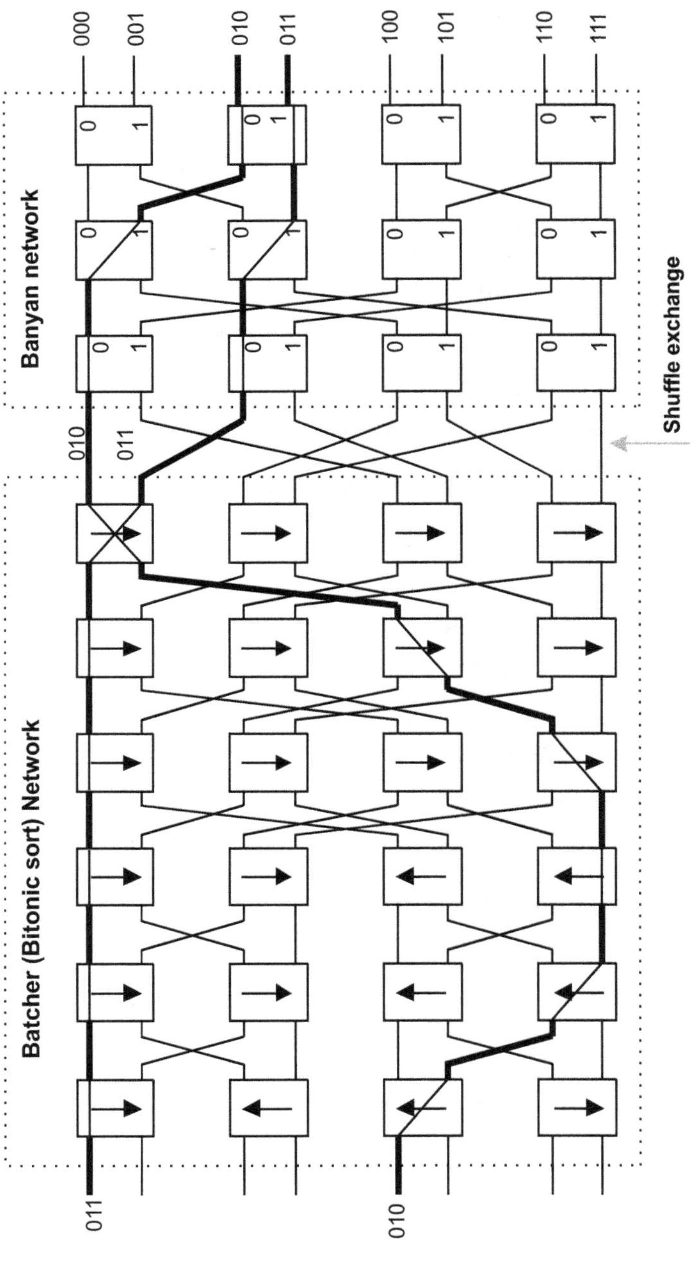

Source: [(Swi H. Ahmadi) © 1989 IEEE]

Fig. 8.15 *Batcher-Banyan network*

8.8.2.3 Scalability

Banyan networks have many desirable features that make their implementation attractive. First, they can be constructed modularly from smaller switches. Second, their regularity and interconnection pattern make their Very Large Scale Integration (VLSI) implementation a good proposition. Still, internal and external blocking severely hamper the throughput of large banyan networks. In fact, the maximum achievable throughput for a moderate size, single-buffered banyan network ($N = 32$) is as low as 0.4 [Swi Fouad A. Tobagi]. This value falls to 0.26 for larger networks ($N = 1024$). The throughput can be improved by increasing the buffer size. Similar performance gains are also observed by introducing distribution networks before the routing network and by decreasing the number of intermediate stages.

8.9 EXAMPLE: SWITCHING IN ATM

ATM has emerged as the preferred transfer mode for high-speed transmission of data. This development is the result of sustained research in the field of switching over a period of one-and-a-half decades (1985-2000). The endeavour is not only to develop switching fabrics capable of switching millions of cells per second, but also to guarantee the QoS promises of ATM technology. After all, efficient switching operations and efficient traffic management of cells hold the key to delivering the QoS promises of ATM.

The fact that ATM is a *hybrid* of traditional circuit switching (for voice transfer) and packet switching (for data transfer) has an important bearing on ATM switching architectures. However, neither packet switching architectures nor circuit switching architectures are directly applicable to ATM switches. The following paragraphs explain why.

Although ATM is connection-oriented, it is significantly different from circuit switching. In circuit switching, dedicated switched paths are set up once a voice channel is established. In contrast, ATM uses a store-and-forward switching that is based on statistical multiplexing. This implies that ATM cells are buffered until the destined output ports are free. While buffering is a key design issue in ATM switches, voice frames are never buffered in circuit switches. Moreover, the speeds at which ATM switches operate are many times higher than the conventional circuit switches. Further, while voice is carried in fixed voice channels (usually 64 kbps), ATM switches are capable of supporting multiple data rates (from as low as few kilobytes/s to as high as few gigabytes/s). Finally, ATM switches do usage control at the ingress point of the network. They also support traffic management features like prioritized transfer (based on multiple classes of traffic), congestion notification and flow control. All these features are not required in conventional circuit switches.

Even compared against packet switching technologies like X.25 and frame relay, ATM switching is different. This is despite the fact that ATM technology is inspired from frame relay. Speed is again a crucial difference. The early packet switches operated at very low speeds (~64 kbps), and thus, software-based processing was possible. Compare this with 150Mbps ATM switches that make hardware-based processing a pre-requisite. Moreover, the low speeds and protocol complexity of X.25 makes it a poor candidate for carrying real-time data like voice and video. Although, as compared to X.25, frame relay is much simpler, but like X.25, it also supports transmission of variable-sized frames. Variable-sized frames introduce unpredictable delay in the switch and hence, increase the jitter. Thus, both X.25 and frame relay are not well-suited for real-time applications. In contrast, ATM provides services like Constant Bit Rate (CBR) and real-time Variable Bit Rate (rt-VBR), which require bounded values of end-to-end transit delay and cell delay variation. This makes buffering of cells a very critical issue.

In essence, switching in ATM is a complicated process because of the following reasons:

- The high-speeds at which ATM switches operate (few megabytes/s to few hundred gigabytes/s)
- The large number of inputs (~1000) that ATM switches must support
- The variety of data rates that ATM switches support
- The traffic management features that ATM switches support (see Chapter 12)

Given the above factors, an ATM switch is more than just an interconnection structure that buffers and routes cells. On the contrary, the control plane functions and traffic management functions of an ATM switch pose significant design complexity. Although this chapter is not an attempt to explore the design issues associated in performing the control plane and traffic management functions, there is a mention of the related functions that an ATM switch must perform.

The functions of an ATM switch are divided according to the three-dimensional reference model. Note that the ATM reference model defines three planes—the *user plane*, the *control plane* and the *management plane*. The following sections describe the user and control plane functions. The management plane functions of an ATM switch include *fault management* and *performance management*. The management plane functions are not discussed any further in this chapter.

8.9.1 User Plane Functions

In the user plane, the primary concern of an ATM switch is to transfer user cells from incoming links to outgoing links. The ATM switch, however, remains transparent to the user information carried in the 48-byte payloads of every cell. The only part of the cell inspected and processed is the 5-byte cell header. The cell processing includes *Virtual Path Identifier/Virtual Channel Identifier (VPI/VCI) translation*, *Header Error Control (HEC) verification* and *few other functions*.

In the context of the user plane, the three core modules (input module, switching fabric and output module) perform three distinct sets of functions. The functions of each of the modules are discussed below.

8.9.1.1 Input Module

The Input Module (IM) of an ATM switch performs several functions. The most important among these is the VPI/VCI translation. The following are the steps associated with the table look-up and VPI/VCI translation:

1. As a cell arrives, the VPI field of the cell header is extracted. This value is used for locating an entry in the translation table associated with each link. The table look-up may be through hashing, search trees, or through

content addressable memories. In order to minimize the table look-up time, only a sub-set of the total VPI/VCI values is allocated. For example, only 12 bits for VCI and 8 bits for VPI may be used. This reduces the search space, and hence the search time of table look-up.

2. The table entry is then used to determine whether the particular VPI value corresponds to a VPC or a VCC, in other words, whether the VPI is a part of a Virtual Path Connection (VPC) or part of a Virtual Channel Connection (VCC). Note that a VPC is switched, on the basis of the VPI value only, whereas a VCC is switched by using the VPI/VCI pair.

3. If the VPI value corresponds to a VPC, then relevant information is fetched from the translation table entry. This information includes the new VPI value and the outgoing link identifier. Additionally, other information like cell priority, delay bound and jitter bound may also be fetched from the table.

4. If the VPI value corresponds to a VCC, another level of search is made by using the VCI value. Again, the appropriate information is fetched from the translation table.

5. The old values of the VPI/VCI are then replaced by the new values (in case of step 3, only the VPI value is replaced). A routing tag is then attached to the cell. This tag contains the information that was fetched from the translation table. Apart from the parameters already mentioned, a multi-cast cell may also carry a bit-map corresponding to the set of outlets for which the cell is destined. The cell is then forwarded to the switching fabric.

The above are the steps involved in table look-up and VPI/VCI translation. Apart from this, the input module also performs the following functions:

- It verifies the Header Error Check (HEC) value of every cell and discards erroneous cells.
- It discards empty cells (i.e. discards cells that carry useful information).
- It identifies the cell boundaries (using the cell delineation procedures).

- It converts optical signals into electrical signals (applicable for SONET/Synchronous Digital Hierarchy (SDH) based interfaces only).
- It separates the user cells from signalling cells and Operation, Administration and Maintenance (OAM) cells. The signalling cells are forwarded to the Connection Admission Control (CAC) module and OAM cells to the management module.
- It does Usage Parameter Control (UPC) and Network Parameter Control (NPC) at the ingress point of the network. This is done to ensure that the user bandwidth utilization is within permitted levels.

8.9.1.2 Switching Fabric

The primary function of an ATM switching fabric is to *transfer cells* from input modules to output modules. Another important function of the switching fabric is to buffer excess cells. Besides routing and buffering, other functions of the switching fabric include.

- Multi-casting/broadcasting cells
- Providing support for fault tolerance
- Traffic concentration and multiplexing
- Traffic expansion and de-multiplexing
- Congestion notification by setting EFCI bit

8.9.1.3 Output Module

The Output Module (OM) is the direct opposite of input module. Thus, the functions of the output module can easily be deduced from the functions of the input module. To summarize, the output module performs the following functions:

- It generates the new HEC value and adds it to the cell header.
- It generates empty cells, as and when required, and sends them (cell rate decoupling).
- It multiplexes user cells with signalling cells and OAM cells.
- It converts electrical signals into optical signals (applicable for SONET/SDH-based interfaces only).

8.9.2 Control Plane Functions

The control plane in the ATM reference model is responsible for the *establishment*, *monitoring*, and *release* of Switched Virtual Circuit (SVC). This control plane function is performed by exchanging signalling messages. As far as signalling is concerned, an ATM switch forwards all the signalling cells received over the reserved VPI/VCI pair (0,5) to the Connection Admission Control (CAC) module. The CAC module processes every connection request, and depending upon the availability of resources, determines whether new connections can be accepted or not. The concept of signalling is detailed in Chapter 10.

8.10 EXAMPLE OF ATM SWITCH ARCHITECTURE: KNOCKOUT SWITCH

The previous section discussed the functions of an ATM switch. This section describes one of the popular types of ATM switch referred to as *knockout switch*.

Typically, output buffering is optimal in terms of throughput, it requires that the memory buffers and switching fabric operate at N times the port speed. Knockout switches reduce this speed-up factor by deliberately introducing some loss probability in the switching fabric. To understand the fundamentals of knockout switching, let us get down to switching basics.

The speed-up factor of N is required because it is quite possible that cells arriving from all the N inlets are destined for the same outlet. Now, if it is decided that only L cells out of the maximum N are buffered in a cell-time, then the speed-up factor is reduced to L. This decision, however, introduces a finite probability of some cells (N-L in the worst case) getting dropped. If, somehow, the cell loss ratio of a connection is maintained within permissible limits, then this approach can easily be adopted. Interestingly, [Swi Y. S. Yeh] have shown that to achieve a cell loss probability of 10^{-10}, a value of $L=12$ is sufficient (this result applies for 90 per cent load and for arbitrarily large values of N). Thus, even for a very large value of N, and even for high load, the speed factor is more or less constant at L.

Besides requiring a very low speed-up factor, knockout switches offer many other desirable features like *self-routing*, *low latency* and *non-blocking transfer*. Moreover, the buffer requirement has a linear growth with N. This is in sharp contrast to many other switching fabrics with output buffering that have quadratic growth with N. All this makes a knockout switch an excellent option for large switching systems.

8.10.1 Components of Knockout Switch

Figure 8.16 depicts a schematic diagram of a knockout switch. Cells arriving at an input are broadcast to all the output ports, using a passive broadcast bus. The broadcast cells are intercepted by a bus interface that precedes each output port. Indeed, the core of the knockout switch resides in the bus interface (see Fig. 8.17). The bus interface first filters the cells destined for its output port and discards the remaining cells. It then queues up cells that are not discarded in a shared buffer. The buffered cells are then served on a first-in-first-out basis.

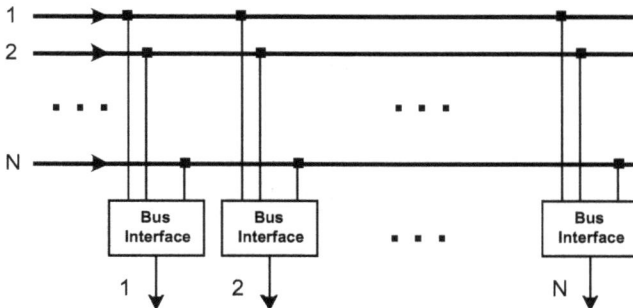

Fig. 8.16 *Basic structure of a knockout switch*

As shown in Figure 8.17, a bus interface has three major components: *cell filter*, *concentrator* and *shared buffer*. Each of these components is discussed below.

8.10.1.1 Cell Filters

Each bus interface has N cell filters, one for each input line. Each filter examines the output port address of the received cell. Irrespective of the input line on which the cell is received,

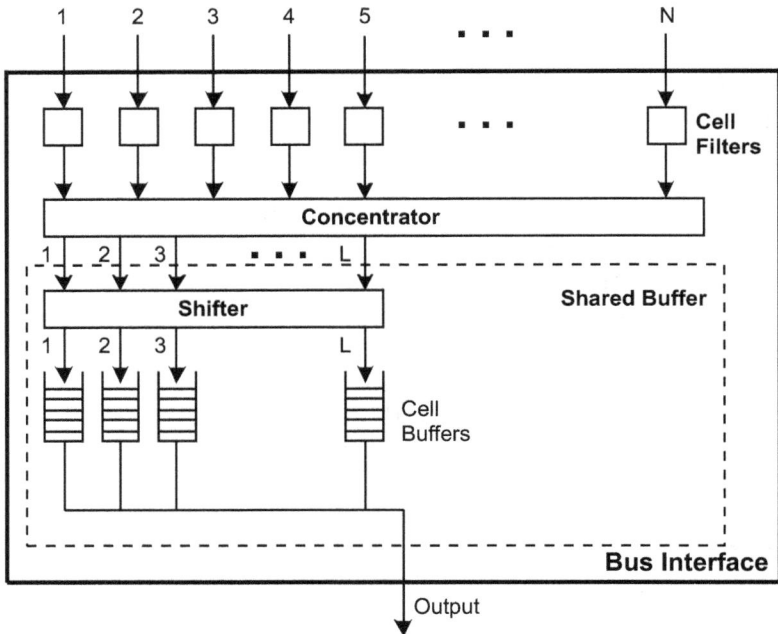

Source: [(Swi Y.S. Yeh) © 1987 IEEE]

Fig. 8.17 *Bus interface of a knockout switch*

a cell filter in the bus interface i allows only cells destined for output port i to pass through. All the remaining cells are blocked.

The implementation of a cell filter is straightforward. A cell carries the output port address as a binary number (with $\log_2 N$ bits). This number is compared bit-by-bit against the output address of the bus interface. If all the bits match, the cell is allowed to go through, otherwise as soon as a compare operation fails, the forward path is blocked for that particular time-slot. If some part of the cell goes through before the path is blocked, then the garbage bits are ignored by the subsequent stages of the bus interface.

8.10.1.2 Concentrator

The cells that successfully cross the cell filter enter the concentrator. The concentrator is an N-input, L-output element that concentrates a number of input lines (N) onto a given number of output lines (L). Specifically, if there are more than L cells for an output port in a particular time slot, only L cells are allowed to pass through, while the remaining cells are dropped. Thus, the concentrator introduces some cell loss probability in the switching process. It can be argued that since cell loss is unavoidable in any practical communication network (due to link failures, transmission errors, buffer overflows, etc.), a switch also has the right to induce some. However, the main issue is whether the Cell Loss Ratio (CLR) guarantees of ATM connection are maintained despite the induced cell loss. If yes, then the logic given for cell loss in knockout switch is justifiable. Now, the typical values of CLR are $\sim 10^{-10}$. It is experimentally proven that even for 100 per cent loading, a value of $L = 12$ (for arbitrarily large value of N) is sufficient to achieve a cell loss of 10^{-10}. Thus, the cell loss of a knockout switch is well with the permissible limits of CLR for ATM connections. It can be further proven that each additional output line of the concentrator (beyond $L = 8$) leads to a decrease in the loss probability by a factor of 10. This is a significant result, because even if the CLR guarantees were to reduce beyond 10^{-10}, with very little modification, the desired result can be obtained. This result also means that the buffer requirement is now a function of $N*L$ (N output ports with L queues each) rather than N^2. This explains why knockout switches have linear growth with N.

The concentrator consists of *contention switches* and 1-bit *delay elements*. A contention switch is a 2×2 switching element, which chooses a *winner* and a *loser*. When cell arrives at only one input port, the arrived cell is automatically declared as the winner. If cells arrive at both the input ports, the winner may be decided on the basis of a static priority or on an alternating priority basis. In case of the former, the input towards left is given a higher priority. The 1-bit delay elements are single-input/single-output delay elements that ensure that the cells leave the concentrator simultaneously.

The peculiar design of the concentrator gives the knockout switch its name. This is because the cells are concentrated by using a *knockout competition*. As stated in previous paragraphs, the winner in a competition may be decided on the basis of a static priority or on an alternating priority basis. The knockout tournament is played across several sections, with one section for each concentrator output. In the tournament, each cell is allowed to compete (and lose) in at most L sections. As soon as a cell loses a match, it is eliminated from the competition for that section. This cell then moves towards the section on its right. There, it competes with other loser cells. A winner cell on the other hand, stays in the competition and fights with other winner cells in the next round. The tournament continues until L winners are chosen and K-L cells discarded. Here, K is the number of cells arriving simultaneously at the inputs of a concentrator. Note that a cell is discarded only if more than L cells simultaneously arrive at the input (that is, $K>L$). For all other cases, there is no cell discard.

Figure 8.18 depicts a 8-input, 4-output concentrator. The eight inputs are fed in pairs in four contention switches. The four winners remain in section 1, while the four losers move towards section two. These four losers are said to be 'knocked

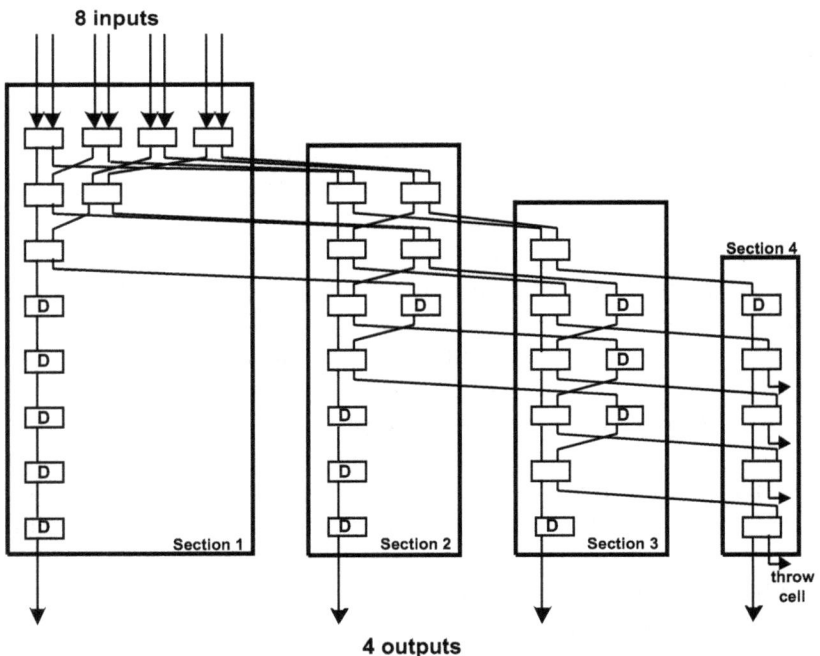

Source: [(Swi Y.S. Yeh) © 1987 IEEE]

Fig. 8.18   *The 8-input/4-output concentrator*

out' of the competition for section 1. However, they are still in the fray for the remaining three sections. The four winners compete in pairs to leave two winners. The remaining two then compete against one another to give the winner of section 1. Similar competition is held in all the remaining sections for choosing a total of L winners (four in this case). The remaining N-L are discarded (again four cells in this case are discarded).

8.10.1.3 Shared Buffer

The cells winning a concentrator output enter the shared buffer, the third component of the bus interface. Note that the speed-up factor requirement for output buffering is reduced significantly because of the N by L concentrator. However, L cells for buffering still requires a speed-up factor of L. To minimize this requirement, L separate FIFO buffers are used.

In the design of the shared buffer, one issue needs to be considered. It is that the cells tend to emerge from the concentrator in left-to-right direction (i.e. the cells first emerge from the leftmost section, then from the second section, and so on). Now, this will lead to overloading of buffers towards the left and underloading of buffers towards the right. In order to prevent this phenomenon, a *shifter* is used. The role of the shifter is to distribute cells in a cyclic fashion such that the L buffers are filled equally. The shifter is designed in a manner such that the total number of cells in each buffer does not differ by more than one cell. This is maintained, irrespective of the number of cells arriving in a particular time slot.

For example, consider the operation of an 8-input shifter. In the first time slot, five cells arrive at inputs numbered 1-5. These cells are passed without any shifting to outputs 1-5. In the second time slot, suppose four new cells arrive. In order to ensure that the cells are distributed equally, the four new cells are shifted to outputs 6, 7, 8, and 1. This cyclic operation of the shifter continues endlessly. Note that the cells will always arrive contiguously, and that too from left to right, because of the peculiar design of the concentrator.

The cyclic distribution of cells and the FIFO-based scheduling ensures that the cells of a particular virtual connection do not go out of sequence. It is left as an exercise for the readers to analyze the cell-ordering problem if either of the two conditions is violated.

8.10.2 Buffering

It follows from the above discussion that a knockout switch uses shared output buffering. The queues are both filled as well as served in a cyclic fashion. The shifter first fills the queues in a cyclic order. Then, the Head-Of-Line (HOL) cell in the leftmost queue is first served. After this, the HOL cell of the next queue is served. This process is repeated cyclically. All the inputs fully share the buffers of L output queues. Cell loss occurs if all the L queues are filled. In order to prevent cell loss at the output queues, the dimension of the queue must be chosen appropriately. Experimentally, to achieve a cell loss probability of nearly 10^{-10} for 90 per cent loads and $L = 12$, ten buffer elements (i.e. $B = 10$) are required per queue. This gives a total of 120 buffers per output port.

8.10.3 Multi-casting

Like most other broadcast-based switching fabrics, a knockout switch can realize multi-casting easily. All that is required now is additional control logic in the cell filters. Recall that a cell filter makes a bit-by-bit compare operation with the received output port address to filter cells. To incorporate multi-casting, this scheme has to be changed. One of the solutions is to use a bit-map vector, one bit for each output port. Each filter then makes a check on the corresponding bit to find out whether it is set or not. If it is set, then the cell is allowed to pass through, otherwise it is blocked.

8.10.4 Scalability

A knockout switch can be easily expandable to large values of N (~ 1000) with port speeds of 50 Mbps. This ease of scalability is the result of modular design and linear growth with N. In fact, the knockout switch can grow modularly from $N \times N$ to $JN \times JN$, where $J = 2, 3, \ldots$.

For example, four $N \times N$ knockout switches can be used to realize a $2N \times 2N$ switch with minor modifications. In the new approach, each concentrator has L additional inputs, with the inputs being received from the concentrator outputs of the previous stage. The operations of the remaining elements remain the same.

For details of various aspects of knockout switch, refer to [Swi Y. S. Yeh].

ⵏ8.11 CONCLUSION

This chapter covered the key issues in switch design and how different architectures tackle these problems. The key issue is that of contention and how packets are buffered when there is a conflict. Different solutions are available for buffering. Further, depending upon the buffering used, there are different switching architectures. A totally different approach is used in space-division architecture in which two examples, crossbar switches and banyan-based switches were discussed. Thereafter, the example of ATM switching were taken up where the key functions of an ATM switch was elaborated. Finally, the knockout switch was considered as an example of ATM switch architectures.

REVIEW QUESTIONS

Q 1. Explain the merits and demerits of input buffering, output buffering and shared buffering.

Q 2. What is the difference between shared-memory and shared-medium architecture?

Q 3. What is a banyan network? What are the key problems in a banyan network and how are they solved?

Q 4. Mention the key requirements and functions of an ATM switch.

Q 5. Why is a knockout switch so named? Explain in the context of the functions of concentrator. Also highlight how the induced cell loss does not lead to deterioration in the performance of the switch.

Q 6. What is the role of shifter in the knockout switch? What happens to cell sequencing if the shifter distributes cells randomly?

FURTHER READING

For further references on switching and the issues raised in this chapter, a number of good papers are available. [Swi M. G. Hluchyj] provides a good description of the issues related to queueing. The switching architectures are discussed in [Swi Fouad A. Tobagi] and [Swi H. Ahmadi]. For more on ATM switching, the reader is referred to [ATM S. Kasera] and [ATM M. Prycker]. For the case study on knockout switches, [Swi Y. S. Yeh] provides a detailed description.

ADDRESSING

9.1 INTRODUCTION

The primary goal of a network is to provide the means for two or more entities to communicate with each other. To communicate, some identifier is required so that one entity can uniquely and unambiguously identify another entity. This requirement is fulfilled by giving an address to the entities. *Addressing* refers to the process of assigning unique identifiers to entities for the purpose of communication.

Here, it is important to stress that addressing and routing concepts go in tandem (i.e. the addressing structure has a direct bearing on routing strategies). The best example in this regard is to consider the problems faced by the Internet community. During the initial phases of Internet development, Internet addresses (also called IP addresses) were distributed without taking into account the effect the address assignment would have on Internet routing infrastructure. The end result was that the routing tables started increasing at a rate that was becoming unmanageable. To prevent a collapse of the routing infrastructure, concrete steps were taken and techniques like *Classless Addressing* and *Classless Inter-Domain Routing (CIDR)* were introduced. The introduction of these techniques alone had a significant impact in stemming the growth of routing tables. In essence, an elegant addressing mechanism generally leads to an efficient routing process and this is why a discussion on routing is preceded by a discussion on addressing. While addressing concepts are discussed in this chapter, routing issues are addressed in Chapter 11.

In order to understand the important aspects of addressing, this chapter first defines addressing and looks at different ways in which addresses are classified. The classification is based on scope, utility, layer and functionality. Then, as examples, three addressing structures are detailed. The most important of the address structures discussed is the Internet address (or IP address) structure. Both classful and classless addressing formats for IP addresses are explained. Next E.164 addresses are explained. E.164 addressing forms the basis for addressing in telecommunication networks. Another important addressing format not covered in this book is the NSAP format. This format is useful in modelling the ATM end-system addresses.

9.2 CLASSIFICATION OF ADDRESSING TECHNIQUES

Addressing refers to the process of assigning unique identifiers to entities for the purpose of communication. In general, there are two different approaches used to assign addresses. In one approach, an address is assigned to an end-system. A good example of this approach is the telephone numbers assigned to telephone sets. In second approach, instead of assigning addresses to end-systems, the interfaces associated with end-systems are assigned addresses. IP addressing is an example of this approach because an IP router with multiple interfaces has multiple IP addresses.

In formal terms, addresses are classified or categorized in a number of different ways. Figure 9.1 illustrates four different methods of categorizing addresses, viz. *scope-wise*, *utility-wise*, *layer-wise* and *functionality-wise* classification. Each method of classification is explained in subsequent sub-sections.

9.2.1 Scope-wise Classification

This classification is based on the *scope* or the *geographical extent* in which a given address is applicable. This form of classification results in the two class of addresses, viz. *globally unique addresses* and *locally unique addresses*.

As the name suggests, globally unique addresses are unique throughout the globe. To provide global uniqueness, these addresses cannot be obtained or assigned arbitrarily. Rather, assignment of globally unique addresses is controlled by a central authority. For example, the Internet Network Information Center (INTERNIC) is the central authority responsible for the allocation of globally unique IP addresses to organizations joining the Internet. Such globally unique IP addresses assigned by the INTERNIC are also called *public* IP addresses. The addresses are assigned as a contiguous block. This contiguous block is also referred to as address space or address range. Note, however, that how the assigned address space is used, is beyond the control of INTERNIC, as this is an internal matter of the organization.

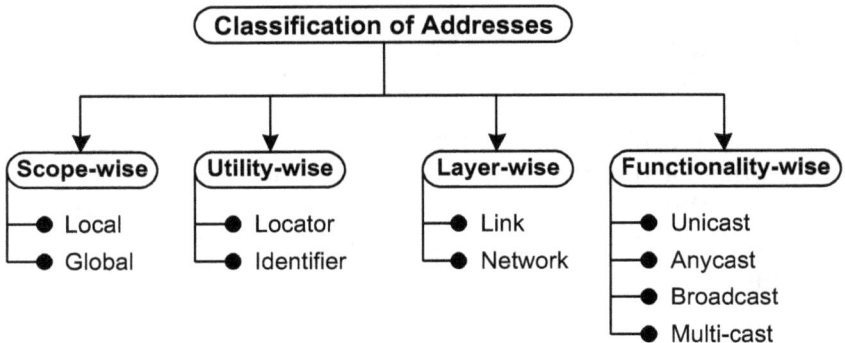

```
                    ( Classification of Addresses )
                                  |
        ┌─────────────────┬───────┴────────┬──────────────────────┐
        ▼                 ▼                ▼                      ▼
  (Scope-wise)      (Utility-wise)     (Layer-wise)      (Functionality-wise)
    ● Local           ● Locator          ● Link             ● Unicast
    ● Global          ● Identifier       ● Network          ● Anycast
                                                            ● Broadcast
                                                            ● Multi-cast
```

Fig. 9.1   *Classification of addresses*

A locally unique address has several connotations. For one, it may refer to an address that is unique within an organization, but is not used outside it. For example, an organization may use IP addresses to number the hosts and routers of its network, without obtaining such addresses from the INTERNIC. These addresses are referred to as *private addresses,* because they are used privately within an organization. Hosts bearing private addresses cannot directly communicate with external hosts. To do so, they require some other mechanism, whereby private addresses are converted to public addresses before packets are sent to external hosts. One such mechanism is called Network Address Translation (NAT) and is detailed in [RFC 1631].

Another interpretation of locally unique addresses is seen in the case of telephone numbers. Within the premises of a city or town, the telephone numbers are unique. However, the same numbers can be (and actually are) used across different cities. Thus, these telephone numbers are locally unique. At the same time, if the country code and the city code are specified alongside the local telephone number, the numbers become globally unique.

Yet another example of a locally unique address is the link level Virtual Channel Identifier (VCI) that identifies a virtual channel. Since the scope of these identifiers is limited to the link between two nodes, such identifiers may also be viewed as locally unique addresses.

9.2.2 Utility-wise Classification

This form of classification is based on the *utility* or *purpose* that an address serves and leads to two classes of addresses—*locators* and *identifiers*.

A locator helps to locate the entity being addressed by the locator. The location here either refers to the geographical location or the topological location. As an example of the first case, a telephone number of the form +91-11-258XXXXX indicates that the number belongs to a resident of India (ISD code 91) who lives in New Delhi (STD code 11) and who is served by 258 telephone exchange. In the second case, the locator is used to identify the location of the entity in the underlying network topology. For example, each entity in the Internet is identified by using IP addresses. The IP address of an entity, with the help of routing tables, is used to route the IP datagrams destined to it. Thus, an IP address, coupled with routing infrastructure, provides the topological information of an IP host. In essence, addressing for locator address follows a certain order and hierarchy. This hierarchy provides a framework for hierarchical routing.

Unlike locators, identifier addresses do not provide any information about the location of the entity being addressed. The best example of identifier addresses is the Ethernet address. An Ethernet address provides no information about the location of the host bearing that address, but uniquely identifies a host machine attached to a network.

9.2.3 Layer-wise Classification

Another form of classification seen in networks, which follow the OSI-RM model, is the *layer-wise* classification. This form of classification results in two broad classes of addresses—*link layer addresses* and *network layer addresses.*

The link-layer address, also known as the hardware address or MAC address, is associated with a Network Interface Card (NIC) that a network entity uses to attach itself to the network.

A network layer address refers to addresses used by entities to communicate with other entities that reside outside their own network/subnetwork. The IP address is a good example of a network layer address that is used for communication between hosts that reside in various parts of the world.

Another set of identifiers, though not strictly addresses, but can be viewed as application addresses, is the transport layer port numbers. As discussed in chapter 2, they are used to identify the application residing above the transport layer.

9.2.4 Functionality-wise Classification

The final form of classification is based on the *number of entities* being addressed, and leads to the following classes of addresses, viz. *unicast*, *multi-cast*, *broadcast*, and *anycast* addresses.

A unicast address uniquely identifies a particular entity in the network and is used for normal, point-to-point communication.

A multi-cast address refers to an address that identifies a group of unicast addresses. The group of addresses is also known as a *multi-cast group.* A packet sent on a multi-cast address is directed to all the entities that are a part of the multi-cast group.

A broadcast address is used to forward packets to all entities residing on a network. In general, broadcasting increases network traffic and thus, should be used with care. However, there are many scenarios where broadcasting is unavoidable, address resolution being one of them.

An anycast address is used when it is acceptable or desired to reach any member out of a group. Here, the group is referred to as an *anycast group*, and the addressing mechanism is called anycast addressing. The difference between anycast and multi-cast is that in former case the packet is sent to only one member, while in the latter it is sent to all the members of the group.

9.3 EXAMPLE: ADDRESSING STRUCTURE IN INTERNET

The Internet Protocol [RFC 791] defines a 32-bit structure for Internet addresses (or IP address). The 32-bit address is divided into two parts—*network identifier* and *host identifier*. The network identifier, also referred to as network prefix, uniquely identifies a network, while the host identifier identifies a particular host within the given network. The division of IP address into network identifier and host identifier leads to a two-level hierarchy, which is essential for routing. Outside the network, a host identifier is generally not visible and routing is done solely using the network identifier. Within a network, routing is done using host identifiers, or a group of host identifiers. In the latter case, the group of identifiers forms a sub-net (a small network within a network), where a sub-net is identified by a network identifier and a subnet number.

In order to meet the requirement of having a varied size network, RFC 791 defines three classes of IP addresses, which are defined below:

- **Class A:** Class A address is designed for use by very large networks. This is the reason why only seven bits were allocated for network identifier (meaning a possible of 128 networks). However, out of the 128 numbers, 0.0.0.0 is used for specifying a default route and 127.0.0.0 for loopback. Thus, only 126 Class A addresses are possible. The Class A address ranges from 1.XXX.XXX.XXX to 126.XXX.XXX.XXX. Figure 9.2 shows the structure of a Class A address. As shown in the figure, the most significant bit is set to 0, implying that a Class A address follows. This bit, along with the 7-bit network identifier, forms the network prefix. The remaining 24 bits are available for addressing individual hosts (implying that a possible of 16,777,216 hosts can be addressed). Here again, two host addresses ('all zeroes' and 'all ones') are not used for host identifiers. 'All zeroes' in the host identifier part specifies "this network" and hence, is not used. Similarly, 'all ones' specifies a broadcast address and is again not used. In effect, for all networks, the number of possible

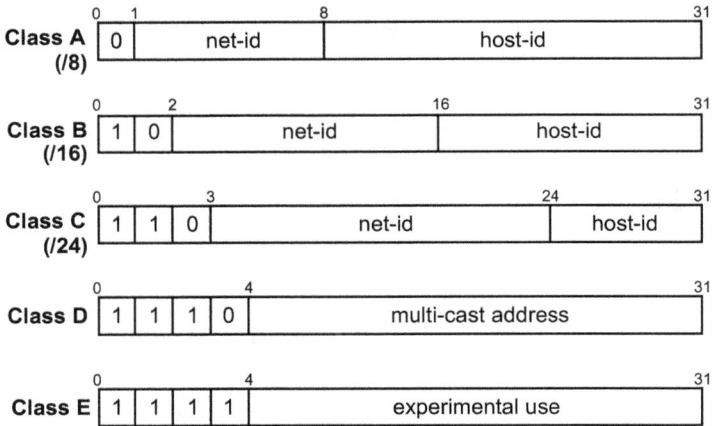

Fig. 9.2 *Classful Internet address structure*

host identifiers is reduced by two because of these two exceptions.

- **Class B:** Class B address is designed for use by moderate-sized networks. The network is identified by a 14-bit number, followed by a 16-bit host identifier. The 16-bit host identifier, can be used to address a total of 65,534 hosts. The first two bits (10) identify that a given address is a Class B address. These two bits, along with the 14-bit network number, form the network prefix. The Class B address ranges from 128.0.XXX.XXX to 191.255.XXX. XXX.

- **Class C:** Class C address is designed for use by small networks. The network is identified by a 21-bit number, followed by an 8-bit host identifier. The 8-bit host identifier can be used to address a total of 254 hosts. The first two bits (110) identify that a given address is a Class C address. These three bits along with the 21-bit network number form the network prefix. The Class C address ranges from 192.0.0.XXX to 223.255.255.XXX.

Besides the above three classes, two more classes have been subsequently defined for multi-casting and other experimental purposes. The structures of these two classes—Class D and Class E—are also depicted in Fig. 9.2. Since addressing is based on classes, this mechanism of addressing is termed *classful addressing*.

It is customary to refer to a Class A network as /8 network, because it has 8-bit in its network prefix. The same holds true for Class B and Class C networks, which are referred to as /16 and /24 networks, respectively. In general, a network with N bits as its network prefix is referred to as /N network. This notation also stems from the migration from classful addressing to classless addressing.

In *classless addressing*, a network prefix and a network mask are used instead of Class A, Class B and Class C addresses. The network prefix can be of arbitrary length. The number of significant bits in the network prefix is denoted by the number of 1's in the network mask. For example, a network mask of 255.255.254.0 denotes a /23 network, because the first 23 bits of the mask is 1. A network of arbitrary size can be represented by using this notation.

Classless notation is preferred over classful addressing because of its inherent advantages. First, classless notation results in the efficient use of IP addresses. This is not the case in classful addresses, where Class C addresses are too small for a moderate-sized network, and Class B is too large. Since most networks opt for the Class B address, this leads to the huge wastage of a precious public resource. Second, classless addressing facilitates address summarization (i.e. ability to represent a group of addresses through a single address), thereby resulting in lesser routing traffic and smaller routing table size.

9.4 EXAMPLE: ADDRESSING STRUCTURE IN TELECOM NETWORKS

The address numbering plan for fixed and mobile terminal networks is based on the E.164 address structure, published in the ITU-T recommendation [ITU-T E.164]. An E.164 number uniquely identifies user-network interfaces [e.g. Integrated Services Digital Network (ISDN) or Public Switched Telephone Networks (PSTN) interfaces], mobile stations [in which it is called mobile subscriber ISDN (MSISDN) numbers] and individuals utilizing specific global services. The E.164 is compatible with the numbering plan for international telephone services.

E.164 defines three different structures for the international public telecommunication number—for *geographical areas, global services* and *networks*. This section explains the public telecommunication number structure for geographic areas. For other structures and associated information, the reader is should refer to [ITU-T E.164].

Figure 9.3 depicts the E.164 international public telecommunication number structure for geographic areas. As

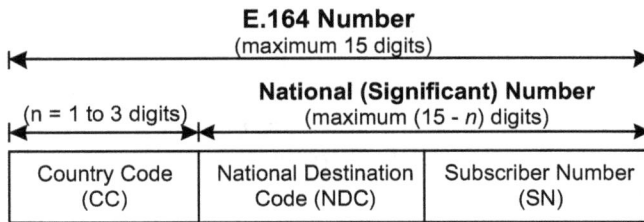

E.164 Number
(maximum 15 digits)

	National (Significant) Number	
(n = 1 to 3 digits)	(maximum (15 - *n*) digits)	
Country Code (CC)	National Destination Code (NDC)	Subscriber Number (SN)

Fig. 9.3   *E.164 structure for geographical areas*

shown in the figure, the number consists of two fields, viz. Country Code (CC) and National (Significant) Number N(S)N. The CC specifies the destination country. Here, country refers to a specific country, group of countries in an integrated numbering plan or a specific geographical area. The country code is followed by the N(S)N, whose function and format is determined nationally. The N(S)N consists of a National Destination Code (NDC) and Subscriber Number (SN). The NDC is an optional field and is used to select the destination network and/or trunk code. For example, consider a telephone number of the form +91-11-2588XXXXX. Here, '+' is used as an international prefix (i.e. the number following this is an international number). 91 stands for the country code of India, followed by 11, which represents the NDC or area code of New Delhi. The code 258 represents a particular local exchange of New Delhi, followed by the number associated with a particular local loop, from the exchange office to the subscriber's residence.

9.5 CONCLUSION

This chapter covered the need for an addressing structure and different ways of its classification. Four different ways of categorizing addresses, viz. scope-wise, utility-wise, layer-wise and functionality-wise, were discussed in this chapter. Thereafter, two popular addressing structures, viz. IP addressing structure and E.164 addressing structure, were discussed. The IP addressing structure is of phenomenal importance and the shortage of IP addresses has led to the standardization of new standards for IP (i.e. IPv6). Just as IP is popular in data communication networks, the E.164 is a popular addressing structure in telecommunication networks.

REVIEW QUESTIONS

Q 1. What is the difference between locator and identifier? Explain with examples.

Q 2. Give examples of addresses at different layers of the OSI reference model including link layer, network layer and transport layer.

Q 3. Compare classful and classless addressing in the Internet. Explain how classless routing can be used for more efficient routing.

Q 4. Compare routing of packets in the Internet using IP address with routing of calls in the telecommunication networks using E.164 addresses.

FURTHER READING

To know more about E.164, the reader should refer to [ITU-T E.164]. For addressing in the Internet, the reader should refer to [RFC 791]. The paper by 3COM [Addr 3COM IP Addressing] also provides a very good description of IP addressing.

SIGNALLING

10.1 INTRODUCTION

Literally speaking, signalling is an act whereby signals are used to exchange information and/or to convey instructions. For example, traffic lights are used as signals to maintain the efficient and smooth flow of road traffic. These traffic signals are used by drivers to determine their right of way. Similarly, in a communication network, signalling is used between the user and the network, or between two network elements, to exchange control information of different kinds. In order to appreciate the need for signalling, consider the following questions: How are connections established and released dynamically? For a virtual circuit-based network, a user must know the virtual circuit identifier of the connection before the data transfer starts. How is this knowledge obtained? A user can specify the traffic and service parameters for a connection. How does the network obtain these connection parameters? If the network does not have adequate resources to support a new connection, how is the connection request rejected?

The answers to all the above questions lie within the realm of signalling. *Signalling* is used to dynamically establish, monitor and release connections. Here, a 'connection' is an overloaded term. In a telecommunication network, connection refers to a physical connection established during a telephone conversation. In a virtual circuit-based network like ATM and frame relay, connection refers to a dynamic virtual connection which is established by using the signalling procedure. Note that in a virtual circuit-based network, there are two types of connections—*static connections* and *dynamic connections*. The static connections are configured and do not require signalling. In contrast, dynamic connections require signalling and it is these types of connections that are within the purview of this chapter.

Besides connection establishment/release, signalling is used to exchange *connection-related information* between the communicating entities. Connection-related information refers to the parameters that define a connection. In the simplest example of a connection (also referred to as a call), that of a telephone call, the telephone numbers of the callee and the caller together is connection-related information. Going a step ahead, for dynamic virtual connections, the virtual circuit identifiers, the service parameters and the traffic descriptors constitute connection-related information. The protocols used during data transfer, the maximum size of the protocol data units (PDUs) and the protocol parameters, are also classified as connection-related information.

10.2 SIGNALLING COMPLEXITY IN DIFFERENT NETWORKS

The complexity of the signalling procedure is determined by the flexibility that a network provides. In this regard, the differences in the signalling capabilities of telecommunication networks, virtual circuit-based networks (e.g. frame relay and ATM) and datagram networks (e.g. IP) make for an interesting study.

A telecommunication network is designed primarily to carry voice. Thus, signalling in such a network is restricted to the establishment of a voice channel in order to allow telephonic conversation. The act of signalling takes place right from the time the dial tone is heard, till the time the phone is kept on the hook after the conversation is over. Intermediate signalling procedures include dialing of the destination's number and hearing of the ringing sound. The exchange of signalling is at a bare minimum and is limited to dialling of the phone number. With the introduction of value added services (e.g. call waiting and tele-conferencing), there is corresponding increase in the signalling complexity,

In virtual circuit-based networks, signalling is not mandatory. It may or may not be required. If *Permanent Virtual*

Circuit (PVC) is established, no signalling is required. As the name suggests, PVC is established statically and no exchange of signalling information takes place. The subscriber of a PVC connection requests for a particular type of connection and puts forward his/her service requirements. Depending upon the resource availability, the service provider accepts or rejects the user's request. For *Switched Virtual Circuit (SVC)*, on the other hand, signalling takes place by using well-defined signalling protocol. The complexity of signalling is dependent upon the underlying technology. For example, signalling in ATM is significantly more complex than signalling in frame relay, though both are virtual circuit-based technologies. The reason for this is the greater capability and flexibility that ATM provides, which complicates the signalling mechanism.

Datagram networks, in general, do not require signalling. This is because by its very definition, a connectionless network does not entail a connection set-up. A typical example in this regard is the Internet Protocol (IP). Since the datagram forwarding in IP is done by using a best-effort model, neither is there any signalling involved nor is there any resource reservation. This best-effort nature of IP makes supporting Quality of Service (QoS) in IP-based networks very difficult. In order to circumvent this problem, IETF has formed different working groups in areas like Multiprotocol Label Switching (MPLS), DiffServ and Resource Reservation Protocol (RSVP). Among these working groups, RSVP and MPLS protocols mandate some form of signalling message exchange and resource reservation. Without going into the details of these protocols, the essence of the matter is that even in connection less networks, some signalling (and associated resource reservation) may be required. It may be added here that Transmission Control Protocol (TCP), which provides reliable service over unreliable IP, uses signalling to establish TCP connections between communicating peers.

10.3 CLASSIFICATION OF SIGNALLING TECHNIQUES

Signalling techniques are classified or categorized in a number of different ways. Figure 10.1 illustrates the different methods

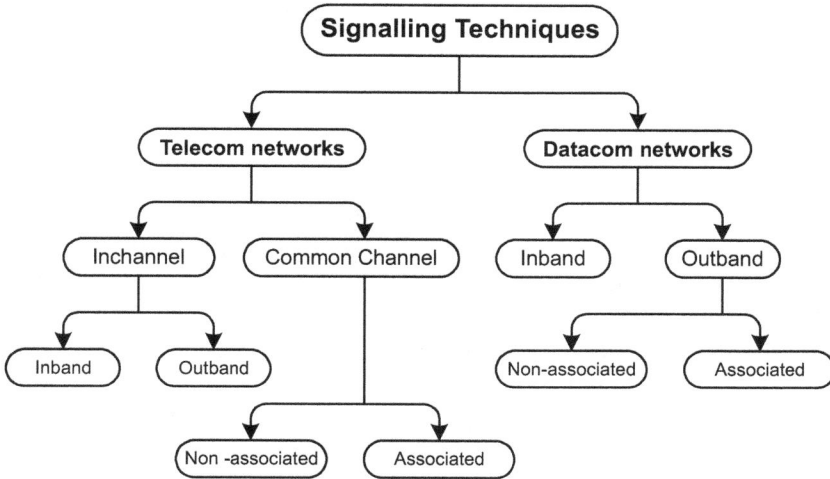

Fig. 10.1 *Classification of signalling techniques*

of categorizing signalling techniques on the basis of a number of factors. This categorization is explained as follows:

- **Inband versus outband signalling:** In a data communication network, *inband signalling* refers to the use of the same virtual channel to carry signalling information as that used to carry data. In contrast, in *outband signalling*, the signalling information and data are carried on different channels. In a telecommunication network, *inband signalling* refers to the use of the same voice frequency band to carry signalling information as that used to carry voice (i.e. 300–3400 Hz). In contrast, *outband signalling* refers to the use of frequencies above the voice band (but below the upper threshold of 4000 Hz) to carry signalling information. Both inband and outband signalling have their limitations. Inband signalling is susceptible to emulation of signalling codes by speech, which can lead to undesirable consequences like unwanted line-disconnect. Outband signalling also suffers from limited signalling bandwidth (i.e. 3400 Hz – 4000 Hz) and the need for additional electronics to support signalling, thus making it unpopular.
- **Inchannel versus common channel signalling:** Both inband and outband signalling techniques in telecommuni-

cation networks come under the purview of *inchannel signalling*. In inchannel signalling, the same physical channel carries signalling information as well as voice/data. In contrast, *common channel signalling* uses a separate channel for solely carrying signalling information for a number of connections. Thus, to some extent, inchannel signalling and common channel signalling in telecommunication networks are analogous to inband signalling and outband signalling of data communication networks, respectively.

* **Associated versus non-associated signalling:** In a telecommunication network, common channel signalling is again classified into two categories: *associated signalling* and *non-associated signalling*. In associated common channel signalling, the signalling channels and the data paths pass through the same network elements, that is, the signalling channels closely follow the data paths. However, unlike in inchannel signalling, these signalling channels do not share the same physical channel. In non-associated common channel signalling, there is no correspondence between signalling channels and data paths. In data communication network, the meaning of associated and non-associated signalling is technology-dependent. For example, ATM provides support for both channel-associated signalling as well as channel non-associated signalling. In channel associated signalling, all the signalling messages for each virtual path is exchanged on a fixed virtual connection (VCI=5) of that virtual path. In channel non-associated signalling, all the signalling messages of all the virtual paths are exchanged on fixed virtual path and connection (VPI=0 and VCI=5). Typically, the channel non-associated signalling mode is used and thus all signalling messages are carried on VPI=0 and VCI=5.

Apart from the techniques explained above another technique—*metasignalling*—finds mention in various signalling standards. Metasignalling refers to the process of establishing signalling channels by using signalling procedures. The signalling channel thus established is then used to establish channels for data transfer.

⚓10.4 SIGNALLING ISSUES

There are various issues that govern the design of an efficient and effective signalling protocol. These issues arise primarily because a signalling entity has to manage multiple connections and the underlying cannot be fully reliable. Due to these points, the following signalling issues must be considered:

- Acknowledgements
- Timer protection
- Handshaking
- Connection identification
- Finite state modelling of calls

The subsequent sections discuss each of these issues in detail.

10.4.1 Acknowledgements

The transmission media, by nature, are unreliable. Thus, the receipt of a signalling message needs to be *acknowledged*. However, is a single acknowledgement sufficient? Suppose an end-system 'B' receives a message from end-system 'A' and sends an acknowledgement for it to A. After sending the acknowledgement, B can rest assured that A will also come to know that B has safely received the message. But what if the acknowledgement gets lost? For B to ascertain that A has received the acknowledgement, A must also send an acknowledgement for B's acknowledgement. But the acknowledgement's acknowledgement can also get lost and then, A cannot be sure whether B has received the acknowledgement's acknowledgement or not. This means that B should also send an acknowledgement for the acknowledgement's acknowledgement. However, this is a never-ending process!

It can be shown that an unreliable link makes it impossible for a communicating end-system to successfully acknowledge the receipt of a message. However, under most conditions, two-way or three-way handshaking is sufficient. Improvements in the transmission systems and sharp fall in error rates further justify this statement.

10.4.2 Timer Protection

Timers are used at the end which originates a signalling message in order to avoid inordinate delays in case the signalling messages get lost or corrupted. In order to protect a message through a timer, the timer is started after the transmission of that message. In case that message is lost or discarded in transit, the timer at the originating end expires and the message is re-transmitted. On the other hand, if the message reaches safely and its receipt acknowledged by the peer (terminating end), the timer is stopped. Generally, only those signalling messages that require a response from the peer are timer-protected. In other words, messages which are sent as a response to some other message, are not timer-protected. The idea behind this is to avoid signalling complexity while keeping the sender protected against data loss/data corruption.

Choosing the correct *time-out period* is an important design issue. If this value is too small, then timers will take time-out very frequently. This will lead to duplicate packets in the network. If a very large value is chosen, then it may defeat the purpose for which the timers are maintained in the very first place. Typically, a value little more than the round-trip propagation time is chosen. This value tries to a strike a balance between the two extremities. On one hand, this value ensures that the peer gets sufficient time to respond to a message. On the other hand, it limits the wait of the sender to a reasonable limit.

10.4.3 Handshaking

Handshaking refers to the process of fixing the connection parameters through the exchange of signalling messages. Through handshaking, certain parameters can also be negotiated. The extent of parameter negotiation is limited by the number of handshakes in the signalling protocol. Signalling protocol generally adopts either a *two-way* or a *three-way handshaking* mechanism. In a two-way handshake, parameter negotiation is limited to the extent that the terminating end can either accept or reject the parameters specified by the originator of the connection. The terminating end has no means of specifying its own choice of parameters.

The *two-way handshaking* protocol has typically three types of messages for connection establishment, viz. ESTB_REQ, ESTB_REQ_ACC and ESTB_REQ_REJ. Figure 10.2 depicts the two-way handshaking process by using these messages. In Fig. 10.2(a), the terminating end accepts the connection request of the originating end. In contrast, Fig. 10.2(b) depicts a case when the terminating end rejects the connection request. In both cases, the originating end sends its choice of parameters in ESTB_REQ message. On receipt of this message, the terminating end decides whether it can accept the request or not. Accordingly, it either sends an ESTB_REQ_ACC message or an ESTB_REQ_REJ message back. It may be noted that different signalling protocols choose to name these three generic messages differently. It is also possible that the ESTB_REQ_ACC and ESTB_REQ_REJ are a single message where the result of the request is returned in some 'result' field, but these are protocol design issues.

The *three-way handshake* provides a greater flexibility for signalling. Strictly speaking, only a three-way handshake provides scope for parameter negotiation. Two-way handshaking provides no negotiation; it is merely a means to inform the terminating end of connection parameters, and the ability/inability of the terminating end to support connection parameters. In three-way handshaking, the originator of the connec-

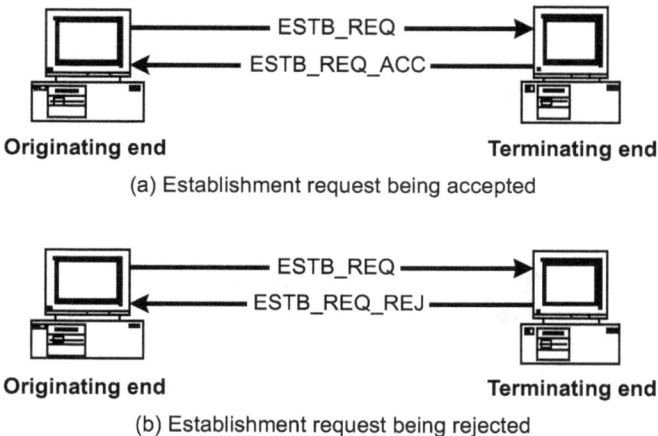

Originating end   Terminating end

(a) Establishment request being accepted

Originating end   Terminating end

(b) Establishment request being rejected

Fig. 10.2   *Two-way handshakes*

tion specifies its parameters in ESTB_REQ message and sends this message to its peer (i.e. the terminating end). The peer has three options. First, it can reject the request straightaway. This case is then reduced to the case depicted in Fig. 10.2(a). Second, it can accept the parameters specified as it is and send the ESTB_REQ_ACC message back. The originator then sends an acknowledgement for the acknowledgement, thereby completing the three-way handshake [see Fig. 10.3(b)]. Third, the terminating end can modify the parameters to declare its choice of parameters and then send the modified message back. The originator then has the option of either accepting or rejecting the modified parameters by sending the ESTB_RSP_ACC message [see Fig. 10.3 (b)] or ESTB_RSP_REJ message [see Fig. 10.3(c)], respectively. Again the names of these messages are specific to the protocol under consideration.

10.4.4   Connection Identification

In order to establish a signalling connection, various levels of addressing are used. At the network layer, the network layer

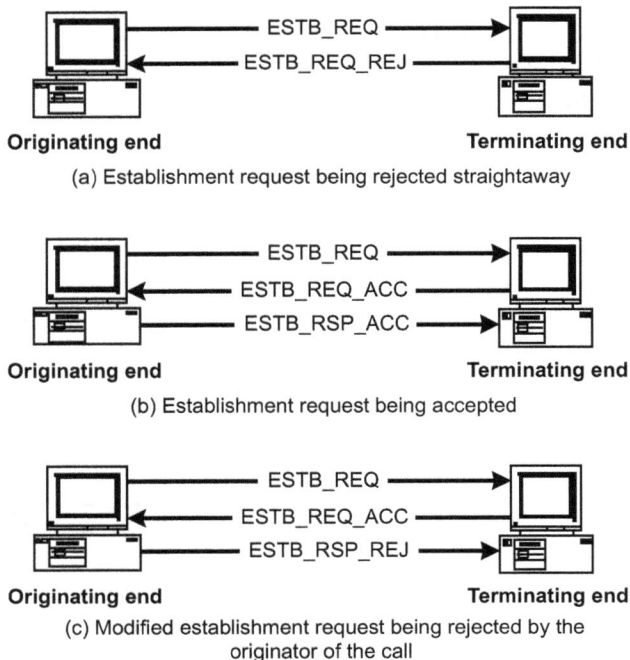

Originating end   **Terminating end**

(a) Establishment request being rejected straightaway

Originating end   **Terminating end**

(b) Establishment request being accepted

Originating end   **Terminating end**

(c) Modified establishment request being rejected by the
originator of the call

Fig. 10.3 *Three-way handshakes*

addresses are used to identify the terminating end. At the terminating end, many instances of the signalling protocol can run. It is also possible that two end-systems can have more than one active connection simultaneously. In such a scenario, the 'end-system address' alone is not sufficient to uniquely identify a connection. In order to understand the implications of this statement, consider the following steps of an event:

1. An end-system 'A' sends a connection establishment request to 'B'.
2. Soon after the first request is sent, another request is send by 'A' to the same end-system 'B'.
3. 'A' then receives a reply from 'B'. How does 'A' identify to the request to which 'B' replied to?

This is a very common issue in the design of signalling protocols. Different technologies adopt different strategies to tackle this problem. However, the essence of all the solutions remains the same. Every time a new connection X is established, a unique number, say uniq_num(X), is generated by the originator and sent along with the connection request message. Subsequently, the response for this request contains the same uniq_num(X) that was originally sent in the request message. This helps to identify the request for which the response is received. Going back to the example presented above, in step(1) end-system, 'A' sends uniq_num(1) along with the first request. The number is changed to uniq_num(2) in the second connection establishment request [i.e. in step(2)]. Now, the reply from 'B' must either contain uniq_num(1) or uniq_num(2). This number is used to identify the request for which the response is received.

Although the above explanation makes things very simple, the generation and allocation of these unique numbers raises a few interesting points. Firstly, the number of bits allocated for uniq_num's puts an upper bound on the number of active connections at a given time. A very large value of uniq_num means significant overheads in every signalling message sent. A very small value can severely restrict the number of active connections at a given time. Thus, one to four bytes are considered optimum for these uniq_nums.

Secondly, it is advisable to avoid the immediate re-use of a uniq_num if the corresponding connection has been just released. The immediate re-use of uniq_num can lead to unpredictable protocol behaviour. Consider a situation where a connection-clearing request is sent to the peer but the message is lost, and the uniq_num(X) for this connection is then used for establishing a new connection. The peer of the previous connection, realizing that the connection is idle for a very long time, sends a connection clearing message with the same uniq_num(X). This connection-clearing message will clear the new connection, though the old connection was already cleared. In order to preclude such possibilities, the general rule is to prevent the immediate re-use of these unique numbers.

Thirdly, in order to avoid collisions in the allocation of such numbers, a bit is reserved to identify whether this is the originating end or the terminating end. Thus through the received message and through this special bit of the uniq_num, an end-system can determine the further course of action.

Another design innovation followed in newer protocols is that both end-systems allocate their own uniq_nums. This helps in database optimization (refer to one of the review questions to understand how).

10.4.5 Finite State Modelling of Calls

As stated earlier, a signalling entity has to manage multiple connections. Thus, it has to model a connection as a *finite state machine*. In order to understand what finite state modelling of calls is, the different phases of a connection must first be explained. In general, a connection goes through the following three distinct phases (see Fig. 10.4):

- **Connection establishment phase:** In this phase, the communicating end-systems try to establish a connection between them. This phase is marked by a two-way or a three-way handshake, and parameter negotiation.
- **Data transfer phase:** As the name suggests, this phase is one in which the actual data transfer takes place. The duration of this phase depends upon the amount of data to be transferred.
- **Connection clearing (releasing) phase:** In this phase, the connection is cleared and all resources allocated for the connection are freed-up. Like the connection establishment phase, this phase can also have a two-way or a three-

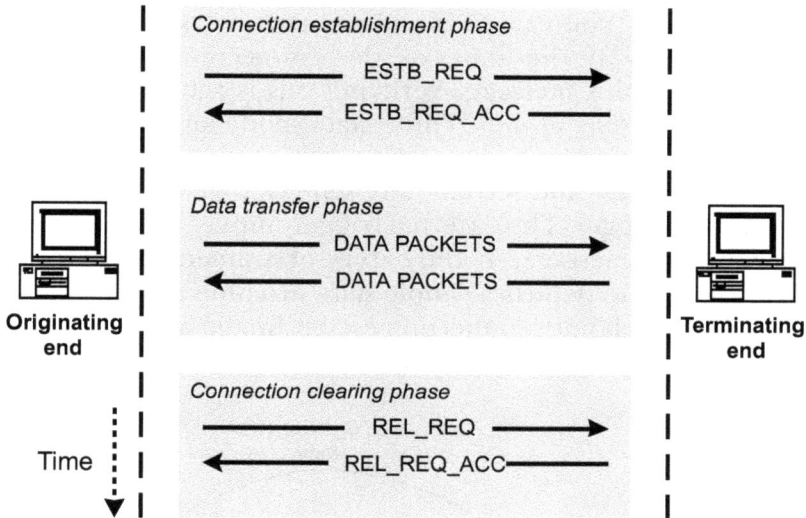

Fig. 10.4   *Three phases of a connection*

way handshake. However, typically, the connection-clearing phase involves a two-way handshake only. There are a couple of reasons for this. First, connection clearing does not involve any parameter negotiation, and thus, a two-way handshake is sufficient. Second, during connection establishment, the terminating end can choose to either accept or reject the request. But this is not the case when an end receives a connection-clearing request. It is mandatory for it to clear the connection and also to free the resources allocated for the same (thus again, the two-way handshake is sufficient).

Now, consider an end-system attached to a network element (i.e. a switch or a router), with the latter routing many calls for the former. Since different calls will be made in different phases, it is important for the network element to take actions on the basis of the state of the connection. For example, if a connection is in the connection establishment phase, then data cannot be accepted by the network element for transfer. As another example, if an ESTB_REQ message is already sent to the peer, then even if the end-system re-transmits this message, the network element should not re-transmit the message. Thus, a network element should maintain a state machine for each connection it routes. This state machine should accept messages only if the state of the connection permits the acceptance of this message. Verifying this is achieved through *finite state modelling of calls*. Finite state modelling here implies two things. First, each connection can exist only in a finite number of states, and second, any state change is the result of an external trigger. The external trigger implies the sending or receiving of a message, or the expiry of a timer.

Figure 10.5(a) depicts a simple state machine at the originating end in which both connection establishment and connection clearing procedures adopt a two-way handshaking scheme.

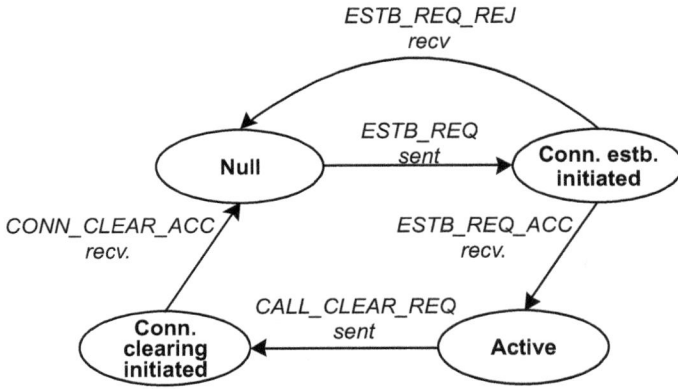

(a) Finite state machine at the originating end

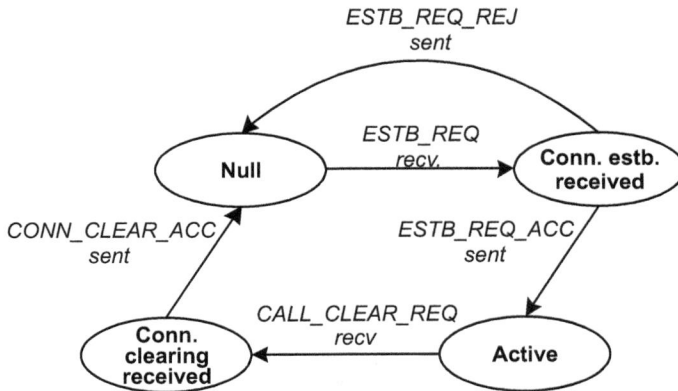

(b) Finite state machine at the terminating end

Fig. 10.5 *Finite state machine of a connection*

Initially, all calls are in the *null* state. When the ESTB_REQ message is sent, the connection state is changed to *connection establishment initiated*. Now, there are two possibilities. First, the peer can accept the request, and send a ESTB_REQ_ACC message. In this case, the connection-state is changed to *active*. An active connection state implies that the connection has been successfully established and is ready for data transfer. Alternatively, the peer can send an ESTB_REQ_REJ message, in which case, the connection state is again set to *null*.

All these events are a part of the connection establishment phase. After the data transfer phase, connection clearing is initiated by sending a CONN_CLEAR_REQ message and the state is changed to *connection clearing initiated*. On receipt of the CONN_CLEAR_ACC message, the connection state is re-set to *null*. It may be reiterated that it is obligatory for a node to accept a connection-clearing request. As another illustration, Fig. 10.5(b) depicts the state changes at the terminating end.

10.5 SIGNALLING MODELS

Quite often, the definition of signalling is restricted as a connection establishment or release procedure between two end-systems. This restriction, though valid under most circumstances, fails to capture the scenarios wherein a multi-partite communication is going on. Distant learning, distributed audio/video conferencing and video broadcasting are all examples of multi-partite communication. Multi-partite communication demands a signalling model that is significantly different from the traditional, two-party signalling model. Thus, depending upon the number of parties involved in communication, signalling models can be classified as: *point-to-point signalling*, *point-to-multi-point signalling* and *multi-point-to-multi-point signalling*. As the name suggests, point-to-point signalling defines procedures for connection establishment and release between two end-stations. In contrast, point-to-multi-point provides connections between a root and multiple leaves. The multi-point-to-multi-point model provides communication between various entities without any specific root entity. The following section provides details of the point-to-multi-point model. The multi-point-to-multi-point model is not discussed in this chapter.

⚡ 10.6 POINT-TO-MULTI-POINT SIGNALLING

Point-to-multi-point signalling is used to establish and release connections between a *root* and *multiple leaves*. This form of signalling is mainly used for applications that require multi-casting or broadcasting of data. Typical applications of point-to-multi-point calls include distant learning, distributed audio/video conferencing and video broadcasting.

Figure 10.6(b) depicts a simple, point-to-multi-point topology. Each point-to-multi-point connection has a single root and multiple leaves. Typically, a point-to-multi-point connection is preceded by a point-to-point connection. Once a point-to-point connection is established, subsequent leaves (also called parties) are added either by the root or by the leaves.

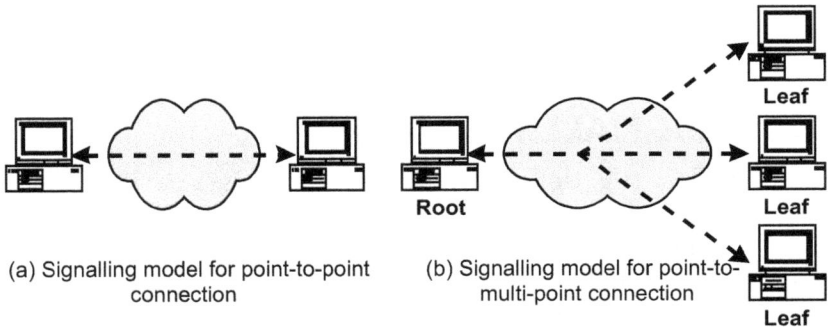

(a) Signalling model for point-to-point connection

(b) Signalling model for point-to-multi-point connection

Fig. 10.6 *Signalling models for point-to-point and point-to-multi-point connections*

Related to point-to-multi-point connection are the following issues that must be addressed:

1. Who starts a point-to-multi-point connection, the root, or the leaves?
2. How are subsequent parties added? Can these parties modify the parameters of the original connection?
3. How are the parties removed from a point-to-multi-point connection?
4. During the data transfer phase, can the leaves send data? If no, why? Moreover, do all the parties get the same data sent by the root?

5. What are the advantages and disadvantages of having a point-to-multi-point connection?

The following sub-sections answer each of these questions.

10.6.1 Starting a Point-to-multi-point Connection

A point-to-multi-point connection is generally started by the root. The root may take this step voluntarily or it may do the same after receiving an explicit request from a leaf. The leaf can send the connection establishment request to the root, either through the signalling channel or through other means. Consider a subscriber of a video broadcast service calling the service provider and requesting the latter to start a particular movie. Here, the telephone connection is an example of 'other means'. Irrespective of the means, once a root receives the connection establishment request, it follows procedures that are identical to establishing a point-to-point connection. The parameters of the connection are fixed during this period and the same parameters are used for the entire lifetime of the connection. Hence, it is not easy to change parameters midway, because the resource allocation for the entire connection is based on the initial parameters.

10.6.2 Adding Parties to a Point-to-multi-point Connection

Once a point-to-multi-point connection is established, subsequent parties are added to the connection by the root. The question that now arises is how does the root know when to add a party? This knowledge is obtained by the root in the same manner as it obtains the information about starting the point-to-multi-point connection, that is, either through a signalling message, or through some 'other means'. Subsequent parties have no say in determining the parameters of the connection, as it has already been fixed (by the root and the first leaf).

10.6.3 Dropping Parties and Releasing the Point-to-multi-point Connection

A leaf of a point-to-multi-point connection can drop itself out of the connection by sending a message to the root. It is manda-tory for the root to entertain this request, and to drop that particular party. However, if the root drops itself out of the connection, the whole connection has to be cleared and all parties have to be dropped.

10.6.4 Nature of a Point-to-multi-point Connection

By definition, a point-to-multi-point connection is one in which one point (root) is connected to multiple points (leaves), and the data flows from the root to the leaves, that is to say, flows in point-to-multi-point calls are unidirectional in nature. It may be argued that, theoretically, nothing precludes bi-directional data flows in point-to-multi-point calls. However, if leaves are allowed to send data to the root, there is a multipoint-to-point connection along with the point-to-multipoint connection. If the leaves are also allowed to send data to other leaves, then there is a multipoint-to-multipoint connection. Within the purview of the current discussion, these possibilities are not to be discussed. This implies that the root is only entitled to send data, and every leaf gets the same data (from the root) as any other leaf.

10.6.5 Analysis of Point-to-multi-point Connection

The above discussion only elaborates upon general concepts for the establishment and release of point-to-multi-point calls. Implementations are usually technology specific, and can vary. The main advantage of having a point-to-multi-point connection is the resultant saving in network resources. This is explained through Fig. 10.7. As shown in the figure, there is only a single data flow on the dashed links. Had there been two point-to-point connections (root-leaf 1 and root-leaf 2), the data would have traversed the dotted links twice. In contrast, for point-to-multi-point calls, data is duplicated on a link only at the *branching point* (or the *bifurcation point*). The branching point refers to a switch that duplicates data because the leaves are on different interfaces (see Fig. 10.7).

Since the data is duplicated as late as possible, all common links carry data only once. This results in significant saving of network bandwidth. The percentage saving depends upon the breadth and depth of the point-to-multi-point tree. If the tree

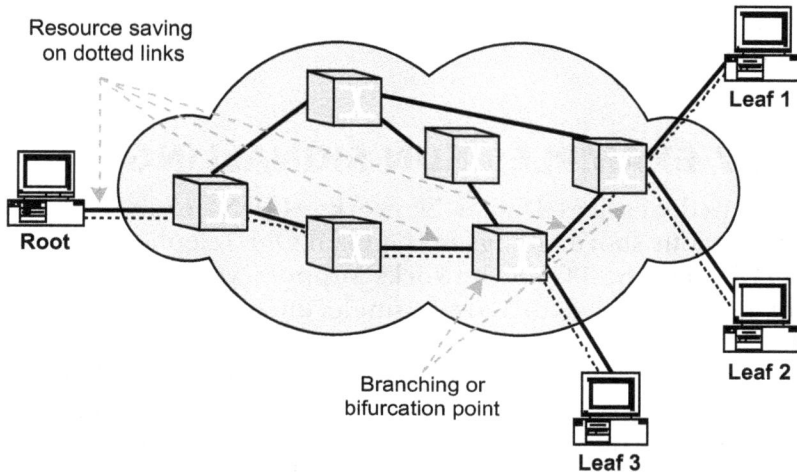

Fig. 10.7 *Resource saving in a point-to-multi-point connection*

is very deep, the number of common links will be more, and subsequently, there will be more saving. If the tree is very wide, the number of branching points will be more, and the gain will be adversely affected. Besides saving bandwidth, point-to-multipoint connection also results in saving the capacity of switch buffers and reducing the consumption of link layer addresses (i.e. virtual circuit identifiers). Thus, having a point-to-multi-point connection is a better option than having multiple point-to-point connections.

Consequently, it is important to assess the disadvantages of point-to-multi-point calls. First, these connections are difficult to establish, manage and release. It is not easy for the leaves to indicate to the root to start a connection. This master-slave relationship severely restricts the scope of signalling procedures. Second, the unidirectional nature of point-to-multi-point calls means that those applications that demand bi-directional flows cannot be supported. It is like asking a teacher to teach a lesson through a virtual classroom, but restricting the students from asking any question!

Third, it is important to realize that the point-to-multi-point calls do not result in an optimum signalling procedure, as it still uses the basic point-to-point signalling methodologies. Finally, a very wide point-to-multi-point still tree may not result in any saving, which was the initial motivation for having the point-to-multi-point calls.

⚡ 10.7 EXAMPLE: ISDN SIGNALLING

Integrated Services Digital Networks (ISDN) evolved due to some serious shortcomings in the Plain Old Telephone Systems (POTS). Firstly, POTS networks support voice-only services and it was desired to have a single integrated network that could support all types of services, namely, *voice-based services*, *data services*, *video-based services* and *multimedia services*. Secondly, the last mile (i.e. the local loop) in POTS being analog led to poor quality transmission. ISDN was developed to overcome these two prime shortcomings of POTS networks.

ISDN is an end-to-end digital network, which supports voice, video and data services in an integrated fashion. One of the biggest advantages of having a digital network is that data from digital devices (like computers) can be directly exchanged, without requiring any transformation from digital to analog form and vice-versa. If the transmission media is analog, as in POTS, it requires an intermediate modulation/de-modulation step by using a *modem*. The following sections discuss the basics of ISDN in a little more detail, after which, signalling in ISDN is discussed.

10.7.1 ISDN Protocol Stack

ISDN technology is the brainchild of the ITU-T standardization body. ITU-T has defined an entire I-series of recommendations that are devoted to integrated networks (ISDN and Broadband-ISDN, or B-ISDN). Figure 10.8 depicts the protocol stack layering in ISDN. Although only *user plane* and *control plane* are shown in the figure, the ISDN protocol stack is normally depicted as a three-dimensional model, consisting of an additional *management plane*. However, for the purpose of our discussion, the management plane is not required, and is hence not introduced.

At the physical layer of the ISDN, two interfaces are defined, namely Basic Rate Interface (BRI) and Primary Rate Interface (PRI). As depicted in Fig. 10.9, the BRI interface, as defined in [ITU-T I.430], is a 144 Kbps channel, consisting of two bearer

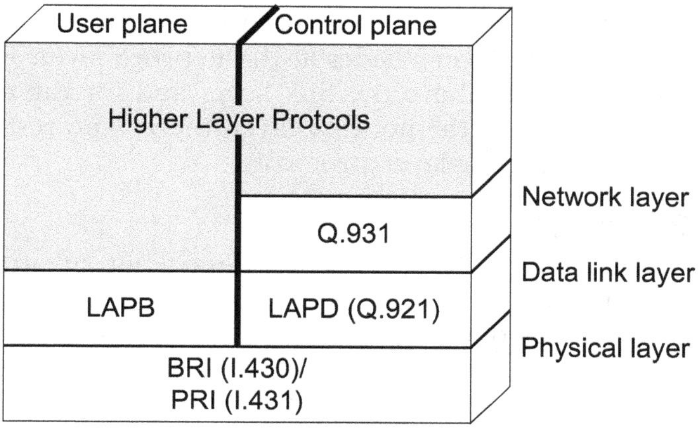

Fig. 10.8 *Protocol layering in ISDN*

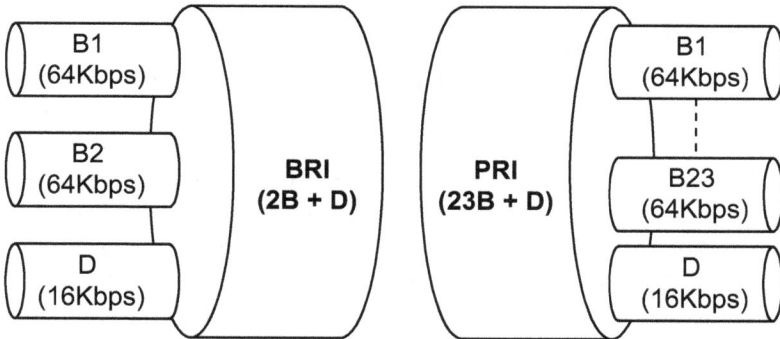

Fig. 10.9 *B and D channels in ISDN*

channels (2B) and a signalling channel (D). The bearer channels are used for information transfer at 64 Kbps each, while the D channel (16 Kbps) is used for signalling and a part for user data (9.6 Kbps). The PRI interface (also depicted in Fig. 10.9) is defined in [ITU-T I.431] and has two variants, one for the US and the other for Europe. In the US, PRI interface is a 1534 Kbps channel, consisting of 23 bearer channels and one signaling channel. In Europe, PRI interface is a 1984 Kbps channel, consisting of 30 bearer channels and one signalling channel. Note that the figure depicts both the BRI and the PRI for the US.

At the link layer of the ISDN protocol reference model, Link Access Procedure–Balanced (LAPB) and Link Access Procedure for D channel (LAPD) is used. In the control plane, [ITU-T Q.931] signalling layer resides at the network layer. For the user plane protocols above the link layer, and for the control plane protocols above the network layer, there is no restriction imposed on the higher layer protocols.

10.7.2 Services

When compared to POTS, ISDN offers a lot of attractive services. For example, consider the case of Internet access. Through an analog dial-up modem operating over a POTS line, the maximum speed of an Internet connection is about 56 Kbps. However, if one uses a BRI instead, a 128 Kbps channel can be obtained, implying a quadruple improvement in speed. Other services of ISDN include voice, fax, video-conferencing and LAN interconnection.

10.7.3 Signalling in ISDN

Signalling in ISDN networks is based on the specification [ITU-T Q.931] (mentioned in section 10.7.1) and the ISDN User Part (ISUP) protocol [ITU-TQ.761-Q.764]. ISUP is a simplified form of the ISDN Q.931 signalling, which was introduced for inter-exchange signalling in the Public Switched Telephony Networks (PSTN). Unlike Q.931, which uses LAPD as the link layer protocol, ISUP uses the SS7 Message Transfer Part (MTP) layers as the lower layers. In fact, ISUP can be used both in ISDN as well as in non-ISDN (e.g. PSTN) networks.

Figure 10.10 depicts two ISDN Digital Terminal Equipments (DTE) connected to their own respective local exchange switches via an ISDN interface. The local exchange switches are themselves connected to each other via a PSTN or an ISDN network. A DTE in some respect can be considered as an ISDN modem, which can be connected to any digital communication device (e.g. computer).

Fig. 10.10 *ISDN DTE connected via PSTN/ISDN*

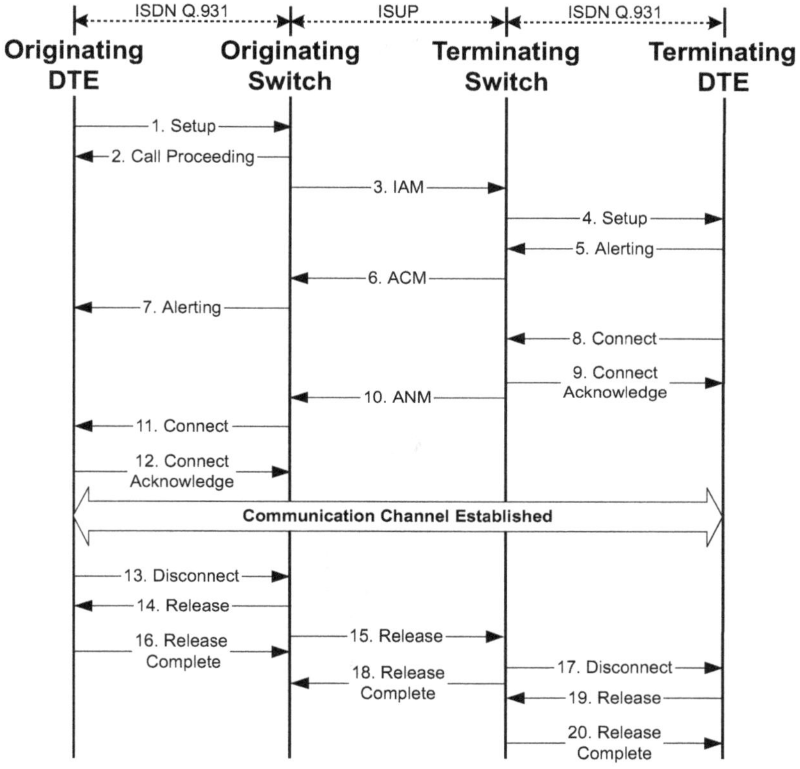

Fig. 10.11   *ISDN signalling sequence diagram*

Assume that DTE-A (Originating DTE) initiates a connection establishment procedure towards DTE-B (Terminating DTE), via the switches at local exchange A (Originating Switch) and local exchange B (Terminating Switch). Figure 10.11 depicts the sequence diagram for the ISDN signalling procedure. The ISDN Q.931 signalling protocol is used between the DTEs and the switches at the local exchange. Within the PSTN/ISDN network, the ISUP signalling procedures are used instead. Collectively, the Q.931 and the ISUP procedures are called the *ISDN signalling procedures*.

ISDN signalling involves the following sequence of steps:

1. The originating DTE initiates connection establishment by sending a *set-up* message to the originating switch over the ISDN D-Channel. The set-up message includes the *called*

and the *calling party number*, apart from an indication of the *bearer capability* (data rate, etc.) that is desired for the user plane communication.

2. The originating switch analyses the received set-up message. If it determines that the set-up message has complete information to route the call, and if the outgoing call is allowed, it sends back a *call proceeding* message to the originating DTE. The call proceeding message is an indication to the originating DTE that the call is in progress, and no more information related to the set-up of the call is required.

3. The originating switch then sends an Initial Address Message (IAM) towards the terminating switch, via the intermediary network. Much of the information in the IAM message (e.g. called and calling party number) is obtained from the set-up message.

4. On receipt of the IAM message, the terminating switch initiates connection establishment towards the terminating DTE by sending a set-up message.

5. The terminating DTE then sends back an *alerting* message, indicating that the called user alerting has been initiated. This is an indication that the called user is able to hear a ring-tone, or any other tone, indicating an incoming call.

6. The terminating switch then sends an Address Complete Message (ACM) towards the originating switch, indicating that the call has reached the called user, and called user alerting has been initiated.

7. On receipt of the ACM message, the originating switch sends an alerting message to the originating DTE, indicating it to play the ringing tone to the calling user.

8. If the called user accepts the incoming call/connection, the terminating DTE sends a *connect* message towards the terminating switch.

9. The terminating switch then locally sends back an acknowledgement to the connect message using the connect acknowledgement message. At this point in time, the user plane bearer is connected, and the charging begins for the called user.

10. Next, the terminating switch sends an Answer Message (ANM) toward the originating switch, indicating that the end user has answered/accepted the call/connection.

11. On receipt of the ANM message, the originating switch sends a connect message towards the originating DTE, and connects the user plane bearer to the call. At this point, the charging begins for the originating/calling user.
12. The originating DTE acknowledges the receipt of the connect message by sending a connect acknowledge message. The user plane communication channel is now established, and the two users can communicate over this user plane bearer.

Assume that after communication, the calling user decides to terminate the connection. The following sequence of steps is followed to release the user plane bearer:

13. The originating DTE sends a *disconnect* message towards the originating switch, requesting for a release of the connection.
14. The originating switch then releases the local resources reserved for the connection, and sends a *release* message towards the originating DTE. This is an indication to the DTE to locally release the resources it has reserved for the connection. At this point in time, charging for the calling user is stopped.
15. The originating switch also forwards the request to release the connection towards the terminating switch, by sending a release message.
16. In the meanwhile, the originating DTE acknowledges the receipt of the release message by sending a *release complete* message.
17. On receipt of the release message, the terminating switch sends a disconnect message towards the terminating DTE, requesting it to release the resources for the connection.
18. The terminating switch also sends a release complete message towards the intermediate network to indicate receipt of the release message.
19. After the terminating DTE has locally released all resources reserved for the connection, it sends a release message to the terminating switch.
20. The terminating switch then releases all resources for the connection. It stops the charging for the called user, and sends a release complete message towards the terminating DTE. This completes the connection release procedure.

🌡10.8 CONCLUSION

Signalling protocols constitute the key to understanding any given technology and this chapter introduced the basic concepts. Various types of signalling for telecommunication and datacommunication networks were explained. Then, the key signalling techniques including handshaking, timer protection, connection identification and modelling of a connection as a finite state machine were discussed. The different models of signalling, in particular the point-to-point and point-to-multi-point model, were also discussed. As an example, the connection set-up and release in ISDN was explained.

REVIEW QUESTIONS

Q 1. Why does signalling complexity differ in different networks? Discuss this with reference to the PSTN network, IP network and ATM network.

Q 2. What are the important issues in the design of signalling protocol? Discuss three of these issues.

Q 3. Define a technique for database optimization when both end-systems issue their own uniq_nums for connection identification. [Hint: Assume that an end-system manages a database for all active calls. Further, it can initiate connection requests as well as get call requests. Now, if the uniq_nums is issued by peer entity, how will this affect the database search?]

Q 4. Referring to various issues in point-to-multi-point, analyse how this may apply in multi-point-to-multi-point model.

Q 5. Describe the protocol stack for ISDN.

Q 6. What are the key message flows for ISDN signalling?

FURTHER READING

For signalling related concepts and their applicability in ATM networks, the reader is referred to [ATM S. Kasera] and [Sig R. Onvural]. For the ISDN case study presented in the chapter, [ITU-T Q.931] and [ITU-T Q.761-Q.764] describe the ISDN signalling and the ISUP signalling, respectively. The broadband ISDN protocol reference model is described in [ITU-T I.321].

ROUTING

⚡ 11.1 INTRODUCTION

Routing is the act of forwarding packets from the source to the destination using intermediate network elements called *routers*. Routers are devices that use the destination address carried in the packet header to appropriately route packets to their respective destinations. For the purpose of routing, routers maintain routing tables that contain a list of possible destinations and the next hop for each possible destination. Figure 11.1 shows a simplistic view of a routing table for router R1. The circles in the figure depict routers and the rectangles depict end-systems. As shown in the figure, a routing table contains at least two columns. The first column lists the possible destination addresses reachable through the router. For each destination address, the row entry in the second column provides the next hop (outgoing interface) to which a packet containing this destination address is to be forwarded. For example, to forward a packet to A3, router R1 uses R2 as the next hop. Some end-systems reside in the same sub-net as that of the router. For such end-systems, the adjoining router marks the next hop as 'directly reachable', implying that the packet is to be forwarded on the local sub-net.

The above example presents a simplified view of things. In practice, instead of keeping an entry for every destination, a router maintains an entry for every reachable network. Thus, the column for the destination address in the routing table is replaced by the destination network prefix. When a packet arrives, the routing table is searched for the closest possible

Routing table at R1

Destination	Outgoing If.
A3	R1-R2
A9	R1-R2
A6	R1-R3
A2	R1-R3
A1	Direct
A7	Direct

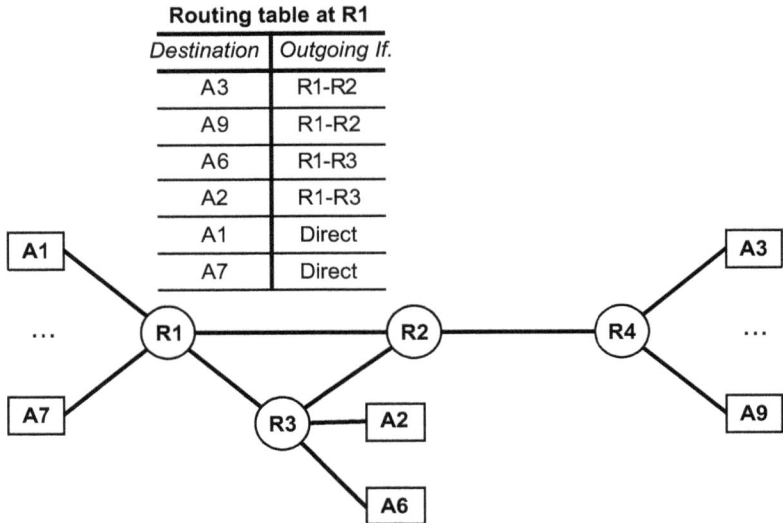

Fig. 11.1 *Routing table structure*

match corresponding to the destination address. The entry thus found provides the most specific route, which is then used to forward the packet. This process of finding the closest match is also referred to as the *longest prefix matching*.

The routing process is divided into two basic tasks. The first task is to exchange routing information so that every router in the network knows about the reachable destination (or set of destinations). In order to obtain this information, routers use routing protocols. Examples of routing protocols include the Open Shortest Path First (OSPF), Routing Information Protocol (RIP), Border Gateway Protocol (BGP), Exterior Gateway Protocol (EGP), Interior Gateway Routing Protocol (IGRP), Enhanced Interior Gateway Routing Protocol (EIGRP) and Intermediate-System to Intermediate-system (IS-IS), among others. The routing information collected by using the routing protocols is then used to build routing tables.

The second step in the routing process is to use the routing tables to forward packets based on the destination address. This act is also referred to as packet forwarding. Routing table look-up, longest prefix matching and determining the next hop, are all a part of packet forwarding.

11.2 DESIGN REQUIREMENTS FOR A ROUTING PROTOCOL

All routing protocols are guided by numerous design requirements. Some of the important design requirements are explained below:

- **Optimality:** One of the primary goals in designing a routing protocol is that the routes found using the protocol must be optimal in nature. By optimality, one implies that the routes calculated must be the best in terms of a given metric. For example, an optimal path in terms of delay is one in which the delay encountered in reaching a given destination, is minimal. Similarly, there can be optimal paths in terms of other metrics (e.g. the number of routing hops).

- **Reducing routing information exchange:** It is observed that a significant percentage of network traffic comprises control information like routing updates. These routing updates are referred to as *management overheads*, because they are merely overheads involved in making a network function, i.e. they do not carry user traffic. Thus, one of the requirements in designing a routing protocol is to minimize the flow of routing information so that more bandwidth is available for carrying user traffic.

- **Reducing processing load on routers:** The previous goal seeks to minimize the flow of routing information so that more network bandwidth is available. On similar lines, another design requirement is to reduce the burden of processing routing information on the routers, so that more processing power is available to forward the packets.

- **Faster convergence:** Another important parameter in routing protocol design is the time it takes for a network to converge after a change has taken place. *Convergence* is the process by which routers achieve a consensus on optimal routes in the network. In order to understand convergence, consider what happens if a link between two routers, *A* and *B*, breaks. *A* and *B* will immediately come to know about the link failure. However, far-off routers will come to know about this change only when the information propagates to them through *A* and *B*. Thus, till the information of a change reaches all the routers in the network, different routers will have different views of the network and hence, different routing tables. This will lead to *inefficient routing*, *misrouting*, and in an extreme case, *routing loops* (a situation in which a packet loops around two or more routers in a cyclic fashion and is unable to reach its destination). Thus, it is important that the time it takes for a network to converge (i.e. convergence time) is the lowest possible.

The convergence time depends upon the following factors:

- The distance of a router (in number of hops) from the point of change.
- The type of protocol used (i.e. static or dynamic. In case of static routing, the change has to be manually configured, which can be inefficient and time-consuming), and
- The mechanism used by routers to notify a change to neighbouring routers.

- **Scalability:** In the context of routing, scalability refers to the ease with which a routing protocol can be extended to cover a larger network area. Roughly speaking, the scalability of a routing protocol is inversely related to the size and frequency of routing table updates and the processing load on routers. However, this is not necessarily true, because there are other considerations as well. For example, though RIP protocol is simple and less-CPU intensive, it puts a limitation that the hop count cannot exceed 15. This implies that RIP cannot be used for very large networks where the number of intermediate routers between two end-systems exceeds 15.

- **Robustness:** Robustness refers to the ability of a routing protocol to tackle unforeseen or unusual circumstances. In other words, the routing infrastructure should not degrade or break in the wake of unexpected changes in network topology.

As generally observed, the above design goals are orthogonal in nature. For example, a simple routing protocol that minimizes the processing load on routers may not provide optimal routing paths for all path metrics (like delay and bandwidth). On similar lines, a robust protocol that handles all eventualities, may require a router with significant processing power.

11.3 CLASSIFICATION OF ROUTING PROTOCOLS/TECHNIQUES

There are a number of ways of classifying routing protocols and techniques. Figure 11.2 illustrates four different methods of categorizing routing protocols, viz. *nature-wise, control-wise, scope-wise classification,* and *classification based on a number of destinations.* Each method of classification is explained in the subsequent sub-sections.

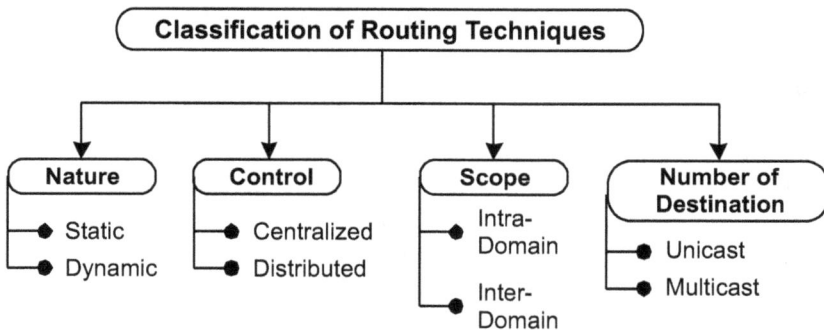

Fig. 11.2   *Classification of addresses*

11.3.1 Nature-wise Classification
This criterion decides the extent to which dynamic *changes in the network topology* get reflected in the routing tables. In general, this criterion leads to the two categories of routing—*static routing* and *dynamic routing.*

Routers engaged in static routing do not exchange any routing information. Rather, routes are static (i.e. fixed) and configured manually. Routing takes place using manually-configured routes or using default routes. (Default routes are routes used when no entry corresponding to the destination address exists in the routing table). *Interim Inter-Switch Protocol*

(*IISP*), a primitive routing protocol for ATM networks, is an example of a static routing protocol.

Although not apparent, static routing has a few advantages. It imposes very little overheads on the routers in terms of memory and CPU requirements. For very small networks, static routing is quite efficient because it does not require the exchange of routing information, and thus does not consume network bandwidth. Static routing is also useful when there is only one path to a given destination. Moreover, static configuration is desirable when the network administrator wants the manual configuration to over-ride the routing table entries computed using dynamic protocols. Note that static configuration is not the same thing as static routing. Even in a dynamic protocol, the costs or various other attributes of links/routes can be configured manually as deemed fit by the network administrator.

However, for large networks or for networks having multiple paths to a given destination, manual configuration of large routing tables is very cumbersome and leads to mis-configuration and inefficient routing. During network failures, it becomes very difficult for static routing to perform routing correctly. In such situations, dynamic routing is highly desirable.

In dynamic routing, routers exchange routing information to maintain the latest information about network topology. The exchange of routing information takes place either periodically or when a change has occurred. In the first case, routers exchange routing information after every 'T' seconds. The value of T is protocol-dependent and its typical value ranges between 10 to 90 seconds. In environments where the network is relatively stable, a periodic update leads to avoidable management overheads. Moreover, such updates consume the routers' processing time, which can otherwise be utilized for the basic task of routers, i.e. forwarding packets. To avoid these drawbacks, some routing protocols send updates only when a change has occurred. This reduces management overheads and does away with the unnecessary processing of routing updates. However, a case may arise wherein a router's neighbouring router may be dead and this router may wrongly assume that no change has occurred. In order to ensure that a dead router is not wrongly assumed to be active, a router periodically polls neighbouring routers using 'Hello' packets. 'Hello' packets (or *Hellos*) are used for the following purposes:

- To discover and verify the identity of a neighbour
- To determine the status of links with neighbour nodes
- To exchange state information with its neighbours.

Exchange of routing information only on a change and the exchange of *Hello* packets constitute a common feature of most of the routing protocols including the likes of Internet routing protocol Open Shortest Path First (OSPF) and ATM routing protocol Private Network Node Interface (PNNI).

11.3.2 Control-wise Classification

This form of classification is based on the *control* (or power) that resides at a router to compute routes and leads to two major categories of routing—*distributed routing* and *centralized routing*.

In centralized routing, a *central server* also called the *route server* collects state information from each router in the network and calculates the routing table using the information collected. The server then distributes the computed routes to all the routers. When an event or a change occurs, the information is conveyed by the affected router to the route server. The route server re-computes the routing tables and distributes them to all the routers. A smart route server can optimize by sending routing information only to routers affected by the re-computation. Centralized routing is very difficult to implement in a large network, because the load on the route server can be excessive. Also, centralized routing is vulnerable to single-point failure, preventing which requires redundant route servers.

On the other hand, in distributed routing, the path is computed in a distributed fashion. *Adjacent routers* exchange state information, which are then used to compute optimal paths. In the wake of a change, a routing update message is generated and sent by the affected router(s). This updated information is propagated through the network, reaching all the affected routers, after which routing tables are re-computed. The distributed routing is the preferred form of routing.

11.3.3 Scope-wise Classification

In the context of the Internet, based on the *scope* of the routing protocols, all routing protocols are classified as either *intra-domain protocol* or *inter-domain protocol*. This form of classification uses the concept of Autonomous Systems (AS). An *autonomous system* is defined as a collection of networks and routers, which is administered by a single administrative authority. Routing within an autonomous system is achieved through intra-domain protocol (or interior protocol). Open Shortest Path First (OSPF) and Routing Information Protocol (RIP) are two most commonly used intra-domain protocols.

Routers belonging to two different autonomous systems exchange routing information by using inter-domain protocol (or exterior protocol). Border Gateway Protocol (BGP) and Exterior Gateway Protocol (EGP) are two of the most commonly used inter-domain protocols.

Figure 11.3 depicts the difference between intra-domain and inter-domain routing protocols. The figure shows two autonomous systems, AS1 and AS2. Within the autonomous system, AS1 and AS2 use an intra-domain routing protocol like OSPF or RIP. Between them, AS1 and AS2 use an inter-domain routing protocol like BGP.

11.3.4 Classification Based on Number of Destination End-systems

Depending upon whether a packet is routed to a *single end-system* or to a *group of end-systems*, the routing protocol thus used

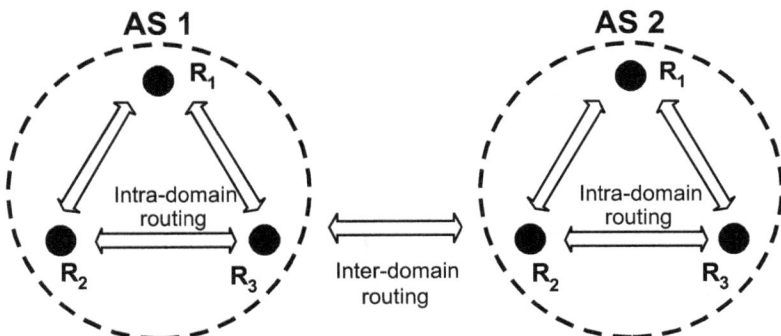

Fig. 11.3 *Intra-domain and inter-domain routing protocols*

is classified as *unicast routing protocol* or *multi-cast routing protocol*. Most of the routing protocols support unicast routing only. Multi-cast routing protocols are specialized in nature and require provisions for the addition to and the removal of end-systems to a multi-cast group. Examples of multi-cast routing protocol include *Protocol Independent Multicast (PIM)*, *Multicast OSPF (MOSPF)*, an extension of OSPF, and *Distance Vector Multicast Routing Protocol (DVMRP)*.

11.4 CORE ROUTING CONCEPTS

This section reviews the core routing concepts that are central to the understanding any routing protocol. These concepts are:

- Route Summarization
- Routing Hierarchy
- Routing Information Flows and Route Computation
- Path Selection
- Route Parameterization

The subsequent sections describe each of these issues.

11.4.1 Route Summarization (or Aggregation)

Route summarization refers to the process of using a single network prefix to represent a collection of nodes/sub-nets that begin with the same prefix. By summarizing routes, instead of sending routing information of individual nodes/sub-nets, a single summarized route is advertised to the external world. This significantly reduces the amount of routing information exchanged and also the number of routing table entries required. For example, consider an enterprise network which is assigned a class B address (141.34.0.0), and which contains 250 sub-nets, each containing 250 hosts. If an entry is to be maintained for each node of this network, a total of 62,500 routing table entries are required. Now, instead of maintaining information about individual nodes, if routes to individual sub-nets are maintained, the routing table entries required reduces to 250 entries. For routers external to the network, if only the class B network prefix is advertised (141.34.0.0), only one entry is required in the external routers.

The above example proves that route summarization is a very effective mechanism for condensing routing information. Besides the benefits accruing from it is quite logical that a network does not advertise its internal topology to the external world (for security or for other reasons). This is because once a packet destined for an internal node reaches the network's border router, it is the responsibility of routing protocols running inside the network to ensure that the packet reaches the destination safely. There is no apparent need for routers outside the network to maintain this internal routing information.

Route summarization is possible only when the network addresses are derived from contiguous address space. For example, if a network has some addresses starting with the address prefix 110.34.0.0 and a few others with prefix 112.39.0.0, then a single prefix is not sufficient to identify nodes belonging to this network. In this case, two routing table entries are required. In general, the more dis-contiguous the address space, the higher is the number of routing table entries required. In the worst case when there is no ordering in address assignment, an entry is required for each end-system in the network.

11.4.2 Routing Hierarchy

The basic aim of creating *routing hierarchy* is to reduce the flow of routing information and to allow routers to operate effectively without actually obtaining complete information of every end-system/link in the network. Thus, in some sense, hierarchical routing and route aggregation seek to satisfy similar objectives. In order to understand routing hierarchy, revisit the route summarization example presented in the previous sub-section. If there is a completely flat network, a router needs to maintain routes for individual end-systems. However, by maintaining a single network prefix for the set of end-systems in the sub-net, the routers can reach all the end-systems in the sub-net. This is the simplest case of creating routing hierarchy. Extending the concept further, a set of subnets can be aggregated to form a network prefix corresponding to the enterprise network. Moving one step further, the network prefix of various enterprise networks (i.e.

subscribers) can be combined to form a network prefix corresponding to the regional service provider.

Figure 11.4 shows how a routing hierarchy can be created by using route aggregation and proper assignment of IP addresses. The classless notation based on *Classless Inter Domain Routing (CIDR)* is used to show route aggregation. At each level, only a single routing table entry in the form of <network prefix, network mask> is sufficient to route any destination that lies below in the routing hierarchy. The *network mask* is also represented as in notation, where n is the number of bits in network prefix.

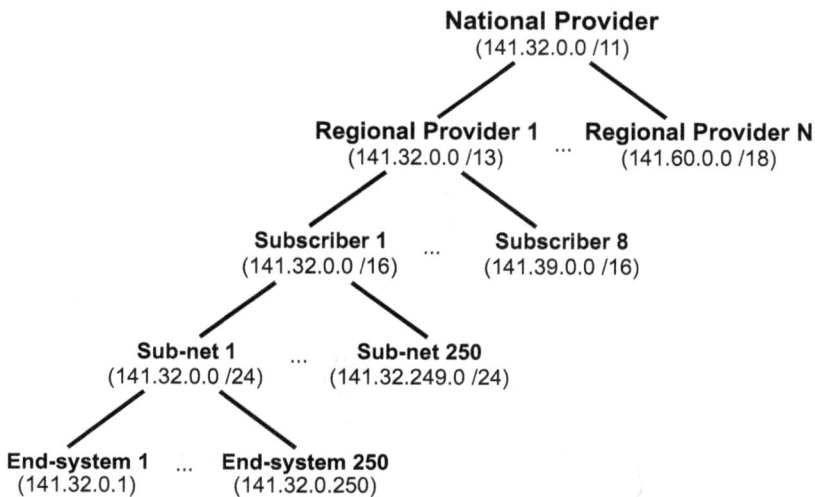

National Provider
(141.32.0.0 /11)

Regional Provider 1   **Regional Provider N**
(141.32.0.0 /13)   ...   (141.60.0.0 /18)

Subscriber 1   ...   **Subscriber 8**
(141.32.0.0 /16)   (141.39.0.0 /16)

Sub-net 1   ...   **Sub-net 250**
(141.32.0.0 /24)   (141.32.249.0 /24)

End-system 1   ...   **End-system 250**
(141.32.0.1)   (141.32.0.250)

Fig. 11.4 *Creating routing hierarchy using route aggregation*

Figure 11.4 is an example of creating routing hierarchy by using the addressing structure. There is another method of creating routing hierarchy by building a hierarchical network topology as used in OSPF. The OSPF network topology is discussed later in this chapter.

The essence of creating a routing hierarchy in the above-mentioned fashion is to limit the flow of routing information and to provide a mechanism to aggregate routes. Efficient route aggregation is possible only if a router can aggregate all possible network prefixes of its domain into a single network prefix. If this is possible, the routers of other areas domains need to maintain only a single entry corresponding to that area. Otherwise, each network prefix that cannot be aggregated forms a separate entry in the routing tables of other routers.

11.4.3 Routing Information Flows and Route Computation

Besides concepts like hierarchical routing and route aggregation, another important routing issue pertains to the method to be used to calculate the optimal routes. In general, there are two different methods to compute optimal routes. The choice of method leads to two major categories of routing protocols—ones that follow the *distance-vector method* (e.g. RIP [RFC 1058] and [RFC 2453]) and others that follow the *link-state method* (e.g. OSPF [RFC 2328]).

In the distance vector method, neighbouring routers exchange *distance vector tables*. A distance vector table is a list of pairs (V, D), where V identifies a destination (or a vector), and D the distance to that destination. The distance vector table of a router provides the router's view of the least cost paths to all reachable destinations. Upon receiving the distance vector table from a neighbouring router, the router adds the weight of its link with the neighbour to each entry in the distance vector table. The sum gives the cost to reach each destination in the distance vector table from this router, via the neighbouring router. If the cost of the path thus found is less than the cost of the best path currently available, the entry in the routing table is changed, and the next hop for the destination is marked as the neighbouring router, otherwise the entry in the routing table is left unchanged. Then, the next entry in the distance vector table is inspected and the process is repeated till all entries have been examined.

In link-state protocols, each node maintains the *status of its links* with each of its neighbours and periodically propagates this information to other neighbours. On the basis of the link-state information received from all routers in the network, a router computes the shortest path to each destination by using the *Dijkstra's shortest path algorithm*. The process is equivalent to finding the shortest path between two nodes of a graph, when the weights of all the edges are given.

11.4.4 Path Selection

Assuming that each intermediate router in the network has the capability to find a path to a given destination, there exist two basic techniques to select a path between a source and a destination. These techniques are *hop-by-hop routing* and *source routing*.

In hop-by-hop routing, each router in the path to the destination inspects the destination address and makes a forwarding decision based on this address. In source routing, the *source end-system* or the *ingress router* specifies the complete path to the destination. A complete path here refers to the set of routers that a packet traverses to reach its destination. Thus, in source routing, the intermediate routers in the path make no routing decision; they only forward the packet based on the next hop address carried in the packet.

Both the path selection techniques discussed above have their own merits and demerits. Hop-by-hop routing is better equipped to handle transient network changes and link failures, but suffers from the demerit that routing decisions have to be made at so many places. Another undesirable outcome of hop-by-hop routing is the possibility of routing loops. Routing loops result from database inconsistency; in particular, loops occur when a change in topology does not reach all the nodes in the network. Thus, a router may send packets to another router that has have no other option but to send the packet back and the packets then go round and round in circles.

In order to prevent routing loops, it is imperative that all nodes in a network implement the same route-computation algorithm and have the same view of the network. This is a severe restriction and stifles innovation. Moreover, in a heterogeneous environment where different parts of a network are administered by different bodies, this may not be a feasible solution.

Source routing is better than hop-by-hop routing in many aspects. In source routing, since the sender or the ingress router specifies the complete path to the destination, the possibility of routing loops is completely ruled out. Moreover, source routing also restricts the processing related to path selection to one entity (sender or ingress router). This reduces the load on the intermediate routers. However, source routing also suffers from a few drawbacks. In IP routing, the use of source routing implies that each and every datagram carries the source route. This increases the per packet overhead. For a packet with a large source route, which can be quite significant for a packet with a large source route.

Source routing also warrants that the source node maintain a large amount of topology information, which is not always possible, especially when the internal topology of one autonomous system is hidden from the other.

Another serious drawback of source routing is its inability to handle transient link failures. For example, consider a particular path comprising of intermediate routers $<N_1, N_2, N_3$ and $N_4>$ chosen by the ingress node N_1 between end-systems A and B (see Fig. 11.5). Suppose, by the time a packet reaches node N_3, the only link between N_3 and N_4 breaks down. In such a scenario, the source routing technique fails to find a path to the destination, even though alternate path $<N_1, N_2, N_4>$ exists to reach end-system B. In such situations the request is routed back till a new route can be found till the destination.

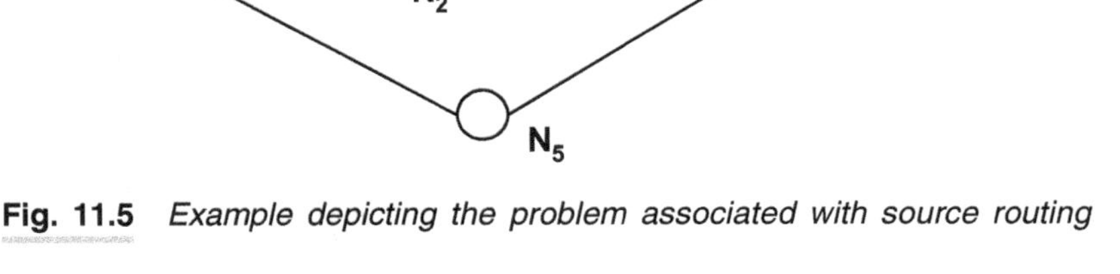

Fig. 11.5   *Example depicting the problem associated with source routing*

11.4.5 Route Parameterization

In order to define optimal paths, one must parameterize the routes/paths on which the criterion of optimality is based. *Route parameterization* is done by associating one or more metrics with every route in the network. In case only a single metric is used, route computation becomes relatively easy. However, when more than one metric is involved, the individual metrics are combined to produce a hybrid metric. Individual metrics are either constants (configured by network administrator) or vary with changing network conditions. Some of the metrics used in routing protocol, are explained below:

- **Hop count:** One of the simplest ways to find optimal paths is to find the number of intermediate routers encountered to reach a given destination. The route having the lowest hop count is then chosen as the optimal path. Although simple, hop count is not the most accurate measure of finding optimal paths. A 3-hop route may traverse a low-quality, low-speed leased line, whereas a 4-hop route may use a high-speed, highly-reliable fiber link. In essence, the hop-count may not be the best indicator of delay or reliability of the path.

- **Bandwidth:** Some routing protocols use the bandwidth of links to find optimal paths. For example, in OPSF, the cost of a link is inversely proportional to the bandwidth of a link. A 100 Mbps link is assigned a cost of 1. High-speed links are assigned lower costs and vice versa. This is because more packets can be sent at a higher bandwidth link (thereby incurring lesser cost) than a link with lesser bandwidth.

- **Delay:** Another parameter used as a metric is routing delay. The delay refers to the time required to forward a packet from a source to a destination. In the simplest case, routing delay is a function of link bandwidth. In more advanced protocol, the delay can also take into account the transient load in the network and the level of congestion.

- **Reliability:** This refers to the extent of packet loss on a given path. A more reliable link is one that has a relatively lower packet loss. Generally, the reliability is configured by the network administrator, taking into account the general nature of physical links.

- **Load:** Load refers to the degree to which a router or a link is utilized. A more loaded link is likely to cause greater delay, and in the extreme case, even packet loss. Thus, a lightly loaded but longer link may be preferred over a shorter but heavily loaded link.

In general, all routing metrics cannot be satisfied at a given time. A route with the shortest hop count may have poor delay and reliability measure. Alternatively, a very reliable link may induce significant delay. Thus, a routing protocol may choose to maintain separate routing tables according to the type of metric optimized.

⁑11.5 EXAMPLE: ROUTING PROTOCOLS IN INTERNET

As discussed in section 11.3.3, routing protocols are categorized as intra-domain and inter-domain protocols, depending upon their scope of operation. Within the Internet, *Routing Information Protocol (RIP)* and *Open Shortest Path First (OSPF)* are the two most commonly used intra-domain protocols. For inter-domain routing, the *Border Gateway Protocol (BGP)* is the most popular protocol in the Internet. The following sections describe these three protocols (RIP, OSPF and BGP) in greater detail.

11.5.1 Routing Information Protocol (RIP)

The Routing Information Protocol is a distance-vector protocol, which is defined in the IETF [RFC 1058]. The general concepts related to distance-vector protocols were discussed in section 11.4.3. Just like any other distance-vector protocol, routers implementing the RIP protocol periodically exchange route updates with each other. These route updates contain information about the cost to reach each destination, via the router sending the route update. Besides periodic updates, these route updates are also exchanged in case of changes in the network topology, which results in a change in the distance-vector table at a router. Normally, routers implementing the RIP protocol exchange the routing information with a period of 30 seconds.

11.5.1.1 Routing Updates in RIP

In RIP, the metric used for route parameterization is the hop-count. As discussed in section 11.4.5, hop-count is one of the simplest mechanisms for route parameterization. Upon receiving a route update from its neighbouring router, a router simply adds 1 to the hop-count for each entry in the route update. This is done to account for the extra hop required to reach the neighbouring router, when transmitting any packet towards the destination via the neighbouring router. Within its route table, each router only maintains the best path to reach a destination. Thus, when the router receives a route update from its neighbour, it updates its route table conditionally, as follows:

- If the route update from the neighbour advertises a route to a destination, and the router does not already know the destination, the router updates its route table with an entry for this new destination. The number of hops to reach this destination is marked as the hop-count in the route update, plus 1. The next hop to reach this destination is marked as the neighbouring router sending the route update.
- If the route update from the neighbour advertises a route to a destination, and the router already knows of a path to the destination, the router updates its route table conditionally, as follows:
  - If the currently known best path towards the destination is via the same neighbour sending the route update, the route table is updated with the new hop-count information.
  - If the currently known best path towards the destination is via some other neighbouring router, the router compares the hop-count of the currently known best path, with the hop-count advertised in the received route update. Only if the hop-count (plus 1) in the received route update is less than the already known best path, the router updates its route table.

11.5.1.2 Loop Prevention with RIP

A common problem with all distance-vector protocols is the possibility of forming routing loops. Since the only information carried in the route updates is the cost to reach a destination, routers may not realize if a routing loop is formed. This is because the route contains no information about the path to the destination, and only describes the cost to the destination, and the next hop, which is implicitly advertised as the node sending the route update.

Consider, for example the situation depicted in Fig. 11.6. As depicted in part a) of the figure, router R2 has a direct route to node A. Using the RIP protocol, router R2 advertises the route to A, with hop-count 0, to both R1 and R3. On receipt of this route update, both R1 and R3 will update their route tables to indicate the path to A, via next-hop R3, and of hop-count 1. After updating their routing tables, both R1 and R3 will advertise this route to each other too, with hop-count 1. The figure depicts the route update sent by router R3 to R1. Since

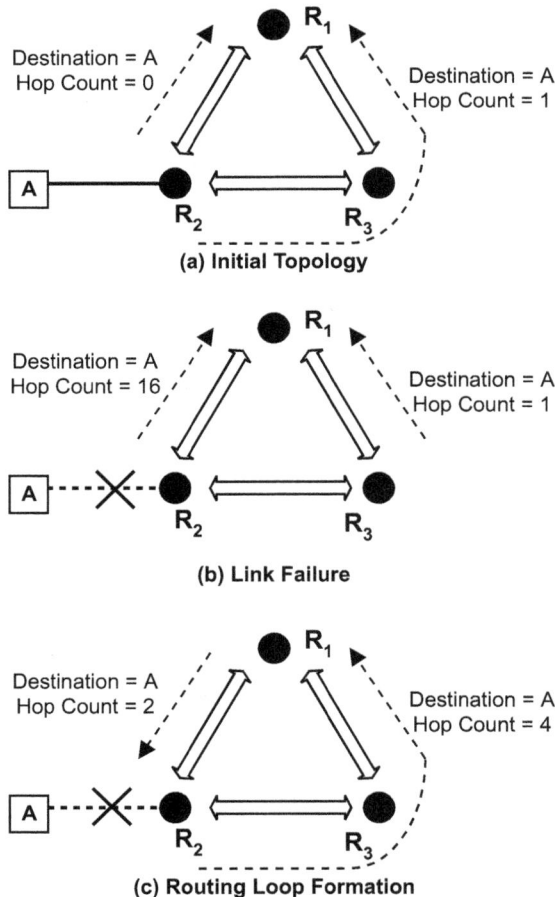

Fig. 11.6 *Routing loops in distance-vector protocols*

R3 is unaware of the fact that R1 has already received the advertisement from R2 for destination A, R2 also advertises the destination with hop-count 1. However, R1 discards this route update, since it knows of a better path to A, i.e. via R3. Thus, in stable state, both R1 and R3 will know of a route to A via R2 and of hop-count 1.

Next, consider what happens when the link between router R2 and node A is broken (see Fig. 11.6 b). In this case, router R2 will send out route updates, indicating hop-count 16 to destination A, to both R1 and R3. The value 16, as we shall later see, is considered as being equal to infinity in RIP. Also, consider that before this route update reaches router R3, router R3 has sent out a periodic route update to R1, advertising a path to A via hop-count 1 (this is as per its old information). Now, when router R3 receives the two route updates, one from R2 and the other from R3, it would be misled to believe that the best path to A is via R3, and will update its routing table accordingly.

Next, router R1 will now start to advertise a route to destination A to R2 with hop-count 2 (see Fig. 11.6 c). Since the route updates do not describe the path to the destination, router R2 will not know that the route advertised by R1 is incorrect. Router R2 will also update its routing table, to have a route to A, via R1, with hop-count 2. In turn, R2 will advertise this new route to R3, which will advertise it back to R1, and the cycle will go in. In each new hop added in the cycle, the hop-count will continue to increase by 1. This is a typical case of how routing loop is formed, with each node assuming that the route to a destination is via another node.

In order to overcome this limitation, RIP defines the value 16 for the hop-count to correspond to infinity. Thus, in the above problem, as soon as the hop-count reaches 16, the routing loop will be detected. All routers will then flush their entries to destination A.

A shortfall resulting from this definition of a maximum hop-count of 16 is that RIP networks cannot have a diameter of more than 16 hops. This limits the size of the network in which RIP can work. However, this is not a very big limitation, since RIP is anyway only used as an intra-domain routing protocol.

11.5.1.3 RIP Packet Format

Figure 11.7 depicts the RIP packet format. The RIP packet consists of the following fields:

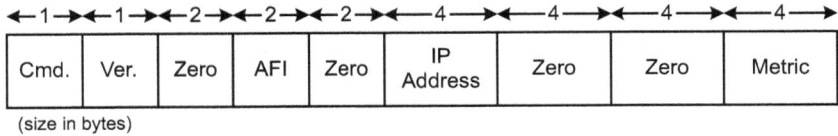

Cmd.	Ver.	Zero	AFI	Zero	IP Address	Zero	Zero	Metric

(size in bytes)

Fig. 11.7 *RIP packet format*

- **Command:** A one-octet command field is used to indicate whether the RIP packet is for a request or a response. The request command is used to ask a neighbouring router to send all or part of its routing table. The response command is used either to respond to a request, or to send an unsolicited routing update.
- **Version number:** A one-octet version number field is used to indicate the version of the RIP protocol being used.
- **Zero:** Two two-octet and two four-octet zero fields exist in the RIP packet format. These are added solely for backward compatibility to some proprietary implementations of the RIP protocol, prior to its standardization. The default value of this field is zero, and hence the name.
- **AFI:** A two-octet Address Family Indicator (AFI) field is used to specify the address family. This field is required since RIP can be used to carry routing information for many different protocols. The AFI value for the Internet Protocol is 2.
- **IP Address:** This four-octet field specifies the IP address of the router sending the RIP packet.
- **Metric:** This four-octet field is used to specify the hop-count to a particular destination. Valid values for a reachable destination are between 1 and 15.

Each RIP packet can consist of up to 25 occurrences of the AFI, IP Address and Metric fields. For larger routing tables, multiple RIP packets are used to convey the information.

11.5.1.4 RIP Version 2

The RIP protocol has two inherent limitations, other than the limitation of maximum hop-count. One of these limitations arises from the fact that RIP does not support sub-nets, since it has no field to carry network masks. The other limitation of the RIP protocol is that it requires each router to periodically broadcast the routing table. This period is typically 30 seconds for RIP networks. This introduces a lot of routing traffic on the network, which is a big overhead.

In order to overcome these problems, a newer version of the RIP protocol was designed, which is backward-compatible with the earlier RIP protocol. This protocol came to be known as the RIP Version 2 protocol. RIP Version 2 is defined in the IETF [RFC 2453]. RIP Version 2 solves the two problems of the earlier RIP protocol. One, RIP Version 2 only requires routers to broadcast parts of the routing table, which have either been modified, or are added new. The complete routing table is not broadcast, leading to a decrease of the routing traffic in the network. Secondly, RIP Version 2 supports sub-nets by including a field for carrying network masks (see Fig. 11.8).

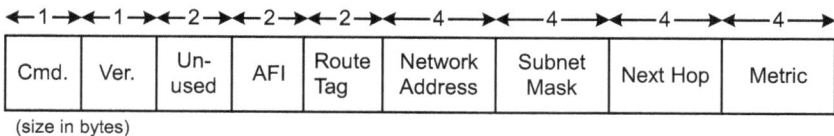

←1→	←1→	←2→	←2→	←2→	←——4——→	←——4——→	←——4——→	←——4——→
Cmd.	Ver.	Un-used	AFI	Route Tag	Network Address	Subnet Mask	Next Hop	Metric

(size in bytes)

Fig. 11.8 *RIP version 2 packet format*

Besides the fields in the RIP packet format, RIP Version 2 packet format defines the following additional fields:

- **Route tag:** This two-octet field is added in RIP Version 2 to distinguish between routes learned by RIP itself from routes learned from some other external protocols. These other protocols can be BGP, for example, which is normally used as an inter-domain protocol.
- **Sub-net mask:** This four-octet field is used to support subnets. If the route does not include a subnet, this field is set to zero.

- **Next hop:** The next hop field specifies the next hop to which packets destined for the advertised destination should be forwarded. While in RIP, the next hop was always implicitly assumed to be the router sending the route update, this is not so in RIP Version 2, where an explicit next hop can be specified.

11.5.2 Open Shortest Path First (OSPF) Protocol

Despite the introduction of RIP Version 2, there still exist some limitations in the RIP protocol. Amongst these limitations is the limitation on the maximum hops that are possible in any network supporting RIP. RIP networks (even RIP Version 2 networks) cannot support more than 15 hops between any two nodes. Another limitation of the RIP protocol is the implementation of the route metric. As discussed in section 11.5.1.1, routing metrics in RIP are a simple hop-count. In RIP, the best path to a destination is considered as the one that has the least number of hops. Thus, RIP does not accommodate some of the other routing metrics, which were discussed in section 11.4.5.

While RIP is still very popular in the Internet despite its limitation, an alternative and more scalable routing protocol was desired. This led to the design of the Open Shortest Path First (OSPF) protocol. OSPF is an intra-domain routing protocol, which is highly scalable in terms of the number of nodes and the diameter of the network, and also provides support for several other concepts like route aggregation and *Classless Inter-domain Routing (CIDR)*. OSPF is a link-state routing protocol, with all the features of a link-state protocol that were discussed in section 11.4.3.

11.5.2.1 OSPF Link State Protocol

In OSPF, each router maintains the status of each of its links with its neighbouring routers. It then broadcasts this information within the OSPF network in the form of *Link State Advertisements (LSA)*. Using the LSAs received from all the routers in the network, each router builds its view of the network topology. It is expected that barring a few aberrations, each router in an OSPF network would have the same view of the network, which is formed as a result of receiving the LSAs.

Once each router knows the network topology, it independently calculates its routing table. The route to each destination in the network is calculated by running the Dijkstra's *Shortest Path First (SPF)* algorithm. The SPF algorithm, also called the Dijkstra's algorithm, after the name of its creator, is used to calculate the shortest path between any two nodes in a network, given the network topology. The shortest path can be computed by using any metric for route parameterization, i.e. either hop-count, or bandwidth, or any other metric as discussed in section 11.4.5.

Thus, unlike RIP where route table updates are broadcast in the network, in OSPF, only the link state is broadcast. The route table is instead independently computed by each router in the network using its view of the network topology, built using the link state advertisements. In OSPF, each router uses the Hello protocol (refer section 11.3.1) to determine its neighbours and the status of its links with the neighbours. In case of change in the status of any link or neighbouring node, the router broadcasts an LSA in the network with the updated link information. This leads to each router independently re-computing its routing table.

11.5.2.2 OSPF Network Hierarchy

The OSPF network topology is designed to consist of multiple hierarchical areas (see Fig. 11.9), which are routing domains created in order to limit the flow of routing information (LSAs) within the network. All areas within an OSPF network are connected to each other via another special area, known as the backbone area. The hierarchical structure of an OSPF network is maintained by using three different types of routers, which perform well defined but different roles in propagating the routing information within the network. These different OSPF routers are as follows:

- At the bottom of the hierarchy is an internal router, whose scope is limited to an area. Formally, an internal router is a router whose attached networks lie within the same area. An internal router has the complete view of the routers and networks present in its area, but has only a partial view of networks and routers present in other areas (*view* here refers to the topological information gathered).

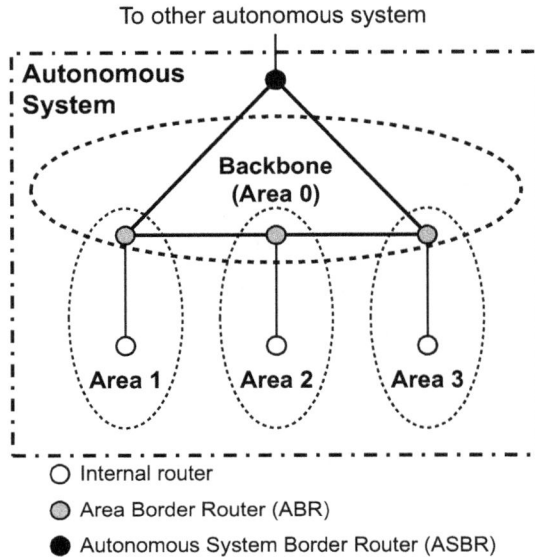

Fig. 11.9 *Routing hierarchy and type of routers in OSPF networks*

- This partial view is provided by *Area Border Routers (ABRs)*. ABR are routers connected to multiple areas. An ABR summarizes the routing information of its area and floods it to ABRs of other areas through the backbone area. Other ABRs, on receiving this summarized/aggregate information, propagate this information to internal routers of their respective areas. This is how internal routers obtain the partial view of other areas.
- The third category of routers is the *Autonomous System Border Routers (ASBRs)*. An ASBR is a router that is attached to routers of other autonomous systems. ASBR collects routing information of other autonomous systems and injects this into its own autonomous systems. To obtain routing information of other autonomous systems, an ASBR runs an inter-domain routing protocol like the BGP. Alternatively, ASBR can use statically-defined routes to inject external routing information into its own autonomous system.

By using a hierarchical network topology, OSPF limits the routing information required to be maintained by each router in the OSPF network. Since each router in an area only needs to maintain a partial view/aggregate routes for inter-area routing, the number of routing table entries and LSAs to be managed by each router are restricted. This makes the OSPF protocol scalable to larger networks.

11.5.2.3 OSPF Packet Format

The OSPF packet format is depicted in Fig. 11.10. An OSPF packet consists of the following fields:

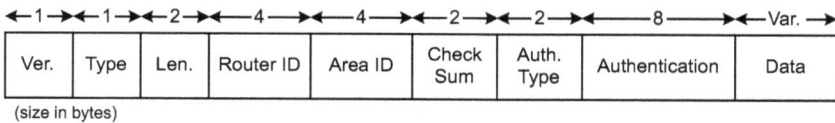

Ver.	Type	Len.	Router ID	Area ID	Check Sum	Auth. Type	Authentication	Data

←1→←1→←2→←4→←4→←2→←2→←8→←Var.→

(size in bytes)

Fig. 11.10 *OSPF packet format*

- **Version number:** This one-octet field is introduced to identify the version of the OSPF protocol used.
- **Type:** This one-octet field is used to identify the type of the OSPF packet. The type field can take on the following different values:
  - **Hello:** Used to establish the neighbour relationship using the Hello protocol
  - **Database description:** This message type is used after the Hello protocol has established its adjacency with neighbours. This message type is used to exchange the topological database.
  - **Link-state request:** This message type is used to request the neighbouring router to send the topological database. This message can be used the first time a router comes up, or after it discovers that a part of its topological database is obsolete.
  - **Link-state update:** This message type is used to respond to a Link-State Request message, and also for unsolicited periodic broadcast of LSAs.
  - **Link-state acknowledgements:** This message type is used to acknowledge the receipt of a Link-State Update message.

- **Length:** This two-octet field is used to specify, in bytes, the size of the OSPF packet.
- **Router ID:** This four-octet field is used to identify the router sending the LSA.
- **Area ID:** This four-octet field is used to identify the area to which this LSA belongs, and to which it is restricted. The LSA is not allowed to go out of the area boundary.
- **Checksum:** This two-octet field is used for the error detection and correction of the OSPF packet.
- **Authentication type:** This two-octet field is used to identify the authentication scheme used. In OSPF, all protocol exchanges are authenticated.
- **Authentication:** This eight-octet field is used to carry the authentication information for authentication purpose.
- **Data:** This variable-octet field contains the actual OSPF information.

11.5.3 Border Gateway Protocol (BGP)

The Border Gateway Protocol (BGP) is an inter-domain or inter-autonomous system protocol, which is widely followed in the Internet. Normally, private customer networks use the services provided by an Internet Service Provider (ISP) to access the Internet. Within the private network, an intra-domain routing protocol like the RIP or the OSPF is employed. This protocol is used to also exchange routes to nodes in the customer network with the ISP. The ISP subsequently advertises these routes to other ISPs using an inter-domain protocol like the BGP.

BGP, defined in IETF [RFC 1771], is a path-vector protocol. This means that the BGP routing protocol only advertises the availability of paths to destination. It does not propagate any cost information for the paths. Thus, the decision to choose between the multiple available paths is left for the network administrator to decide.

11.5.3.1 Route Aggregation in BGP

BGP is a very robust and scalable routing protocol. The scalability of the protocol to extremely large networks comes from the ability of BGP to aggregate routes. On learning intra-autonomous system routes, a BGP router aggregates these routes before advertisement to neighbouring networks. This limits the number of routes that each router needs to maintain in its routing table.

BGP supports Classless Inter-domain Routing (CIDR) for route aggregation. This is done by accompanying each network address with a network mask. Consider for example, an ISP network, which has offered 256 Class C addresses to its customers. Let these addresses be in the range from 195.10.0.* to 195.10.255.*, and assigned to 256 different customers. In the IP address notation, the "*" denotes a wild card, meaning all values in the range 0 to 255. Without route aggregation, an ISP BGP router would advertise 256 Class C addresses. However, with CIDR and the consequent route aggregation, the ISP BGP router only advertises a single route, that of 195.10.*.*.

11.5.3.2 BGP Path Selection

As mentioned in section 11.5.3, BGP is a path-vector protocol, and advertises only path availability, without any associated cost information. Network administrators define policies in order to help routers choose one of the multiple paths to a destination. The policy itself is defined on the basis of multiple BGP route attributes. Some of these route attributes include:

- **Local preference attribute:** This attribute is used to define a preferable exit point from the local routing domain/autonomous system. Thus, if multiple exit points exist via different Autonomous System Border Routers (ASBRs), the route via the preferred ASBR is chosen.
- **Origin attribute:** The origin attribute defines how BGP learned of a particular route. The origin attribute could be either an intra-domain routing protocol, or an inter-domain protocol. Normally, a route learned via an intra-domain protocol is preferred over a route learned via an inter-domain protocol.
- **AS path attribute:** This attribute is used to choose a path that takes the minimum number of hops via different Autonomous System (AS). Thus, a path with minimum number of AS hops can be chosen as the preferred path.

Multiple other route attributes are also possible, which are defined in the [RFC 1771]. A network administrator can define policies for path selection on the basis of these different route attributes. For a detailed description of the BGP routing protocol, the reader should refer to the BGP [RFC 1771].

11.6 CONCLUSION

Routing, along with addressing and signalling, governs how packets are exchanged within a network. For connectionless networks, every packet is routed from the source to the destination. For connection-oriented networks, routing is performed for setting up connections. This chapter looked at routing issues and solutions. The case study of routing within the Internet was also taken up with special reference to intra-domain protocols including Routing Information Protocol (RIP) and Open Shortest Path First (OSPF), and inter-domain routing protocol, Border Gateway Protocol (BGP).

REVIEW QUESTIONS

Q 1.   What are the desirable characteristics of a good routing protocol? Which of these characteristics are satisfied by OSPF?

Q 2.   What is route aggregation and summarization? Why is it so important in routing?

Q 3.   Describe the problems in the RIP routing protocol, and how RIP Version 2 solves these problems.

Q 4.   Compare the OSPF protocol with the RIP protocol. How does the OSPF protocol solve the problems of the RIP Version 2 protocol?

Q 5.   What is the need for BGP? What are its peculiar characteristics that are absent in RIP or OSPF?

FURTHER READING

[Gen A. Tanenbaum] and [Gen S. Keshav] are two good texts on getting a general understanding of routing concepts. For the case studies presented in this chapter, [RFC 2328], [RFC 2453] and [RFC 1771] describe the OSPF, RIP and BGP protocols respectively. [TCP/IP D. Comer] has dedicated chapters for the RIP and OSPF routing protocols. Also Cisco Press published [Rou M. Sportack] describes the RIP, OSPF and BGP protocols in relatively simpler form.

- Traffic Management

- Network Management

- Security Management

Part 3

ADVANCED CONCEPTS
OF COMMUNICATION
NETWORKS

The previous part of the book focused on the information flows within the network. Thus, concepts like addressing, bridging/routing, switching, signalling and multiple-access were elaborated in it. The essence of all discussion is how information is transferred from one node to another. Thus, while concepts like bridging are applicable within a network (at network layer or layer 2), routing is appropriate for information flows across networks (at network layer or layer 3). Signalling is applicable for connection-oriented networks for purposes of resource reservation and exchange of connection-related information. For all these concepts, addressing remains an integral part of the need to identify a node.

This part of the book raises a few more questions and seeks their answers. In particular, it considers the following questions:

- How is user traffic defined? And how is the service provided to the user parameterized? What are the mechanisms to manage traffic flows within a network? What if the resource requirements exceed the available resources? What are the means to derive optimum throughput?
- What are the mechanisms to ensure that the network is operating properly and giving optimum performance? What are the means to detect faults in the network? What are the means to exchange information between the managed objects and those that are managing them?
- What are the security threats in the network? What are the means to tackle these threats?

The aforementioned questions touch upon three advanced concepts in the field of networking. This part of the book is an attempt to elaborate upon these concepts to provide readers a further insight into the field of communication networks. For this purpose, the following concepts are defined:

- **Traffic management:** This refers to various mechanisms for managing network traffic, providing service guarantees to user connections and ensuring the optimal utilization of network resources. This topic is covered in Chapter 12.
- **Network management:** This refers to the act of monitoring and modifying network operations with the purpose of detecting and solving problems, improving performance and collecting reports and statistics. This topic is covered in Chapter 13.

- **Security management:** This refers to techniques like authentication, confidentiality and data integrity for the purpose of safe and secured information exchange. This topic is covered in Chapter 14.

TRAFFIC MANAGEMENT

12.1 INTRODUCTION

Communication networks are all about providing means to communicate (i.e. about exchanging user information). The user information can be viewed as user traffic akin to the traffic on the road that moves across the road network. This chapter touches upon the concept of traffic and what it takes to manage it.

Theoretically, *traffic management* is the act of managing network traffic, providing service guarantees to user connections and ensuring the optimal utilization of network resources. This definition of traffic management encompasses the following three basic concepts:

- Concept of traffic
- Concept of service
- The mechanisms to ensure optimal utilization of the network resource.

Accordingly, this chapter first explains the twin concepts of *traffic* and *service*, and then discusses the mechanisms for ensuring optimal resource utilization. These mechanisms are also referred to as elements of traffic management. This chapter elaborates upon six basic elements of traffic management—*traffic contract management, traffic shaping, traffic policing, priority control, flow control* and *congestion control*.

⚡12.2 CONCEPT OF TRAFFIC

What does one mean by traffic? On the roads, what constitutes traffic? Obviously, it is the cars, buses, trucks and two-wheelers moving from one place to other that constitute road traffic. Similarly, in a communication network, the data packets exchanged between communicating entities define a traffic flow. It is a customary to refer to it as a *flow,* because data packets are not static but are moving from the source to the destination.

In order to allocate resources for user traffic, it is necessary to accurately model traffic flows. Associated with traffic modelling are two important measures, viz. the amount of data carried and the rate at which it is carried. While amount of data defines the volume of user traffic, the rate defines the speed with which the network traffic is serviced.

These two parameters alone, however, cannot define the behaviour of the traffic flow. For example, consider the traffic flows depicted in Fig. 12.1. As shown in the figure, the amount of data carried and long-term Average Data Rate (ADR) is same for both the connections, the values being 120 Kb and 10 Kbps, respectively. However, the behaviour of these traffic flows is markedly different. Traffic flow A is from an application that injects data at a *steady* rate. In contrast, traffic flow B is from a *bursty* source that injects data in an on-off fashion. In essence, this example shows that the amount of data carried and data rate alone cannot accurately model traffic behaviour.

In order to accurately model traffic flows, two sets of parameters are necessary, one set that measures the *short-term* changes and another that measures the *long-term* average. The short-term measures include the following:

- **Maximum Allowable Data Rate (MADR):** This is the maximum data rate that a user is allowed at any instant of time. In the example presented above, traffic flows A and B have a maximum allowable data rate of 10 Kbps and 20 Kbps, respectively.
- **Burst Duration (BD):** This is the time duration for which a user is allowed to send data at the maximum allowable rate.

Besides these two parameters, another parameter is required which measures the long-term average of the connection. This parameter, termed as Average Data Rate (ADR), is the average rate for the duration of the connection.

The short-term and the long-term measures collectively form what is referred to as *traffic parameters*. Traffic parameters are necessary in order to determine the resource requirements for a given connection. For example, consider a traffic flow that has equal values of ADR and MADR. This connection is obviously similar to traffic flow A of Fig. 12.1. Such a connection requires steady bandwidth, which is equal to the ADR. Therefore, if a network has a total bandwidth of W, it can support W/ADR connections of such kind.

On the other hand, there are also connections that are inactive most of the time, but send data at MADR for a brief period of time (similar to traffic flow B of Fig. 12.1). In this case, it has to be statistically computed as to how many such connections can be supported. A simple solution is W/MADR connections. But this would lead to inefficient resource utilization. If it is assumed that the period of bursts of multiple such connections is spread over time, the network can grant W/ADR connections instead of W/MADR connections, thereby benefiting from the statistical distribution of traffic bursts. However, the correct approach is to support connections somewhere between W/ ADR and W/MADR. Since a network supports both types of traffic flows simultaneously, it is customary to allocate connec-

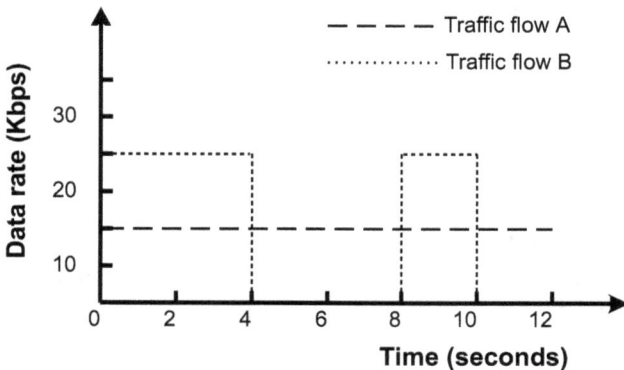

Fig. 12.1 *Traffic flows*

tions with steady rate and then allocate bandwidth to traffic flows like *B* from the bandwidth that is left over.

12.3 CONCEPT OF SERVICE

The previous section highlighted the concept of traffic. This concept is incomplete in itself because it only provides information about the offered load. From a user's point of view, it is important by the user to know how the network handles the offered load. For this, the user must have the means to quantitatively measure how the offered load is serviced by the network. Such quantitative measures include the time taken for packets to reach their destination (delay), the variation in this delay (jitter), the average loss encountered during transit and the extent of data corruption. These aspects translate into four basic parameters, viz. *delay, jitter, loss ratio* and *error ratio*.

- **Delay:** Delay is measured between a pair of boundaries. It is the measure of time taken by a data unit to reach one boundary from another. Specifically, it is the difference between the time when the first bit of a data unit crosses the first boundary and the last bit of the same data unit crosses the second boundary. It is customary to refer delay as *transit delay*. The transit delay can be either measured hop-by-hop or end-to-end. Hop-by-hop transit delay is a measure of delay between two consecutive intermediate nodes. The end-to-end transit delay is the summation of the individual, hop-by-hop transit delay. Delays are important in time-sensitive applications (e.g. voice conversation and video conferencing) because after inordinate delay data becomes useless for that application. For example, if a movie frame reaches the destination after its succeeding frames have been viewed, it is of no use and has to be discarded.
- **Delay variation (or jitter):** Jitter is the variation in delay over time. The transient changes in network load cause jittery transmission. For example, when the network is overloaded, there is more delay as compared to the delay in a lightly-loaded network. With oscillating loads, the available capacity in the network varies, resulting in

significant amount of jitter. Jitter is an important parameter in real-time applications. The best example that explains the effect of jitter is voice conversation on the Internet. Since the Internet uses the best-effort mode of transfer (as opposed to dedicated resource reservation model of circuit-switched telephone network), successive voice packets experience highly random delay. This makes the conversation highly jittery.

- **Loss ratio:** This refers to the ratio of total data lost to the total data transmitted. Ideally, no user likes to use a network with a high data loss rate. However, not all applications are sensitive to data losses, the classical example being telephony networks. A loss of a few bits cannot be recognized by humans very easily and this fact is used while building telephony networks. While some data loss is permissible in voice conversation, it is strictly prohibited in data applications (e.g. file transfer protocol or FTP). These applications can tolerate some delays (and even jitter), but cannot tolerate any data loss.

- **Error ratio:** This refers to the ratio of total erroneous data received to the total data transmitted. If the receiver can detect erroneous packets, they are typically discarded. Else, they are forwarded to the intended recipient of the packet. Note that even though loss and error appear to be similar, there is a subtle difference between the two. Loss captures the difference between data received and data transmitted (i.e. loss of data during transmission). In contrast, error ratio refers to erroneous data actually received (and not lost) versus the total data transmitted.

Delay, jitter, loss ratio and error ratio are core parameters that define a service type. These parameters are also referred to as *service parameters* or *service descriptors* because they measure the service attributes of an application. Apart from these basic parameters, sometimes *throughput* is also categorized as a service attribute. Throughput is the traffic injected into the network, minus the lost/corrupted data. Therefore, once the traffic parameters and service parameters (i.e. loss ratio and error ratio) are known, throughput can easily be computed.

Though throughput is an important measure in discussions pertaining to network performance, it does not always give the correct picture. In order to understand this, consider an example in which a sender sends 100 segments, where each segment is divided into 100 small packets. Unfortunately, one packet of each segment is lost in transit. Thus, though the throughput is 99 per cent, the data received at the destination end is of no use at all. This is because the whole segment needs to be re-transmitted.

In order to avoid incorrect appraisal of the situation, as was the case in the above example, another measure *goodput* has been defined. Goodput is the ratio of useful data received and the actual amount of data injected into the network. A data is said to be useful if it is not re-transmitted. Goodput takes into account the possibility of the received data being discarded by higher layers.

12.4 NETWORK CAPABILITIES

Quite often, one comes across statements like "traditional data networks are not suited for voice transfer" or, "ATM network can carry voice, video and data with ease". Since digital information is nothing but a stream of 0's and 1's, why are some networks more suited to carry voice, video and data? In order to answer the above, consider few questions and their analysis.

The first question is whether a network can provide the data rates required by all applications. Simple logic suggests that to support an application, the physical link should be at least capable of carrying the data rate required by that application. Now, link speeds vary from a few Kbps to a few Gbps. Therefore, depending upon the link speed, a network can or cannot support certain applications. For example, a user connected to the Internet using a 33.4 Kbps modem cannot expect to receive high-quality video streams. The reception will be jittery and of poor quality.

To analyse bandwidth requirements further, Fig. 12.2 depicts the bandwidth requirements of various applications. Given the link speeds, this figure provides a rough idea of the ability (or inability) of the network to support an application in terms of its bandwidth requirement. For example, the ISDN connection cannot support applications like medical imaging because the maximum available bandwidth in ISDN (~2 Mbps) is not

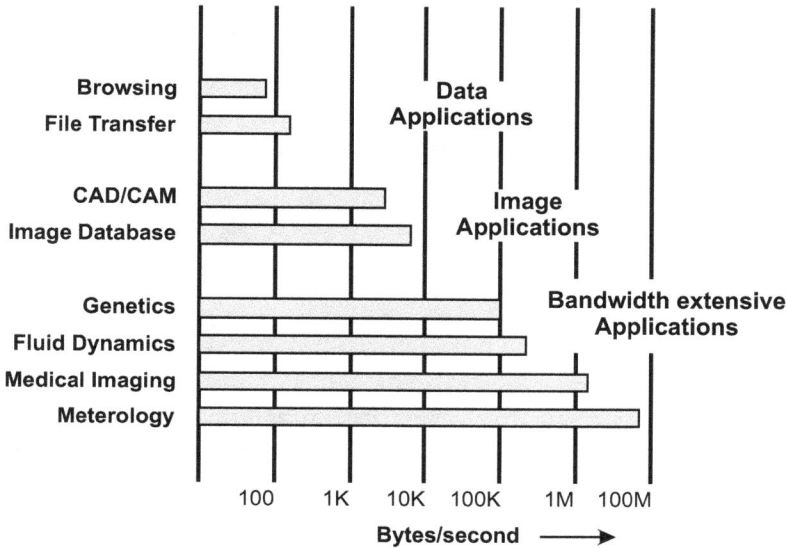

Fig. 12.2 *Bandwidth requirements of different applications*

sufficient enough to meet the demands of these applications. It may be noted that the link speed only provides an upper bound on the available bandwidth; the full bandwidth in not available to a single user because it is generally shared among multiple users.

Consider another question: Can the network provide real-time support, i.e. support real-time applications like voice-conferencing? Now, real-time support is possible only if the network can guarantee appropriate bounds on delay and jitter. However, not all networks are capable of doing this. In fact, as a general rule, Connectionless networks are incapable of providing bounded delay and jitter. For example, consider shared Ethernet networks, which are widely used in local area networks. Since Ethernet uses the CSMA/CD technique, the network cannot provide an upper bound on delay. A packet that collides frequently before finally reaching its destination will have much greater delay than another packet that reaches in the first attempt. Thus, shared Ethernet has no means of ensuring that a packet will reach its destination within a given time and is hence, ill-suited to carry real-time traffic. However, the situation is better for Switched Ethernet wherein there is dedicated connection with each node.

To some extent, the problem with shared Ethernet holds true for the Internet (TCP/IP networks), which provides a best-effort transfer. Since no resources are reserved and no guarantees provided, an IP datagram can be delayed, discarded, lost or mis-delivered. All this essentially means that the Internet is not well-suited to support interactive applications. However, advancements in IP world have made it less difficult for IP to support real-time traffic.

Finally, consider the question: Can the network handle bursty data? In general, the ability of a network to handle bursty data is orthogonal to its ability to support real-time applications. This is because in order to handle large traffic bursts, a network must buffer excess data. The larger the buffer size, the greater is the transit delay and the greater the transit delays, the less suitable is the network for carrying real-time data. This explains why the design requirements of data networks and voice networks are orthogonal to each other. However, the differences are blurring with IP networks carrying voice and the Broadband ISDN technology like ATM carrying data.

It follows from the above that the capability of a network in carrying different data flows does not lie in its ability to carry 0's and 1's. Rather, it depends on whether a network can support the requisite bandwidth or not, whether it can handle bursts or not, whether time-sensitive data is given its due priority or not.

12.5 TYPES OF TRAFFIC

Different applications have different traffic and service requirements. While some applications (e.g. data transfer) are inherently bursty in nature, other applications (e.g. voice conversation) require constant data rates. Depending upon the nature of traffic flows, all networking applications are classified into three categories—viz. *voice, data* and *video*. The characteristics of voice, data and video are explained below.

12.5.1 Voice

Voice transfer is characterized by low data rates. In traditional telephone networks, voice is encoded using the Pulse Code Modulation (PCM) technique in a 8-bit sample at a rate of 8000 samples/sec. This translates into a data rate of 64 Kbps, which is adequate for voice transfer. However, with the advances in encoding techniques, and with new silence suppression methods, voice can even be transferred at rates as low as 4 Kbps. This has leveraged alternative solutions for voice transfer and explains why voice over data networks has emerged as a cheap and viable alternative to voice over circuit-switched networks. However, these developments have not altered the basic requirements for voice communication, that is, the transmission delay should be less than 50 ms and the jitter the value should be in the order of a few milliseconds.

12.5.2 Data

Depending on the type of application, traffic characteristics of data flows show a marked variation. In typical client-server applications, the requests are much smaller than the server responses. The same holds true even for Internet applications like web downloads. In essence, there is a great deal of asymmetry between bi-directional data flows and the data rates may vary from a few kilobytes to a few megabytes.

Data applications are generally insensitive to delays or jitter. However, the desirable values of delay and jitter can vary from application to application. An e-mail transfer, for example, requires the delay to be in the order of a few seconds. Even if this delay is of the order of few minutes, there is hardly any appreciable difference. However, if the delay is of the order of hours, the sender may not receive the response as expected. This may reduce the effectiveness of e-mail application. Unlike delay, jitter is not a significant parameter for e-mail transfer.

For other data applications like net chat, low values of delay and jitter are desirable.

Table 12.1 summarizes the differences between voice and data transfer. While data transfer is inherently bursty, voice transfer is relatively stable. Various other aspects of voice and data transfer are compared in the table.

12.5.3 Video

Although delay and jitter bounds for video is somewhat similar to voice, the bandwidth requirements differ considerably. The bandwidth requirements of video depend on whether

Table 12.1 Differences in Voice and Data Traffic

	Based on	Voice traffic	Data traffic
Service Attributes	Transit delays	Up to 50 ms delay is acceptable for quality voice transfer	For batch data transfer; transit delay is not a concern
	Jitters	Jitter for voice transfer should around a few milliseconds	The affect and significance of jitter be varies from application to application
	Bit/burst errors	Not a serious concern. Does not degrade the quality of voice	Can lead to packet discard and is a serious concern
	Echo cancellation	Is an important issue in voice transfer	Not an issue for data transfer
Traffic Attributes	Nature	Relatively stable. Voice traffic exists as streams of data. Average duration is around 3 minutes	Quite often bursty. Data traffic exists as blocks of data. Duration highly variable
	Data rates	Bandwidth determined by the digitization techniques (generally 64 Kbps full duplex)	Can go up to the link speeds (varies from few Kbps to few Mbps)
	Flow control	Voice does not require any flow control	Data traffic may or may not require flow control

compression is being used or not. An uncompressed video yields a constant data rate stream of a few megabytes per second. Compression significantly reduces the bandwidth required. In compressed video, instead of sending frame by frame, the difference of the current frame with respect to the previous frame is sent. Thus, if the motion picture does not contain rapid scene changes, there is very little information to send and consequently, a low data rate is required. However, for fight sequences and other effects where there are rapid transitions, a significant bandwidth may be required. In essence, compression reduces the bandwidth requirements but makes the data rate variable. Good quality video can be sent using roughly 400 Kbps data rate.

12.6 TRAFFIC MANAGEMENT

The previous sections discussed the twin concepts of traffic and service, albeit independently. However, many questions like the following are left unanswered: What is the mechanism to specify traffic and service parameters? What happens if one or more user connections violate the specified traffic parameters? How is user traffic differentiated, i.e. how is user connection serviced in a prioritized manner? How does the network monitor traffic flows? How is the problem of network overload handled? All these questions come under the realm of traffic management.

Traffic management is the act of managing network traffic, providing service guarantees to user connections and ensuring the optimal utilization of resources. The scope and nature of traffic management depends upon a number of factors like the switching technique used (circuit-switching or packet-switching), the degree of flexibility provided (single-service or multi-service platform), the performance measures provided (simple or complex service guarantees) and the extent of statistical multiplexing done.

In order to provide service guarantees and to ensure optimal resource utilization, several elements of traffic management are used. The important ones among these elements are:

- Traffic Contract Management

- Traffic Shaping
- Traffic Policing
- Priority Control
- Flow Control
- Congestion Control

Each of these elements is explained in detail in the following sections.

12.7 TRAFFIC CONTRACT MANAGEMENT

Bandwidth (or for that matter any resource) in a network is finite and thus, its allocation and use must be monitored. *Traffic contract management* provides the means to allocate network resources to contending user connections and to manage network resources efficiently. Formally, traffic contract management is the act of deciding whether a new connection request can be accepted or not.

As an example, consider the case of telephone networks. When a particular number is dialled the network checks for the availability of trunk lines over which the call can be routed to the destination. Therefore, traffic contract management in telephone network amounts to checking of available switching capacity and trunk lines. If both of these are available, the network establishes a call and traffic contract is made. The traffic contract in a telephone call is defined by the following: the user's traffic information (e.g. PCM samples at 64 Kbps), the user's service requirement (in terms of delay and jitter bounds), the network's ability to service telephone call (based on the availability of switching capacity and trunk lines), the network's service-charges and the users' acceptance to pay the telephone bills.

Note that none of this information is required to make a decision as to whether a new call is to be accepted or not. This is because telephone networks are single-service networks, (i.e. they provide a single service only, that of voice conversation) and the information is implicit. Nonetheless, the information presented above provides insight on the constituent elements of a traffic contract. In general, a *traffic contract* in a network is defined by:

- Type and volume of traffic to be carried (terms of traffic descriptors)
- Requested service parameters (in terms of delays, losses and jitters)
- Network's acceptance to service the user request
- Network's fees for the service and user's acceptance to pay

A traffic contract can be *implicit*, as in the case in telephone networks. This case has been discussed in an example provided earlier. A traffic contract can also be *explicit*, as in the case of ATM and frame relay networks. In explicit traffic contract, the traffic and service parameters are fixed using signalling or other procedures. Parameter negotiation is possible only in explicit contract. The basic requirement for parameter negotiation is that parameters should be easily quantifiable, simple and understandable by both the network and user, and be easily exchanged.

Traffic contract is the agreed place whenever a new connection is established. For establishing a new connection, a user first specifies the traffic parameters (which describe the behaviour of user traffic) and service parameters (which describe the service attributes expected from the network). For these parameters, the network checks whether there are sufficient resources to accept the new connection or not. If it has adequate resources, the connection request is accepted, else, the request is rejected.

Once a traffic contract is negotiated, the network knows the expected user traffic, while the user knows the expected service attributes. Within the terms of the contract, the network is obliged to provide the agreed-upon service parameters throughout the lifetime of the connection. This obligation is not one-sided and it applies to the user as well. That is, the user is also expected to comply with the contract. If the user violates the contract beyond certain limits, the network may take punitive measures and may even terminate the connection. Ensuring that the user conforms to the traffic contract comes under the purview of traffic policing (see section 12.9).

The terms of traffic contract are mentioned in an *Service Level Agreement* (SLA). The SLA defines the roles of network and user, and the penalty that may be imposed if either party violates its part of the traffic contract. The ironical part of the agreement is that if the network violates its part, then it becomes the offender and policing authority at the same time!

12.8 TRAFFIC SHAPING

Many sources on the network emit data in bursts. A bursty transmission is characterized by alternating periods of high and low activity. During the period of high activity, traffic is generated at maximum data rates. This is followed by a period of inactivity when no data is transmitted. If the incoming data from a source is buffered and sent out at a lower rate, a reduction in the maximum offered rate can be achieved. This mechanism of achieving the desired modification of traffic characteristics is known as *traffic shaping*.

Traffic shaping alters the traffic characteristics of the connection, and delivers a more predictable traffic to the network. Traffic shaper is usually applied before the user traffic enters the network boundary (ingress point). Shaping increases the chance of a traffic conforming to the parameters of the *traffic contract*.

There are many schemes used for traffic shaping, the most common being the *leaky bucket technique*. It must, however, be noted that leaky bucket is a generic concept and is widely used for traffic shaping, traffic policing and flow control.

In order to understand how leaky bucket works, consider a bucket (see Fig. 12.3) of volume V units, with a hole at its bottom through which fluid leaks out at constant rate R units/time-unit. Simultaneously, a tap pours fluid into the bucket intermittently. If at any time the tap starts pouring into the bucket at a rate more than R units/time-unit, the fluid will start accumulating in the bucket. If the rate exceeds for a long time, the bucket may eventually fill up and start overflowing. Further, as long as there is any water in the bucket, the rate at which the water leaks out of the bucket is always R units/sec. The rate is zero if the bucket is empty.

This idea of regulating the rate of fluid flow using the leaky bucket can be applied to network traffic as well. Consider the leaky bucket to be a buffer with constant output rate of R units/

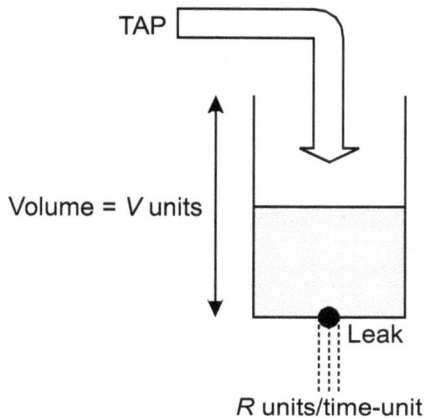

Fig. 12.3 *Leaky bucket*

time-unit. Now, the following operations are permissible:

1. If there is any data in the buffer, the packets are transmitted at a constant rate.
2. If the rate at which packets arrive exceeds the output rate, the buffer will start filling up gradually.
3. If a packet arrives and there is not enough space, the packet is dropped.

Using the three steps mentioned above, the output of the leaky bucket could be controlled. To understand, consider a leaky bucket of volume $V = 50$ Kbits and a leakage rate of $R = 5$ Kbps. Initially, suppose that the bucket is empty and packets arrive from a source at a rate of 25 Kbps for a duration of 2 seconds. Since the amount of data received does not exceed the size of the bucket, there is no overflow and the data is buffered. The buffered data is then transmitted at a constant rate of 5 Kbps for a duration of 10 seconds. This process is shown in Fig. 12.4. As shown in the figure, traffic shaping reduces the data rate from 25 Kbps to 5 Kbps. By appropriately choosing size of the buffer and leakage rate, a traffic profile can be shaped. In this example, shaping requires the ability to delay packets, which may not be always possible. Further, the example here is a very simplistic view of shaping and there can be various complicated algorithms for traffic shaping.

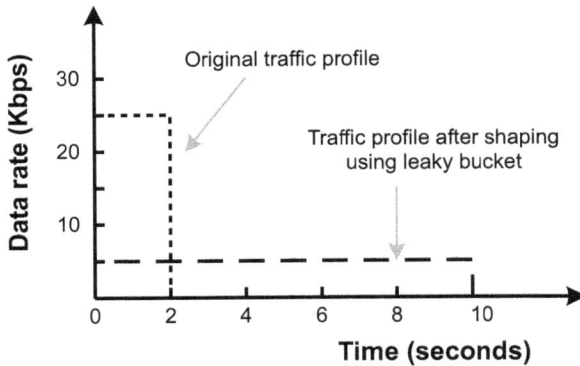

Fig. 12.4 *Traffic shaping using leaky bucket*

12.9 TRAFFIC POLICING

As discussed in section 12.7, a connection establishment is associated with an explicit or an implicit contract. Once the traffic contract is agreed upon and a connection is established, the network has to continuously monitor each connection to ensure that the connection is adhering to the terms and conditions of the contract. This is necessary because the users may specify very low data rates, but send data at substantially higher rates. Such a behaviour on the part of the user may be intentional or unintentional. While specifying connection parameters, the user may not be sure of the characteristics of the application and hence, underestimate the requirements. It is also possible that during the data transfer phase, the end system starts malfunctioning, resulting in violation of the traffic contract.

Irrespective of the cause, the effect of a non-conforming connection can lead to performance degradation for the network. If the traffic of such a connection is allowed to enter the network unabated, the non-conforming connection will consume more than its fair share of resources, thereby degrading the performance of conforming connections. This must be prevented at all costs. The mechanism to ensure that a connection adheres to its traffic contract parameters and that non-conforming connections are penalized is termed as *traffic policing.* Punitive measures taken by the policing function, include the following:

- **Tagging or marking:** This refers to changing the priority of the packets from high to low. Tagging is the first punitive measure that may be adopted for packets belonging to the non-conforming connection. The underlying assumption is that if there are adequate resources, then occasional non-conformance does not disrupt the service guarantees of other connections, and hence, packets are tagged and allowed to go through. If somewhere in the downstream there is overloading, the tagged packets are the first to be discarded.
- **Packet discard:** This refers to discarding packets belonging to a non-conforming connection. This step is adopted when the node under consideration cannot support additional packets.
- **Connection termination:** In the extreme case, a non-conforming connection may be terminated. This step is adopted when the continuous non-conformance of a connection results in disruption of the service guarantees of other connections.

Traffic policing is usually done at the ingress point of the network. Traffic policing can also be done at the boundary of two networks.

There are many features that a good policing function should have. First and foremost, the policing function must be able to identify a non-conforming connection quickly. This is necessary to limit the damage inflicted by the non-conforming connection. Having identified the non-conforming connection, necessary steps must then be taken to prevent degradation of the service of conforming connections. Further, the policing functions must be transparent to the conforming connections, i.e. the traffic characteristics of such connections must not be altered by the policing function.

Traffic policing is most commonly implemented using the leaky bucket algorithm described earlier. As shown in Fig. 12.3, the packets arriving from the source are stored in a bucket and periodically serviced. If the data arrival rate exceeds the rate at which the bucket is serviced, the bucket gradually fills up. If this continues for a long time, the bucket overflows. All overflowing packets are considered to have caused violation of the traffic contract; these packets are tagged or discarded as deemed appropriate by the policing function.

12.10 PRIORITY CONTROL

In elementary terms, *priority control* refers to treating packets of unequal importance unequally. In other words, packets of higher priority are given a preferential treatment as compared to packets of lower priority. Priorities are important in communication networks because the importance of packets differs from packet to packet. As an example, consider a router that needs to process two packets, one an ordinary datagram and the other, a routing update packet. It is possible that if the datagram is forwarded before the routing packet is processed, the datagram will be forwarded to an incorrect or outdated route. Under such circumstances, it is recommended that the routing table update packet be processed immediately, i.e. it be given a higher priority.

The above scenario is just one example where priorities are important. In general, when incorporating priorities in communication networks, a few issues must be considered. First of all, the information content of the packet must be considered. As discussed in the previous paragraph, the information content of packets differs in importance. The routing update packet is much more important than an ordinary datagram. From the network's point of view, as a general rule, processing control messages (like routing table updates, signalling messages and notifications) is more important than processing and forwarding data packets. Hence, the former set of packets must be given higher priority.

Even for datagrams, it is possible that certain packets are coming from or going to certain critical machines. For example, if the IP address of a packet is coming from the computer belonging to the head of a company, it is possible to give it higher priority than some other employees. So source/destination addresses can be used for some form of prioritization.

Another form of prioritization is based on whether the packet belongs to a time-sensitive application or not. As stated in this chapter, time-sensitive applications demand bounded delay and jitter characteristics from the network. Hence, time-sensitive packets cannot be buffered for too long, because otherwise they become useless. Thus, if a choice is to be made regarding buffering of two packets—one that is time-sensitive and the other that is not, the decision must be to buffer the time-insensitive packet. This decision is again a form of prioritization, because time-sensitive packets are given preferential treatment over other packets.

The above is an example where packets are prioritized at the application level. In some cases, packets of an application can require better treatment as compared to other packets of the same application. As an example, consider a *layered video* encoder that produces two types of data streams, an *essential layer* and an *enhancement layer*. The essential layer is sufficient to produce an average quality picture, while the enhancement layer improves the picture quality. The layered encoder is an example where packets belonging to the same application have unequal priority. For such applications, during overload conditions, the packets of the enhancement layer can be selectively discarded. This is again a form of prioritized discard, i.e. discarding low-priority packets first.

In essence, priorities are necessary for providing deterministic delays and losses. As seen from the above discussion, priorities can be incorporated in a number of scenarios. For example, during packet discards, the low-priority packets are dropped first. Or, during switching and buffering, time-sensitive packets are given preferential forwarding treatment.

12.11 FLOW CONTROL

Flow control is a means of synchronizing the sender's sending capacity and the receiver's receiving capacity. The primary concern is to ensure that an entity is not stressed beyond its servicing capability. In other words, flow control is a mechanism to control the flow of data between the sender and the receiver so that the receiver's buffers do not overflow. Flow control is necessary when a fast processor (sender) is communicating with a slow processor (receiver). Even if the processing capabilities of communication entities do not differ too much, flow control is necessary to achieve synchronization between the sending capacity of the sender and the processing capacity of the receiver.

Flow control mechanisms are broadly classified into two categories: *window-based flow control* and *rate-based flow control*. While in window based flow control, memory bottlenecks are tackled, rate-based flow control mechanisms tackle link and processing speed bottlenecks. Both the flow control mechanisms were explained in Chapter 6.

12.12 CONGESTION

Literally speaking, *congestion* is a situation when the demand for a limited resource exceeds its supply. In communication networks, congestion occurs when the available network resources are not enough to meet the demands. Shortage of buffer space, packets discard and packet re-transmission are all indicators of an imminent congestion or an actual state of congestion.

Congestion is one of the most prominent problems in network design. Whenever congestion occurs, the buffers start overflowing. Subsequently, the packets start getting dropped. In order to prevent buffer overflows, it is desirable to have larger buffers. But this leads to other problems. As the buffer size is increased, the transit delay also increases. If a packet is buffered for a very long time, the source assumes that the packet is lost, and re-transmits it. This leads to duplicate packets in the network, which clog the network and further aggravate the congestion. Thus, if not controlled, congestion feeds on congestion.

12.12.1 Causes of Congestion

The following are the main causes of congestion:

- **Over-commitment of shared resources:** By definition, congestion occurs when the demand for a resource exceeds the supply. When a resource is shared, demand for it may come from different quarters. Hence, all shared resources (like memory buffers and trunk lines) are potential points where congestion can occur. For example, consider the telecommunication networks where the total subscriber base is generally much greater than the number of active connections that can be handled at a given time. This is why during peak load conditions, the call requests are blocked. In this example, congestion is caused due to over-commitment of shared resources like trunk lines and switching capacity. On similar lines, statistical multiplexing also causes congestion. This is because, in order to optimize resource utilization, statistical multiplexing is used with the assumption that bursts from different users do not occur simultaneously. But during transient overloading conditions, when the bursts of different users coincide, congestion can occur.

* **Growing disparities in communication link speeds:** With tremendous advancements in the field of fiber-optic transmission, disparities in the speeds of communication links are increasing. This often leads to circumstances wherein incoming links into a router (or a switch) are much faster than outgoing links. The simplest example of such a scenario is a high-speed LAN (e.g. Fast Ethernet at 100 Mbps) connected to a low-speed WAN link (128 Kbps). The router at the LAN-WAN interface in this case receives data at much higher rates than it can possibly send. The end result is buffer overflows and packet discards at the router, both leading to congestion.

* **System breakdown:** Failure of network elements like routers and switches also causes congestion. This is because by the time the failure is detected and appropriate action taken, the transit packets in the network fill the buffers in the routers of preceding stages. The huge traffic jams on highways when a heavy vehicle breaks down, exemplify this point.

* **Non-conformance:** Congestion can also occur if proper policing mechanisms are not in place to control the traffic inflow of non-conforming users.

12.12.2 Congestion Problem in High-speed Networks

It is a general misconception that with improvements in transmission links and end-processors, congestion will cease to be a problem. Contrary to this, in certain situations, high-speed networks can be more susceptible to congestion attacks. The following three factors support this argument:

- **Bandwidth-delay product:** The amount of outstanding data on a link at any given time is the product of bandwidth and delay. The bandwidth refers to the link speed and can vary from anything between few Kbps to few Gbps. The delay refers to the end-to-end transit delay and is in the order of a few microsecond per kilometer. Thus, depending upon link speed, there can be anything between a few hundred kilobytes to a few megabytes of outstanding data on a link. During congestion, even if a signal is sent to the sender to stop data transmission, the outstanding data may still be large enough to overflow switch buffers.

- **Reduced processing time:** The link speeds have greatly outpaced the processing speeds. This has substantially reduced the processing time available to the switches to process incoming packets/cells. For example, at 15 Mbps link speeds and for an average packet size of 1500 bytes, a switch has .8 ms to process a packet. If the link speed is increased to 600 Mbps, the available processing time drops down to 0.02 microseconds. During any transient failure in the switching devices, there can be significant buffering problems, leading to congestion.

- **Outdated flow control methods:** Traditional flow control methods are no longer adequate to handle the heterogeneous traffic carried by high-speed networks. For example, some constant data rate applications like voice or video transfer, do not allow their data rates to be altered. These applications can either run at their specified data rates or have to be terminated. Traditional window-based flow control procedures cannot be applied to them. In such a scenario, flow control mechanisms cannot be used for congestion control.

12.12.3 Effects of Congestion

In general, congestion reduces the throughput and increases the delay of a network. Figure 12.5 shows how the throughput and delay of a network typically varies with an increase in the offered load. During low loads, substantial amounts of network

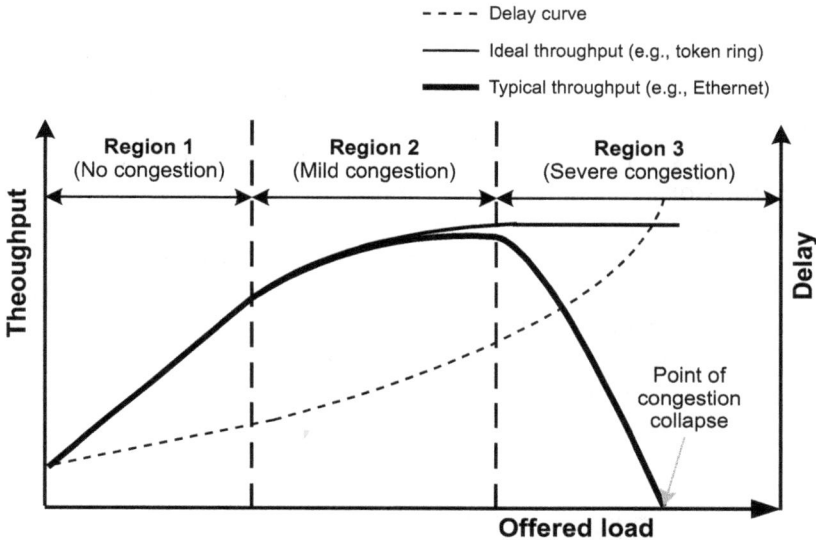

Fig. 12.5   *Effect on delay and throughput with an increase in offered load*

resources remain unutilized or underutilized. Hence, network throughput is low. However, low loads also result in less delay because user data does not have to wait for resources to be available. As the offered load is increased, more and more resources start being utilized and the net effect is an increase in throughput and a proportional increase in delay. However, the linear increase in throughput (region 1) with an increase in the offered load is applicable only till a point. Any further increase in the load results in only a marginal improvement in throughput. This region, where the throughput does not keep pace with load, is the region of *mild congestion* (region 2). Most network managers try to operate their networks in the region of maximum throughput and low delays, i.e. towards the end of region 1, just before the onset of region 2.

Once the offered load is increased beyond region 2, network performance deteriorates exponentially. This is the region of severe congestion where an increase in the load decreases throughput and increases delay. The point where the throughput becomes nearly zero is called the *point of congestion collapse* (see Fig. 12.5). What are the reasons behind this drastic deterioration in performance? If the example of Ethernet networks is considered, then it is obvious that under high loads, the increased collisions severely hamper network performance. For other network architectures also, severe congestion leads to *buffer overflows*, *packets re-transmissions* and *duplicate packets* in the network—all leading to performance deterioration. The re-transmission and duplication happens because even though packets are not lost, they are re-transmitted due to excessive delays. Moreover, if one of the fragments of a higher layer PDU is dropped, then the whole packet is re-transmitted. Under severe circumstances the network carries the same packets over and over again. All these ultimately lead to a point of congestion collapse. Interestingly, token ring networks behave ideally during overloads.

12.13 CONGESTION CONTROL

A lot of terms have been coined in existing literature, which describe congestion control mechanisms: Congestion avoidance, congestion management, congestion recovery, congestion prevention, preventive congestion control scheme and reactive congestion control scheme being a few of them. Depending upon the period when the mechanism is applied, all the mechanisms mentioned above fall under two basic categories: *preventive congestion control* and *reactive congestion control*. The difference between the two is not in 'what they are', but in 'when they are applied'. A preventive congestion scheme, as the name suggests, helps in preventing congestion. On the other hand, a reactive congestion scheme reacts to the problem of congestion once it has occurred. Proponents of 'prevention is better than cure' argue that preventing congestion is preferable. People on the other side argue that prevention leads to wastage and undue caution is unwarranted. They prefer reacting to the problem of congestion. The following sub-sections shed more light upon preventive and reactive congestion control mechanisms.

12.13.1 Preventive Congestion Control

One of the simplest ways to prevent congestion is to build a network that is capable of carrying the worst-case user traffic. The telecommunication network of the United State (US) is a good example in this regard. It is said that the US telephone network is designed in such a manner that even during the heaviest load (which occurs on Mother's Day), the network's resources are still not completely utilized. Although this is a foolproof measure, no one can generally afford the luxury of building such a wasteful network.

A different approach of congestion prevention would be to reserve some portion of the bandwidth (say 10-20 per cent) exclusively for handling transient overload conditions. This is a more practical approach and used in many public frame relay and ATM networks.

Traffic contract management is another congestion prevention technique. Traffic contract ensures that the demand from the existing connections does not exceed available resources. A new connection is accepted only if ample resources are available in the network to service the request. However, this method is applicable only for connection-oriented networks. In connectionless networks, such restrictions cannot be imposed on new connections.

12.13.2 Reactive Congestion Control

In general, congestion prevention techniques result in the under-utilization of resources. Thus, the preferred approach is to react to the problem of congestion than to prevent it. Mechanisms to ease out congestion, once it has occurred, come under the purview of reactive congestion control techniques. Described below are two reactive congestion control schemes, viz. *packet discard* and *congestion notification*.

12.13.2.1 Packet Discard

One of the simplest ways to deal with congestion is to discard excess packets. However, packet discard is not as simple as it may seem. There are many issues that must be taken into account before discarding packets. Some of the issues related to packet discard are:

- **Conformity and fairness:** Consider a case where there are ten applications, each requiring a bandwidth of 100 Kbps and another bandwidth-intensive application requiring 5 Mbps of bandwidth, all sharing a common link. Now, if the congestion is caused due to the non-conformance of 5 Mbps application, it would be unfair to drop packets of other applications. If, on the other hand, all applications are adhering to their traffic contract, the packets must be dropped in proportion to the bandwidth requirements, i.e. congestion control mechanism should be fair.

Further, it is observed that compliant connections quite often suffer because of non-compliant connections. If traffic contract is not stringently enforced, packets are discarded equitably among all connections, thereby benefiting non-compliant connections. Thus, ensuring conformance should assume greater priority than discarding packets.

- **Requirements for maintaining fairness:** Given that fairness is to be maintained, the implementation of this mechanism leads to new problems. This is because then the intermediate node has to then maintain a per-connection state. This would require considerable memory as well as processing capacity. Thus, guaranteeing fairness may become a task in itself.

- **Problems due to packet re-transmissions:** A discarded packet is generally retransmitted by the sender. Thus, random packet discards leads to a situation where the network carries retransmitted packets only, thereby bringing the network performance to a halt. Hence, packet discards must be considered under extreme circumstances.

- **Intelligent packet discards:** Any packet discard strategy must take into account the fact that if a fragment is dropped, then the whole, higher-layer PDU is discarded at the destination end. Thus, it is better to discard all fragments of a single PDU than to discard individual fragments of different PDUs. Such a packet discard mechanism is called intelligent packet discard mechanism and there are many ways to do this. In section 12.15.9 of this chapter, intelligent packet discard mechanisms in ATM are explained as an example.

12.13.2.2 Congestion Notification

Even if packets are discarded, the problem of congestion cannot be solved unless the end systems and intermediate routers take measures to slow down their data injection rates. For them to slow down, they must first be informed of the state of congestion in the network. This information is delivered through *congestion notification* or *feedback*. Notification involves informing—the end-systems and/or the intermediate routers—about the current state of the network. Upon receiving such notifications, it is *expected* that the end-systems and routers reduce their data injection rates.

Although it is assumed that once the end-systems receive a feedback message, they will take appropriate actions, this may not always be true. Depending upon the responsiveness of end-systems, the data flows can be classified as *responsive flows* and *unresponsive flows*.

Responsive flows respond to any notification from the network, *explicit* or *implicit*. Explicit notification involves sending explicit congestion notification messages or involves marking particular bits in data packets indicating impending congestion. Implicit notification is indicated through packet discards. Packet discards by the network are indicators of incipient congestion and suitable actions should be taken by the source upon its discovery. A good example here is TCP which reduces its rate upon detection of a lost packet.

Unresponsive flows are those flows where the source does not react to any congestion notification, explicit or implicit. A good example here is UDP traffic that is quite unresponsive.

Any good notification mechanism should consider various aspects. First of all, it should avoid additional traffic on the network. This is because if mechanisms to alleviate congestion add to the existing burden on the network, then such mechanisms are not desirable. As far as possible, the notification message should be piggy-backed to reduce the additional burden on the network. That is why in many networks (e.g. frame relay and ATM), a single bit in data packets/cells is marked to send congestion notification. Further, for applications having a very small active period, it may not be a very wise decision to send a feedback. It is quite likely that by the time the feedback reaches the source, the application has already quit. Even for other applications, if the delay is considerable, then sufficient damage may have already been done before any recovery starts. In essence, a notification mechanism should inform the source with minimum latency. Lastly, it is not always possible for the source to reduce its data rate, especially in the case of applications like video transfer. In such scenarios, the feedback mechanism is not recommended and other means need to be devised.

There are many prevalent notification techniques, some of which are:

- **Router to source notifications:** These notification messages, also called *source quench messages*, request the source to reduce its current rate of data transmission. Generally, every packet discard may result in a source quench message being sent. To avoid multiple messages, a smart router may monitor flows and send a notification message only to certain flows that are causing congestion. The obvious problem with this kind of notification is that it increases traffic on an already burdened network.
- **Routing table updates:** During congestion, a router may inform its adjacent routers of impending congestion in certain regions of the network. This will result in traffic re-routing and route optimizations, thereby helping in ameliorating the problem of congestion. This scheme also suffers from the drawback as that seen in the previous scheme.
- **Forward and backward notification:** Some networks have fields reserved in their packet headers that are set by intermediate nodes to indicate to the receiver about the impending congestion in the network. Frame relay and ATM networks employ this mechanism. However, while frame relay allows both backward and forward notification, ATM provides forward congestion notification capability only.

12.14 FLOW CONTROL VERSUS CONGESTION CONTROL

In existing literature, there is some confusion regarding the distinction between flow control and congestion control. Some argue that congestion control is a special case of flow control, while others argue that flow control is one of the means to tackle congestion. In this chapter, the two have been treated as logically distinct concepts, with the understanding that one may have a bearing on the other.

While flow control is limited between two entities (i.e. a *local problem*), congestion control is a network-wide phenomenon (i.e. a *global problem*). In flow control, a sender is throttled from sending at a rate more than what the receiver can receive, whereas the congestion control mechanism prevents network collapse due to traffic overload. Flow control involves two cooperative users, whereas congestion control involves many users (not necessarily co-operative). The distinctions between flow control and congestion control is summarized in Table 12.2.

Table 12.2   *Flow Control Versus Congestion Control*

Flow Control	*Congestion Control*
Throttling a sender from sending at a rate more than what the receiver can receive	Mechanism(s) to prevent network collapse due to traffic overload
It is a bipartite agreement	It is a network-wide control
Generally, the two parties involved in flow control are co-operative	The numerous parties involved in congestion control may not co-operate
Fairness is not an issue in flow control	Fairness is an important design issue
It solves the problem of resources at the destination being a bottleneck	It solves the problem of routers, switches and links being a bottleneck
Proper flow control is a necessary but not a sufficient condition for congestion-free network	Congestion control can have some bearing on the flow control

12.15 EXAMPLE: TRAFFIC MANAGEMENT IN ATM

This section highlights the concepts explained in this chapter earlier through the example of traffic management in ATM networks. To do so, first the ATM traffic and service descriptors are defined. These two sets of parameters form the very important notion of Service Categories in ATM. Five important service categories are discussed. Thereafter, various techniques for traffic management in ATM are discussed.

12.15.1 ATM Traffic Descriptors

In ATM, several *traffic parameters* are defined. Some of the important parameters are:

- **Peak Cell Rate (PCR):** This is the maximum rate at which a user is allowed to inject data into the network. Specifically, PCR defines an upper bound on the traffic that can be submitted by an ATM source. The inverse of PCR gives the 'minimum inter-arrival time' of cells for a given connection. The maximum value of PCR is bounded by the link rate. This is comparable to Maximum Allowable Data Rate (MADR) discussed earlier.

- **Sustainable Cell Rate (SCR):** This is the measure of the long-term average of user traffic. Specifically, SCR is an upper bound on the long-term average of conforming cells for an ATM connection. An ATM source may, at times, send at a rate greater than SCR. However, the average rate of the connection must remain less than SCR. This is comparable to Average Data Rate (ADR) discussed earlier.

  Once PCR and SCR are defined, it is important to ask, "For how long can a user send cells at PCR and still ensure that the long-term average is less than or equal to SCR?" Putting it differently, how much data can a user send at its peak cell rate without overshooting the average rate of the SCR? This is defined by the MBS and BT as explained below.

- **Maximum Burst Size (MBS):** This is the amount of data that an ATM source can send at its peak cell rate. Specifically, MBS is the number of cells an ATM source can send at the PCR. If MBS and PCR are known, then the maximum burst duration is determined by using the relation $T_{MBS} = \text{MBS/PCR}$, where T_{MBS} is the duration for which a user can send data at the peak cell rate.

- **Burst Tolerance (BT):** BT is a measure of the interval between consecutive bursts during which cells are sent at PCR. In other words, BT is the time interval after which an ATM source can again send data at PCR without violating the long-term average of SCR.

- **Minimum Cell Rate (MCR):** This parameter is defined for low-priority applications involving best-effort transfer. Specifically, MCR is the minimum cell rate that the network must provide to a connection. Its value can even be zero. This parameter can be viewed as the minimum guarantee that the network is willing to provide to a low priority application.

All the aforementioned parameters collectively define the source traffic descriptor. PCR, SCR and MCR are expressed in terms of cells per second. MBS is expressed in terms of cells, and BT, in terms of seconds.

Besides the traffic descriptors, there is another parameter in ATM called the *Cell Delay Variation Tolerance (CDVT)*. The relevance of CDVT is explained as follows: a cell stream arriving at the user network interface experiences some cell delay variation while entering the network. The reasons for this include multiplexing of several incoming cell streams or due to the variable processing delays in switches at the network ingress point. The delay variation may also result from insertion of physical layer overhead cells or management cells. This delay variation or randomness affects the inter-arrival time (i.e. 1/PCR) between consecutive cells of a connection as monitored at the UNI. Due to this, a cell stream injected at PCR may suffer some delay variation, resulting in the rate exceeding PCR. Cell delay variation tolerance or CDVT represents an upper-bound on the delay variation (i.e. jitter) of cells at a UNI. Thus, the measurement of cell rates (PCR/SCR) at the network ingress point is done keeping in account the CDVT for that interface.

12.15.2 ATM Service Descriptors

In ATM, the *service parameters* are commonly referred to as Quality of Service (QoS) parameters. The ATM Forum recommendation [ATMF TM4.1] specifies six QoS parameters. Depending upon the capability of the network and the requirements of individual connections, a network may offer one or more QoS parameters. The six QoS parameters defined are:

- **Maximum Cell Transfer Delay (maxCTD):** It is the transfer delay of a (major) portion of cells. This parameter captures the delay component.
- **Peak-to-Peak Cell Delay Variation (peak-to-peak CDV or P2P-CDV):** It is the difference between maxCTD and the fixed CTD that could be experienced by any delivered cell

during the entire connection holding time. This parameter captures the jitter component.

- **Cell Loss Ratio (CLR):** It is the fraction of cells that are either not delivered to the destination or delivered after a pre-specified time. This parameter captures the loss ratio discussed earlier.

- **Cell Error Ratio (CER):** It is the ratio of the total number of cells delivered with error to the total number of cells delivered. This parameter captures the error ratio.

- **Cell Mis-insertion Ratio (CMR):** It is the number of cells, meant for some other destination, inserted per second.

- **Severely Errored Cell Block Ratio (SECBR):** It is the ratio of severely errored cell blocks to the total transmitted cell blocks. A severely errored cell block outcome occurs when more than M error cells, lost cells or mis-inserted cell outcomes are observed in a received cell block of N cells.

12.15.3 ATM Service Categories

A unique feature of ATM is the introduction of various service categories. Each *service category* fulfils a particular need and is based on the relevance of a set of traffic and service descriptors. The different service categories in ATM are described below:

- **Constant Bit Rate (CBR) Service Category:** CBR service category is used by applications that require a constant bandwidth allocated to them throughout the life of the connection. This service category places strict upper bounds on cell delay and cell delay variation. Applications using CBR are voice, video and circuit emulation. The bandwidth requirement of the CBR traffic is characterized by PCR only. This is because the average rate and peak cell rate for CBR connections is equal, and so, only one of them is specified. As delay and jitter are critical, CDV and CTD parameters are specified. Loss may also be an important criterion and hence CLR is also specified.

- **Real-time Variable Bit Rate (rt-VBR) Service Category:** rt-VBR service category is intended for applications that have bursty traffic and require strict bounds over CTD and CDV. For rt-VBR traffic, the bandwidth requirements are specified by SCR and PCR. Again, since real-time applications may be sensitive to loss, delay and jitter, CLR, CTD, and CDV are specified for rt-VBR traffic. Applications using VBR are compressed video, voice and other real-time applications with bursty traffic.

- **Non Real-time Variable Bit Rate (nrt-VBR) Service Category:** The nrt-VBR service category is for applications having bursty traffic, but which are not sensitive to cell delay and delay variation. For the same reasons as in rt-VBR, the bandwidth requirement of this service category is characterized by PCR, SCR and MBS. For non-real time applications, since delay and jitter are not important, CDV and CLR are not specified. However, CLR is specified for the nrt-VBR traffic. Typical applications of nrt-VBR are off-line transfer of video, multimedia e-mail and bank transactions.

- **Unspecified Bit Rate (UBR) Service Category:** UBR service category is intended for applications having no requirement, whatsoever, with regards to cell delay, cell delay variation and cell loss. This is the reason why this category does not have any negotiable parameters associated with it. A UBR connection is characterized by PCR. However, the PCR value has little significance because the network never rejects a UBR connection due to unavailability of resources. If the network is unable to carry the offered load, the load belonging to this category is dropped. In essence, this service category is intended to consume the leftover capacity of a network. Example applications of this service category include simple file transfer and e-mail.

- **Available Bit Rate (ABR) Service Category:** ABR service category is used by applications that can adapt to *closed-loop feedback* given by the network. The feedback information contains information about the rate at which the user can transmit data. The ABR service category places no strict bounds on cell delay, cell delay variation, or cell loss. A source adjusting to data rates according to the feedback given by the network is expected to experience low cell loss and receive a fair share of available bandwidth. In other words, although this category does not have any QoS parameters associated with it, compliant users are assured of a low cell loss rate. The traffic is characterized by PCR and MCR. The applications may or may not specify MCR (that is, specifying MCR is optional). A zero MCR can also be specified. In case MCR is specified and non-zero, the network is guaranteed to provide MCR rate throughout the life of the connection. Examples of applications of ABR service category include data transfer and e-mail. Like UBR, this service category tries to utilize the available bandwidth of the network. However, the important difference between the two is that while a minimum value (MCR) is specified in ABR connections that the network must always support, the network is not committed to support the PCR specified for UBR connections. Moreover, unlike ABR, there is no feedback mechanism in UBR that monitors the rate of user data.

12.15.4 ATM Traffic Contract Management

The *traffic-contracting procedures* in ATM are very extensive. A *traffic contract* is defined by connection traffic descriptor (which includes traffic descriptors like PCR and SCR) and QoS parameters (which includes service descriptors like CLR, maxCTD and peak-to-peak CDV).

The contract parameters are specified either explicitly or implicitly. The parameters are explicitly specified when the network uses the values specified by the user. The values are implicitly defined if they are assigned by a network using a default rule (in absence of explicitly-defined traffic parameters contract).

During traffic contract management, the *Connection Admission and Control (CAC)* mechanisms decide whether a call request is to be accepted or not. CAC procedures determine if sufficient resources are available in the network to support the requested call. CAC procedures also ensure that the performance of existing connections is not degraded by accepting the new request. If sufficient resources are not present or the acceptance of connections may endanger the QoS guarantees of the existing connections, the network rejects the call. Thus, CAC ensures that resources are not allocated in excess to the existing capacity. An unchecked admission may lead to a situation where the demand for the resources far exceeds the availability, thereby causing congestion. In this sense, the CAC mechanism acts as a preventive congestion control mechanism.

The CAC algorithm to apply is linked to the ATM service categories. A CBR connection, for example, is characterized by its PCR only (i.e. it does have an SCR). In this case, the peak bandwidth allocation scheme is applicable. This is because cells are transmitted at a constant rate and there is no statistical gain to be achieved. In general, the greater is the burstiness of a connection, the more is the advantage of statistical multiplexing. Figure 12.6 presents a simple CAC algorithm based on the PCR value of CBR connections. The connection is rejected if the PCR value exceeds a certain threshold value. This threshold value can be a percentage of the link rate. Such a clause ensures that no connection occupies a significant portion of the link capacity. The connection request is also rejected if the number of CBR connections exceeds a pre-specified maximum value. The second *if* condition determines if sufficient bandwidth is available for the new connection. Since a switch services the requests of different service categories, the factor %_RESVD_FOR _CBR_CONN measures the part of bandwidth that is reserved specifically for CBR connections.

The peak bandwidth allocation scheme for CBR connections, as mentioned above, does not take into account the QoS parameters (i.e. CLR and CDV) of the request. There are two methods that take CDV into account, viz. *negligible* and *non-negligible CDV methods*, details of which can be found in [TM J. Roberts].

```
simple_cac_algorithm
n: the number of pre-existing connections
PCR_{n+1}: Peak Cell Rate of the incoming call
if((MAX_ALLOWED_PCR < PCR_{n+1}) or (MAX_NO_OF_CBR_CONN < (n+1)))
Reject the call
if (PCR_{n+1} < (%_RESVD_FOR_CBR_CONN * LINK_RATE - Σ_{i=1 to n}(PCR_i)))
    Accept the call
    Increment n
else
    Reject the call
```

Fig. 12.6 *A CAC algorithm*

For VBR connections, the effectiveness of the CAC algorithm depends upon the burstiness of the connection, where burstiness is the ratio of SCR and PCR (i.e. burstiness = SCR/PCR). If the burstiness tends towards 1, then the peak bandwidth allocation is a suitable option. However, as the burstiness value reduces, the link utilization achieved using peak allocation also falls. In the extreme case where SCR/PCR << 1, link utilization can be very low. If instead of PCR, SCR is used to allocate resources, there is a possibility that simultaneous bursts cause severe cell loss. Thus, the preferred option is to use a value between SCR and PCR, where the exact value could be chosen by monitoring the behaviour in an actual ATM network.

The CAC procedure for ABR connections is simple because the only parameter of any significance is the MCR. A peak bandwidth allocation using MCR is a reasonable scheme. CAC for UBR is the simplest, because the network is not obliged to guarantee anything. However, an upper limit on the number of UBR connections may be maintained to provide some bandwidth to UBR connections.

12.15.5 ATM Traffic Shaping

Traffic shaping is typically implemented using the leaky bucket algorithm. In this scheme, bursty data generated by a source is stored in the buffer and sent out at a lower rate, thereby transforming bursty traffic with a high PCR to a more stable traffic with lower PCR. This mechanism is referred to as peak cell rate reduction. Traffic shaping makes the traffic characteristics more predictable and hence, the network can adequately satisfy the requirement of the resources of such a source.

12.15.6 ATM Traffic Policing

In ATM, the mechanism to monitor traffic and enforce the traffic contract is termed as *Usage Parameter Control (UPC)*. Depending upon whether traffic monitoring is performed at UNI or NNI, the monitoring functions are referred to as *Usage Parameter Control (UPC)* or *Network Parameter Control (NPC)* respectively. However, UPC is considered to be a generic concept and includes NPC as well.

There is a minor difference in opinion between the ITU-T and the ATM Forum over the use of UPC/NPC functions. While ATM Forum makes both UPC as well as NPC a network-dependent option, ITU-T recommends the use of UPC while leaving NPC as a network option.

Several algorithms are defined to monitor traffic and to enforce traffic contract. The most commonly used algorithm is the *Generic Cell Rate Algorithm (GCRA)*. GCRA can also be viewed as a 'Leaky bucket' algorithm. The GCRA is used to define conformance of the cells with respect to the negotiated traffic contract. GCRA may be used by the UPC function to enforce conformance. However, the network is not obligated to use this function and may use any other algorithm. The algorithm checks every cell for its *conformance*. A non-conforming cell may be marked (i.e. CLP bit in the header can be set to 1) such that during congestion, such cells can be discarded. Non-conforming cells can also be discarded.

The GCRA is used to define conformance with respect to PCR and CDVT, as well as SCR and BT. GCRA is defined in terms of two parameters: The increment (I) and limit (L), and is denoted as GCRA(I, L). The increment value specifies when the next cell is expected, while the limit specifies the permissible deviation from the expected arrival time. For example, for CBR applications, I is equal to 1/PCR and L equal to CDVT. This form of GCRA is represented as GCRA(1/PCR, CDVT). Similarly, for VBR applications, GCRA is defined in terms of SCR, CDVT and BT.

GCRA checks conformance to the traffic contract, using the twin concepts of *expected* and *actual cell arrival times*. For an incoming cell, the expected arrival time is referred to as the Theoretical Arrival Time (TAT). The conformance of a cell is decided on the basis of its TAT. The cell is said to be conforming if it arrives at time (TAT-L) or after. All cells arriving before time (TAT-L) are marked as *non-conforming cells*.

Figure 12.7 shows the GCRA algorithm. In this algorithm, the first incoming cell initializes the algorithm and sets the TAT of the first cell to ta(1). For the kth cell, the conformance is decided on the basis of TAT(k), according to the following rules (here, TAT(k) is the theoretical arrival time of the kth cell):

1. If the kth cell arrives after time TAT(k), the cell is declared conforming and the TAT for the next cell (i.e. TAT($k+1$))

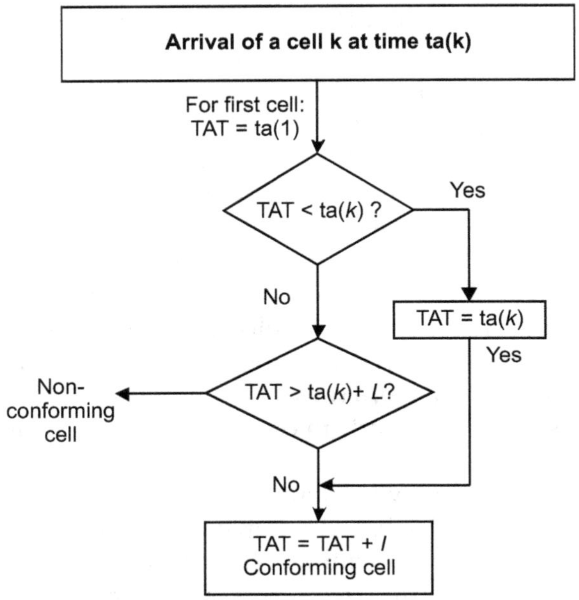

TAT: Theoretical Arrival Time ta(k): Time of arrival of cell k

I: Increment *L*: Limit

Fig. 12.7 *Generic Cell Rate algorithm*

is set to (ta(k) + I) and k is incremented. Here, ta(k) refers to time of actual arrival of kth cell.

2. If the cell arrives between time TAT(k) and TAT(k) – L, the cell is still declared as conforming. However, the TAT of the next cell is still set to (TAT(k) + I) and not to (ta(k) + I). This prevents a smart user from obtaining a data rate of $1/(I - L)$, instead of $1/(I)$. Moreover, the value of k is incremented.
3. If the cell arrives before (TAT(k) – L), the cell is declared as non-conforming and the TAT for next cell is again set to TAT(k). (Note that a cell will not be declared non-conforming if it arrives after its TAT. Thus, even if the TAT of a cell is set to a value much less than the actual arrival time, the conformity of the cell towards traffic contract will remain unaffected).

The three steps mentioned above ensure that the cells transmitted above the allotted rate are declared non-conforming. Consider what happens if a malicious user starts sending cells at a rate slightly greater than the allotted rate (PCR); i.e. at a spacing slightly less than I, say $(I - \delta)$, where $\delta < L$. The first arriving cell initializes the algorithm and sets TAT2 to I. Let the expected arrival times of subsequent cells be as shown in upper part of Figure 12.8 (expected behaviour). The second cell arrives at instant $I - \delta$ and is still conforming. The TAT3 is set to TAT2 $+ I$ by rule 2. The third cell now arrives at $2(I - \delta)$. This cell is conforming if $2*\delta < L$. However, going by the figure, the third cell is also conforming. It is the fourth cell which is declared non-conforming because $3*\delta > L$. In a general case, a user can only send n cells at a rate greater than allotted rate, where n is the largest number such that $n*\delta < L$. For example, if $\delta = 0.4*L$, the user is able to send only two cells, before the third one is declared non-conforming.

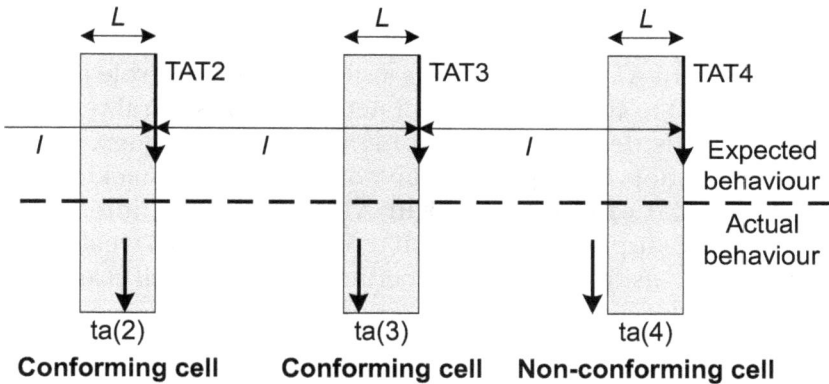

Fig. 12.8 *The cell arrival process*

12.15.7 ATM Priority Control

Although attempts have been made to incorporate priority control features in the ATM network, not much headway has been made in this direction. Knowing well the advantages of priority control in any network, the critics are quick to point out this lacuna in ATM. Incorporating priority in ATM is rather difficult. Complex priority control algorithms will further overload ATM layer functionality. Processing every cell according to its priority will mean additional delays, which might not be acceptable to network users.

Currently, *selective cell discard* is the only priority control function specified in ATM. This function is based on the *Cell Loss Priority (CLP)* bit of the ATM header. The cells with CLP bit set to 0 are considered as high-priority cells, while the cells with CLP bit set to 1 are considered as low priority cells. If the network is forced to drop some cells, the network selectively drops the cells with lower priority.

The cells can be either marked by the user or by the UPC function at the network ingress point. In the first case, the user classifies cells into two priority levels and sends high-priority cells with CLP bit 0 and low-priority cells with CLP bit 1. If there is congestion along the destined path, the network first discards the cells with CLP bit 1. This scheme ensures that high-priority cells are given preferential treatment. As an example, the CLP bit can provide two priority levels for applications like the layered encoder described earlier.

12.15.8 ATM Flow Control

Flow control in ATM networks has a slightly different role to play as compared to conventional data networks, for several reasons. First, ATM is designed to be a high-speed technology. Thus, traditional hop-by-hop flow controls for error checking (as used in X.25) cannot be used in ATM. Moreover, interactive multimedia applications, which communicate in real-time, preclude the use of flow control because a delayed cell is as good as a lost cell.

Currently, flow control in ATM is used only to optimize bandwidth utilization by having feedback mechanisms. Two such flow control mechanisms are specified. One of the mechanisms is *Generic Flow Control (GFC)* that provides a link-level flow control at the UNI. The other mechanism is *ABR flow control* that is used to utilize any leftover bandwidth in the network for connections belonging to the ABR service category. Since GFC flow control is not very popular, only ABR flow control is described below.

For ABR traffic, the source controls its data injection rates according to the loading conditions of the network. At any given time, the current load of the network is conveyed to the source through *Resource Management (RM)* cells. These RM cells carry various pieces of information about the current state of the network. The information includes, among other things, the bandwidth currently available, the state of congestion and the impending congestion.

Although the ABR service does not provide any QoS guarantee, it ensures an efficient means to access the bandwidth available in the network. In order to achieve this goal, a flow control feedback loop is used. For the feedback loop, the source generates an *Resource Management (RM)* cell (called the forward RM cell), which is looped back by the destination and sent back to the source (looped back cell is called the backward RM cell). This end-to-end flow control model is depicted in Figure 12.9.

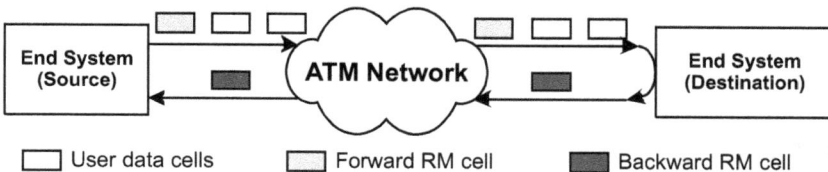

Fig. 12.9 *End-to-end ABR flow control model*

An intermediate node has three options with regard to conveying feedback information to the source. As the simplest of options, a node can directly insert information in the forward or backward RM cells. Alternatively, a node can set the congestion bit of the ATM cell header to convey congestion notification. The destination in this case then updates the backward RM cell according to the received information. Lastly, a node can directly generate a backward RM cell. In this case, the communication no longer remains end-to-end. This option is advantageous, in the sense that it alleviates the latency involved in end-to-end feedback.

12.15.9 ATM Congestion Control

In ATM, both *preventive* and *reactive* mechanisms exist to tackle the problem of congestion. CAC is an example of a preventive mechanism, wherein resources are allocated in such a manner that congestion remains a distant possibility. Among the reactive congestion control mechanisms, three techniques, viz. *selective cell discard*, *frame discard* and *explicit forward congestion indication*, are discussed below.

During congestion, a network may decide to discard cells with the CLP set to 1 and protect the cells with the CLP bit set to 0. This mechanism is referred to as selective cell discard. The basic idea behind this mechanism is to protect high-priority cells while dropping the low-priority ones. Also note that one of the means of traffic policing is tagging the non-conforming cells. Tagging works on the assumption that in case of congestion, the tagged cells are the first ones to be discarded.

In order to understand frame discard, note that in ATM networks, a large IP packet is segmented and carried in a number of cells as the 48-byte payload. Thus, during congestion, if one of the cells belonging to an IP packet is dropped, the remaining cells carried by ATM network become *dead cells* because these cells cannot be properly reassembled at the receiving end. This is because there is no re-transmission at the ATM or AAL level. The carriage of dead cells can aggravate congestion to a great extent.

Depending upon how the dropped cells are bunched, the throughput of the network varies. In the worst case scenario, wherein a cell from each packet is discarded, the throughput can be nearly zero. With the same level of cell discard, the best-case packet discard strategy discards cells that belong to the same packet giving a much better throughput as compared to distributed cell discard strategy discussed earlier (see Fig. 12.10). Thus, it is preferable that if one cell of a packet is

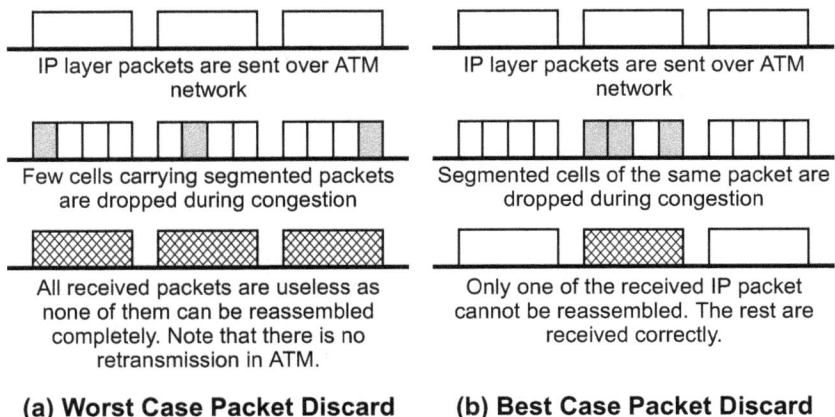

IP layer packets are sent over ATM network	IP layer packets are sent over ATM network
Few cells carrying segmented packets are dropped during congestion	Segmented cells of the same packet are dropped during congestion
All received packets are useless as none of them can be reassembled completely. Note that there is no retransmission in ATM.	Only one of the received IP packet cannot be reassembled. The rest are received correctly.
(a) Worst Case Packet Discard	**(b) Best Case Packet Discard**

Fig. 12.10 *Worst and best case packet discard*

discarded, the rest of the cells of that packet is also discarded. In other words, rather than discarding cells on an individual basis, it is better to discard cells on a frame basis. In order to tackle the packet discard issue, various schemes are currently defined in which packets are discarded at the frame level rather than the cell level. Some of these include Partial Packet Discard (PPD), Early Packet Discard (EPD) and Random Early Discard (RED).

Another reactive congestion control technique is congestion notification, which is sent to the destination using a mechanism known as the Explicit Forward Congestion Indication (EFCI). In this mechanism, the congestion bit in the ATM header is set to 1 to indicate an impending congestion in the network. Upon receiving an EFCI notification, the destination should somehow communicate to the source to slow down the transmission rate because congestion is prevalent in the downstream direction (in the source to destination direction). How the destination actually conveys this information depends upon the protocols running at the end-system. No standard procedures exist in this regard. Moreover, since EFCI is optional for all of the service categories (except ABR), it is recommended that the network should not rely on this congestion control mechanism.

12.16 CONCLUSION

This chapter provided an overview of traffic management. Tightly linked to the concept of traffic management is the concept of traffic (user's view of offered load) and the concept of service (network's parameters to define the quality of service provided). The key traffic parameters are average data rate, maximum data rates and burst tolerance. The important service parameters are delay, jitter, loss ratio and error ratio. After elaborating the traffic and service descriptors, various techniques for traffic management were discussed. These were then elaborated through the example of the ATM network wherein the application of these techniques was examined.

REVIEW QUESTIONS

Q 1. What are the key traffic descriptors? What are the important traffic descriptors in ATM?

Q 2. What are the key service descriptors? Differentiate between loss ratio, throughput and goodput.

Q 3. Why are so many service categories defined in ATM? What is the relation of the ATM service categories with ATM traffic and service descriptors?

Q 4. What are the tools for traffic management? Discuss some of the important techniques with respect to ATM.

Q 5. What is traffic contract management? What are its key elements?

Q 6. What is the difference between traffic shaping and traffic policing? How can shaping avoid policing? Explain with a simple example.

Q 7. Differentiate between flow control and congestion control.

Q 8. Differentiate between preventive congestion control techniques and reactive congestion control techniques? Which ones are used in ATM?

FURTHER READING

[Gen A. Tanenbaum], [Gen W. Stallings] and [Gen S. Keshav] are some of the important references for generic information on traffic management. For details of traffic management in ATM, the reader should refer to [ATM S. Kasera] and [ATM M. Prycker]. The ATM Forum specification [ATMF TM4.1] and [ITU-T I.371] also provides a detailed description of Traffic management procedures in ATM. [TM H.G. Peros], [TM K. Shiomoto] and [TM Raj Jain] are other references on ATM traffic management.

NETWORK MANAGEMENT

13.1 INTRODUCTION

The last couple of decades have seen an explosive growth in the deployment of networks, in both the local area as well as in the wide area environment. However, with the increase in network size, the need to efficiently manage them has also gained significance. Till the mid-1980s, network management was considered to be an ad-hoc exercise, intended to keep all network elements functioning. But this scenario changed after the mid-1980s, when the large size of the networks warranted an exorbitant price in terms of managing them. The heterogeneity of and the incompatibility between networking technologies made network management that much more difficult. At that time, a strong need was felt to have a protocol that could automate the process of network management. Various protocols like Simple Network Management Protocol (SNMP) and Common Information Management Protocol (CMIP) that followed, were the result of this need to have a well-defined management framework. Both SNMP and CMIP are based on the manager/agent/object relationships. As these terms suggest, a manager manages objects with the help of agents. These concepts are detailed in this chapter.

Before these concepts are elaborated, it is important to define network management. Network management, as a term, has many definitions. Put simply, *network management* is the process of managing a network. More formally, network management is the act of monitoring and modifying network operations with the purpose of detecting and solving problems, improving performance, and collecting reports and statistics.

13.2 GOALS OF NETWORK MANAGEMENT

The answer to the question what is inevitably linked with the goals of network management. Some of the *goals* of network management are:

1. To reduce the operational cost
2. To increase flexibility
3. To increase efficiency
4. To increase network availability
5. To provide security.

The goals collectively define the scope of network management.

13.3 FUNCTIONAL AREAS OF NETWORK MANAGEMENT

In order to formalize the scope and functional breakdown of network management, the OSI management framework classifies the network management functions into five major functional areas (refer [ITU-T X.700]). This classification, also referred to as *FCAPS* after the first letter of each functional area, defines the following functional areas:

1. Fault management
2. Configuration management
3. Accounting management
4. Performance management
5. Security management

The next five sub-sections describe each of these functional areas (see Table 13.1).

13.3.1 Fault Management

Faults or failures are a part of everyday life. In communication networks, faults occur in the form of disk failures, system

Table 13.1 *Functional Areas of Network Management*

Functional area	*Definition*	*Functions*
Fault management	Encompasses fault detection, isolation and the correction of abnormal operation.	1. Maintain/examine error logs. 2. Trace and identify faults. 3. Carry out diagnostic tests. 4. Correct faults.
Configuration management	Identifies, exercises control over, collects data from and provides data to open systems for initializing, starting, providing for the continuous operation of, and terminating interconnection services.	1. Set parameters that control routine operation. 2. Associate names with managed objects and sets of managed objects. 3. Initialize and close down managed objects. 4. Change configuration.
Accounting management	Enables charges to be established for the use of resources.	1. Inform users of costs incurred or resources consumed. 2. Enable accounting limits to be set and tariff schedules to be associated with the use of resources.
Performance management	Enables the behaviour of resources and effectiveness of communication activities to be evaluated.	1. Gather statistical information. 2. Maintain and examine logs of system, state histories. 3. Determine system performance under natural and artificial conditions. 4. Alter system modes of operation for conducting performance man-agement activities.
Security management	Provides the means to support security policies of various applications.	1. Create, delete and control security services and mechanisms. 2. Distribution of security-relevant information. 3. Reporting of security-relevant events.

failures, cable faults, loose cable connections, internal hardware failure, etc. Faults lead to inconvenience, irritation, and even financial losses. Thus, network managers are wary of any kind of fault in the network, and give fault management the top priority among all MIS activities.

In essence, *fault management* is the process of detecting, isolating and correcting network-related problems. The fault management procedures are divided into the following areas or components:

1. **Problem detection:** This involves isolation of both the problem and the component in which the problem has occurred. This, however, does not include the root cause analysis of the problem. The emphasis is to take all the necessary steps, after which the problem diagnosis can begin. Problems are usually detected through alarms or event triggers.

2. **Problem diagnosis:** This involves determining the precise cause of the problem and suggesting the actions required to solve the problem.

3. **Problem correction or resolution:** This involves efforts required to solve the problem. Problem resolution can be manual or automated.

4. **Problem tracking and control:** This involves tracking the problem until it is finally solved. Problems are tracked by maintaining problem databases that include problem status reports and status monitoring data.

The aforementioned stages of fault management are comparable to the stages of a person falling ill and recovering from it. First, the person observes certain symptoms of illness (problem detection). Then he/she consults a physician. The physician diagnoses and conducts medical tests to check the validity of his/her hypothesis (problem diagnosis). The physician then prescribes medicines (again a part of problem diagnosis). The sick person takes the prescribed medicines and recuperates (problem correction). Whenever the person falls ill again and consults the physician, the old prescription acts as a problem-tracking tool. The physician remembers the effect of the medicines prescribed earlier and takes actions appropriately.

13.3.2 Configuration Management

Configuration management involves keeping track of the various components of the network. For this, a configuration database is maintained, which contains details of all the resources of the network. The information associated with an entry in the database depends upon the type of resource. For example, for a desktop PC, some of the attributes maintained would be the type of microprocessor, size of the RAM, total number of storage devices (like hard disk) and the capacity of each, type of monitor, total number of serial and parallel ports, type of cards in each slot, etc.

Similarly, for other resources, various attributes are maintained. Configuration management ensures that any configuration change is, captured in the configuration database quickly and accurately. The configuration database is useful for other management functions such as fault management, because it provides the necessary inputs. Configuration management can be further classified into the following functional areas:

* **Change management:** This refers to managing network-related or system-related changes in the network. Another aspect of change management is to minimize the impact that changes in one part of the network may have on the other parts.
* **Topology management:** This entails maintenance of information of the physical, logical and electrical layout of the network. Since a network is essentially an interconnection of communicating entities, it is important to keep track of how the various elements are interconnected. This important aspect comes under the purview of topology management. The scope of topology management depends upon the size and the geographical extent of the network. For a very large and geographically dispersed network, topology management becomes very complicated.
* **Inventory management:** Also sometimes referred to as asset management, this entails keeping track of currently installed components and their spares. The inventory contains information on an array of items like PCs, mainframes, switches, routers, modems, Data Service Units (DSUs), Channel Service Units (CSUs), cross-connects, trunk lines, circuits, vendors, locations, etc.

13.3.3 Accounting Management

Accounting management refers to the process of collecting and reporting information on resource usage. Accounting occupies different meanings in LAN and WAN environments. In a WAN environment, the service provider provides network connectivity, and in turn, charges for the resource consumed. Here, the emphasis is on two things: first, to devise ways to charge the user appropriately and second, to provide means to the user to access accounting information. Charging can be done by using a number of ways, with the common techniques being duration-based billing (e.g. telephone calls and Internet connections) and fixed-rate billing (leased line connections). In a LAN environment, depending upon the management policies, accounting may or may not be required. When required, the endeavour is to maximize resource utilization (because the cost is more or less fixed).

In general, accounting management involves the following steps:

1. **Identifying the cost components of various network elements:** This includes, among various things, the hardware cost, maintenance cost, salaries of employees and charges of facilities.
2. **Establishing charging criteria:** The charging decision is usually based on parameters like duration of connection, total bytes transferred and geographical distance. Once the basis is fixed, the rates are determined accordingly.
3. **Defining procedures for receiving and processing accounting information from various network elements:** The emphasis is upon keeping the procedures simple and stable.
4. **Processing of bills, and sending the same to the respective customers.**

13.3.4 Performance Management

Performance management refers to the set of activities required to maintain the network at optimum performance levels. It includes the following:

- Continuously monitoring the *performance indicators* of network operation

- Verifying how *service levels* are maintained in the network
- Identifying actual and *potential bottlenecks* of the network and
- Preparing *reports and trends* to facilitate better network planning and management.

For the purpose of performance management, a number of performance indicators are defined. These indicators can be classified as *service-oriented indicators*, which measure the quality of service provided by the network, and *efficiency-oriented indicators*, which measure the efficiency and resource utilization of the network. Some of the service indicators include percentage availability of the meantime between failures, the meantime to repair, error rates, total response time and processing delay. Some of the efficiency-oriented indicators include the numbers of calls established per unit time, percentage of CPU usage, percentage of bandwidth utilization and communication overheads.

13.3.5 Security Management

Security management includes providing security to both physical assets (e.g. buildings, computing and machinery) as well as to logical assets (e.g. confidential data, reports and information).

The security threat to physical assets occurs in various forms. The primary threat is from natural calamities like floods, cyclones and earthquakes. In order to minimize the loss of data and operations due to these calamities, one of the solutions is to have a fully functional back-up site. Then there are threats from fire, which can be avoided by taking adequate care. The threat from physical thefts can be tackled by employing security guards, safety locks and automatic entry/exits.

The security threat to logical assets occurs in various forms. These threats can come from malicious users who:

- listen to an ongoing communication, with the intention of deciphering the contents of the message,
- try to guess passwords by decoding the contents of an ongoing communication (for future unauthorized access),
- modify the contents of the sent data, with the purpose of misleading the receiver,

- delete portions of the transmitted information,
- gain unauthorized access with the purpose of destroying stored data, and
- prevent authorized users from accessing stored data.

The primary function of security management is to ensure network access only to authorized users and to prevent security threats from malicious users. The most common way to ensure this is through user identification (using login name and a password). This form of security check, however, has several problems. Passwords are generally disclosed and ensuring authorized access is never possible. Hence, some systems make it mandatory for the user to frequently change the password. Even if passwords are not disclosed, the simple ones can be easily guessed. Thus, some security mechanisms place a restriction on the minimum number of characters and the form of a password. While employees within an organization are less prone to make unauthorized access, the problem is more severe in providing secured remote access. Hackers always try to break through the defence of an organization through devious means. This is avoided by using firewalls at the boundaries of the networks.

13.4 EXAMPLE: SIMPLE NETWORK MANAGEMENT PROTOCOL (SNMP)

Simple Network Management Protocol (SNMP) is an application layer protocol, designed for the exchange of management information between network elements. Although SNMP was developed as a temporary solution to network management problems, it has now become one of the most popular protocols for network management.

Around 1987, three proposals for network management surfaced. These three proposals included the High-level Entity Management System/Protocol (HEMS/HEMP), the Simple Network Management Protocol (SNMP) and the Common Management over TCP/IP (CMOT). Out of these three, the HEMS/HEMP proposal was soon withdrawn, but the latter two remained. In the 1988 meeting of the Internet Activities Board (IAB), it was decided that SNMP be used in the short-term and that CMOT be developed for use in the long-term. The idea behind this move was to use SNMP as a temporary measure only, till CMOT was completely developed. The long-term goal was to use CMOT to merge the Internet management framework around the lines of the OSI management standards. However, owing to the huge success of the SNMP coupled with the excessive delay in the publication of the OSI standards, SNMP became a permanent network management standard widely accepted in the telecommunications industry.

The first version of SNMP (SNMPv1) was released around mid-1990. SNMPv1 is collectively defined in the following RFCs:

- RFC 1157 defines the basic SNMPv1 protocol.
- RFC 1155 defines the Structure of Management Information (SMI) for SNMP.
- RFC 1212 provides guidelines for defining the Management Information Base (MIB) modules.
- RFC 1213 defines the core set of managed objects.

The second version of SNMP (SNMPv2) came three years after the first version was released. The major shortcoming of SNMPv1 was its lack of security features. SNMPv2 was developed to incorporate security features, apart from few other modifications that were done to improve the standard from SNMPv1.

Like SNMPv1, SNMPv2 is also described in a number of RFCs. At the time of its release, SNMPv2 was detailed in RFC 1441-1452. However, some of those RFC's have now become obsolete. Currently, the important RFCs for SNMPv2 include the following:

- RFC 1441 provides an introduction to SNMPv2.
- RFC 1902 provides the Structure of Management Information (SMI).
- RFC 1903 provides the textual conventions.
- RFC 1904 provides the conformance statement.
- RFC 1907 provides the Management Information Base (MIB).

As part of this example, only the first version of SNMP (SNMPv1) is discussed. This is because of the two versions of the protocol are basically similar, with only a few modifications in SNMPv2. Interested readers can refer to the aforementioned RFCs of the SNMPv2 protocol for details.

At the time of writing this book, work was underway for the next generation of SNMP called the SNMPv3.

13.4.1 SNMP Management Model

In order to exchange management information, SNMP framework defines a management model (as depicted in Fig. 13.1). This network management model consists of *agents, network management stations* and *MIBs*.

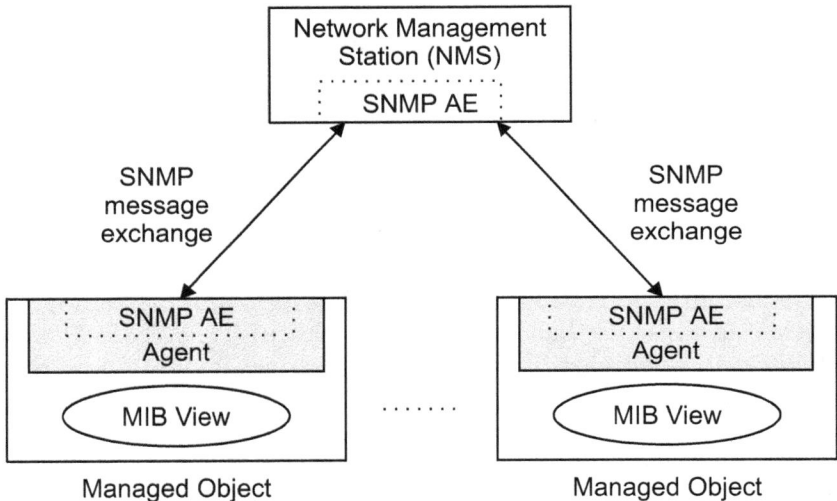

Fig. 13.1 *SNMP management model*

Agents are software modules residing within managed objects. A managed object can be any entity residing on a network that needs to be managed, which includes computers, printers, routers, hubs and bridges. Since the computing power of a managed object can be limited, the agent software must be designed with a knowledge of this limitation. This fact has an important bearing on the design of the SNMP protocol. Not only is SNMP a simple protocol because of its name, it has also been designed to be a simple protocol. All aspects of SNMP reflect this philosophy of being simple.

Agents residing in managed objects provide management information to Network Management Stations (NMS). The fact that agent modules are simple shifts the computing complexity to NMS. An NMS can be viewed as a manager, which monitors the managed objects through exchange of control messages *Get* and *Set*. The NMS queries the state of the managed object through a *Get* operation. It can also alter the value of a managed object by using a *Set* operation.

MIBs, or Management Information Bases, are virtual information stores through which managed objects are accessed. Objects in the MIB are defined by using a sub-set of the Abstract Syntax Notation one (ASN.1) notation as per the guidelines of the Structure of Management Information (SMI). A MIB view of a managed object is defined as the subset of the objects in the MIB that pertains to that managed object.

In Fig. 13.1, besides agents NMS and MIB, another entity called SNMP AE (SNMP Application Entity) is also depicted. An SNMP AE is any entity residing at either the management station or at the managed objects, which communicates with other similar entities using SNMP.

13.4.2 Structure of Management Information

The Structure of Management Information, (SMI), refers to the structures and identification schemes used to define management information. SMI is detailed in RFC 1155. The RFC describes an object information model along with a set of generic types used to describe management information.

According to RFC 1155, the managed objects are represented by using ASN.1 notation. Associated with each managed object are its name, syntax, and an encoding scheme. The details of each of these attributes, as specified by RFC 1155, are explained in the following sub-sections.

13.4.2.1 Name of Managed Object

Names are used to identify managed objects. Each name is represented uniquely, using an *OBJECT IDENTIFIER*. An OBJECT IDENTIFIER can be viewed as a path from the root to a node of a hierarchical tree, where each node in the tree is assigned a number. The assignment of numbers to different nodes is administratively controlled. Figure 13.2 shows the registration hierarchy tree of managed objects. The root of this tree has three children. The left-most child of the root {ccitt,

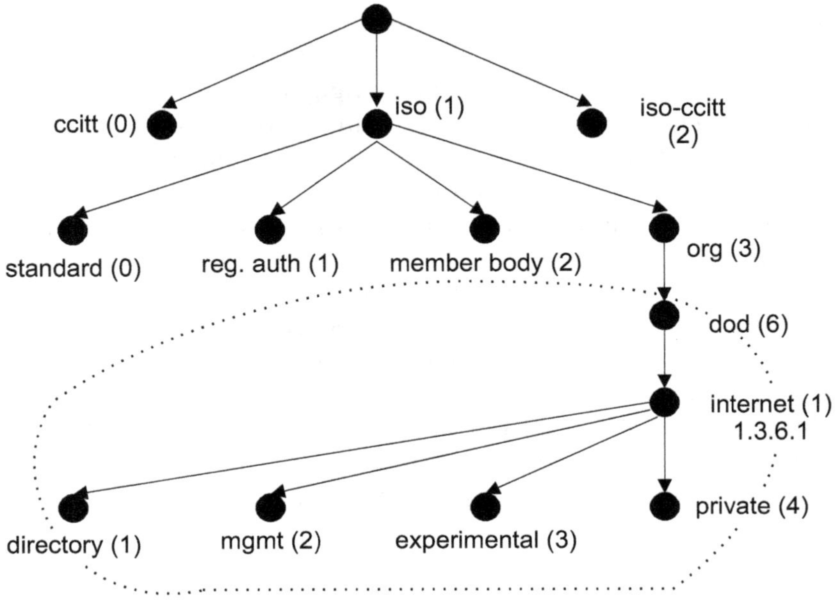

Fig. 13.2 Registration hierarchies

(0)} indicates that the node is managed by CCITT (now known as ITU-T) and has a number 0. The same logic applies for other two children of the root.

Now consider the sub-tree depicted in the figure, which is enclosed within dashed lines. The Department of Defence (DoD) allocated a node for the Internet community. This node is administered by the Internet Activities Board (IAB) and is defined as:

Internet OBJECT IDENTIFIER ::= {*iso(1) org(3) dod(6) 1*},

where '::=' means 'defined as'.

Hence, the Internet sub-tree OBJECT IDENTIFIER starts with 1.3.6.1. The Internet sub-tree contains the following nodes:

- **Directory:** The directory (1) sub-tree is reserved for future use.
- **Mgmt:** The mgmt (2) sub-tree is used to identify objects that are defined in IAB-approved documents.

- **Experimental:** The experimental (3) sub-tree is used for identifying objects used in Internet experiments.
- **Private:** The private (4) sub-tree is used to identify objects unilaterally.

In essence, names of managed objects are represented uniquely, using a hierarchical tree structure, wherein the identifiers associated with each node in the tree is administratively controlled.

13.4.2.2 Syntax of Managed Object

Besides the name of the managed object, the second attribute associated with a managed object is its *syntax*. Syntax refers to the structure used to define object types. All structures are defined using the ASN.1 notation. Table 13.2 shows the built-in types of ASN.1.

Table 13.2 *ASN.1 Built-in Types*

Boolean	Integer	Bit String	Octet String	Null
Enumerated	Real	Sequence	Sequence of	Set
Set of	Choice	Any	Tagged	Character String
External	Object	Object Descriptor	UTCTime	Generalized Time

Built-in types can be viewed as standard data types available for the representation of data. For example, integer, float and char are examples of built-in data types in C programming language. The user is allowed to define his own data type by using built-in types. Similarly, the built-in types of ASN.1 allow the user to define requirement specific data types. In order to keep the SNMP protocol simple, only a sub-set of ASN.1 is permitted. The permitted structures are:

- Primitive types like INTEGER, OCTET STRING, OBJECT IDENTIFIER and NULL
- Enumerated INTEGER
- Constructor types like SEQUENCE and SEQUENCE OF.

Note that there are certain constraints under which the aforementioned structures are permitted. For example, DE-

FAULT and OPTIONAL CLAUSES do not appear in the SEQUENCE definition. This is done to avoid ambiguity in decoding SNMP messages. For details, the reader should refer to RFC 1155.

Besides the structures of ASN.1, RFC 1155 additionally defines the following application-wide types:

- **NetworkAddress:** This represents an address from a particular protocol family. Presently, only the Internet protocol family is supported.
- **IPAddress:** This represents a 32-bit IP address represented as an OCTET STRING of length 4.
- **Counter:** This represents a non-negative integer that increases monotonically until it reaches a maximum value ($2^{32} - 1$) when it wraps around and starts again from zero.
- **Gauge:** This represents a non-negative integer which may increase or decrease, but which latches at a maximum value ($2^{32} - 1$).
- **TimeTicks:** This represents the elapsed time, in hundredths of a second, since the occurrence of a particular event.
- **Opaque:** This type is used to encode arbitrary information as OCTET STRING. However, the SNMP application entities must know a priori the mechanism to parse opaquely encoded data.

The aforementioned types are called application-wide types. All the above application-wide types are either derivable from ASN.1 types or from other application-wide types. The same holds true for application-wide types that will be defined in the future.

13.4.2.3 Encoding of a Managed Object

Once an instance of an object type is identified, its value is encoded by applying the basic encoding rules to the syntax of the object type. Basic encoding rules are applied in the presentation layer before transferring ASN.1 values. Depending upon whether the length of the contents are known or not, the following formats are defined:

- **ILC (Identifier-Length-Contents) format:** This is exactly identical to the TLV (Type-Length-Value) format used for encoding signalling messages. The ILC format stands for Identifier-Length-Contents. The *Identifier* identifies the ASN.1 type, *Length* gives the length of the message (excluding the identifier and the length field) and *Content* carries the information.

- **ILCE   (Identifier-Length-Contents-End_of_contents) format:** The ILCE format, is similar to the ILC format except that, the length field in the former is left unused. In order to distinguish between the ILC and ILCE format, one bit of the length field is reserved. In ILCE format, a two-byte, all-zero field is added after the contents field. This 0x0000 field signifies the end of contents. However, this places a restriction that 0x0000 cannot appear as a valid data in the contents field.

Using one of the two formats, ASN.1 values are encoded. Rules are defined for the encoding of each of the ASN.1 built-in types given in Table 13.2. For example, Fig. 13.3 shows how a BOOLEAN value of TRUE is encoded.

0000 0001 **Identifier**	0000 0001 **Length**	1111 1111 **Contents**

Fig. 13.3 *Encoding of BOOLEAN value (TRUE)*

13.4.3 Management Information Base (MIB)

The Structure of Management Information (SMI), discussed in the previous section, details the elements of an object type. An object type is similar to the data types (like INTEGER) used in programming languages. This section explains how an instance of the object type is defined. An *object instance* is an instantiation of an object type, which is bound to a value. For example, the entries in a particular routing table are instances of object-type routing table entry.

An object-type definition consists of the following fields:

- **Object name:** This refers to a textual name of the object, along with its OBJECT IDENTIFIER. The textual name is called OBJECT DESCRIPTOR. For example, consider example 1 described in Table 13.3. In this example, *atEntry* is the OBJECT DESCRIPTOR and *{atTable 1}* is the OBJECT IDENTIFIER.

- **Syntax:** This refers to the abstract syntax used to represent the object type. In example 1, syntax of *atEntry* conveys that it is a sequence of three variables (*atIndex*, *atPhysAddress* and *atNetAddress*). The item following each variable in the sequence list indicates its data type. For example, *atIndex* is of type INTEGER, *atPhysAddress* is of type OCTET STRING and so on.

- **Definition:** This refers to a textual description of the object type. For the purpose of inter-operability, it is important that objects have consistent meaning across different machines.

- **Access:** This refers to the access permissions available to SNMP application entities while performing management operations. In essence, access fields are similar to the read-write-execute permissions maintained for each file in a UNIX file system. The following access modes are defined for SNMPv1:

  — **Read-only:** In this mode, the value of an object can only be retrieved, it cannot be altered; i.e. only *Get* operations are permitted while *Set* operations are not allowed.

  — **Read-write:** In this mode, both *Get* as well as *Set* operations are allowed on the object.

  — **Write-only:** In this mode, values can only be set. *Get* operation is not allowed.

  — **Not-accessible:** In this mode, neither *Get* operation nor *Set* operation is allowed.

  — **Status:** This refers to status of the object and can assume one of the following: *mandatory, optional* or *obsolete*.

Table 13.3 provides two examples of how object types are defined in the MIB. The object, definition, access and status fields of both the examples are self-explanatory. Regarding the syntax, note that Example 1 contains INTEGER, OCTET STRING and *NetworkAddress* as data types. The first two types (INTEGER and OCTET STRING) are built-in types of ASN.1 notation. The third type, *NetworkAddress*, is an application-wide type defined by RFC 1155. As mentioned earlier, RFC 1155 has defined certain application-wide types, including *NetworkAddress*, to define objects. The motivation behind defining these

Table 13.3 *Examples of Object Type Definition* (source RFC 1155)

Example 1	*Example 2*
Object: atEntry {atTable 1}	**Object:** atTable {at 1}

Syntax:
atEntry ::= SEQUENCE {
   atIndex
   INTEGER,
   AtPhysAddress
   OCTET STRING,
   atNetAddress
   NetworkAddress
   }
	Syntax:
	SEQUENCE of AtEntry

Definition: An entry in the address translation table.	**Definition:** The address translation table.
Access: read-write	**Access:** read-write
Status: Mandatory	**Status:** Mandatory

application-wide types is to facilitate the use of standardized types for objects that will be defined subsequently, as is the case in Example 1, where application-wide types are used to define object atEntry.

13.4.4 Protocol Details and Message Formats

All information exchange between SNMP application entities takes place using *UDP datagrams*. The UDP datagrams carry SNMP messages, which are encoded according to the basic encoding rules of ASN.1. All SNMP messages (except traps) are sent on port number 161. Trap messages (explained below) are sent on port number 162. A default maximum length of 484 bytes is specified by the protocol. Implementations are, however, allowed to support a larger datagram size.

SNMPv1 is essentially a simple request-response protocol. Each SNMPv1 message contains a Message Header and a Protocol Data Unit (PDU). The message header contains a *version* field and a *community name*. The version identifies the version number of the protocol (e.g. for RFC 1157, version-1 is used). An SNMP community refers to a pairing of an SNMP agent with some arbitrary set of SNMP application entities. Thus, the community name in the SNMP message helps in defining an access environment for the set of NMSs using that community name.

Five different types of SNMP PDUs are defined, which include GetRequest, GetNextRequest, GetResponse, SetRequest and Trap. The message formats for each of these messages is depicted in Fig. 13.4. The details of the PDUs are as follows:

- **GetRequest:** This PDU is generated by a protocol entity only at the behest of its SNMP application entity (this also applies for GetNextRequest and SetRequest PDU). The error-status and error-index in GetRequest PDU is always set to 0. This is because error codes are generated only for a response, not for a request. In case no error occurs, the receiver of the GetRequest PDU responds back with GetResponse PDU, which contains the name and the value corresponding to each object in the received variable binding list.

- **GetNextRequest:** This PDU is identical to GetRequest PDU, except that the PDU type is different. In case no

Request-ID	Error Status	Error Index	Variable Bindings

(a)

Enterprise	Agent address	Generic trap type	Specific trap code	Time stamp	Variable bindings

(b)

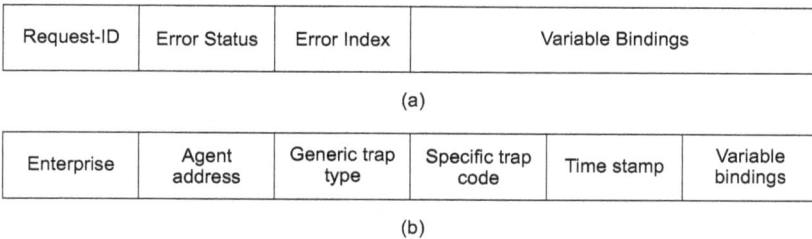

Fig. 13.4 *SNMPv1 message format. (a)* GetRequest, GetNextRequest, GetResponse *and* SetRequest *message format. (b)* Trap *message format.*

error occurs, the receiver of the GetNextRequest PDU responds back with a GetResponse PDU, which contains the name and the value of the immediate successor corresponding to each object in the received variable binding list. *GetNextRequest* is useful in traversing a conceptual table of objects maintained in an MIB.

- **GetResponse:** GetResponse PDU is generated as a response to a GetRequest, GetNextRequest or SetRequest PDU.
- **SetRequest:** This PDU is used to set values of object instances within an agent. In case no error occurs, then for each object named in the variable binding list of the received message, the corresponding value is assigned to the variable. A GetResponse PDU is also generated to inform the sender of the results of the SetRequest.
- **Trap:** Traps are unsolicited message sent by the agents to the NMS. They are sent asynchronously to inform the NMS of the occurrence of some event. *Trap* PDUs do not elicit a response from the receiver.

Except the Trap PDU, the message format for the remaining PDUs is identical (see Fig. 13.4(a)). Each PDU contains the following fields:

- **Request-ID:** This field is used to distinguish between multiple outstanding requests, i.e. it is used to associate an incoming response with an outstanding request.
- **Error status:** A non-zero value of this field indicates that an exception condition has occurred.
- **Error index:** In case of exception condition, this field identifies the variable in the list that caused the exception.

- **Variable bindings:** This refers to the paring of a name of a variable to the variable's value. Note that for some operations (like *Get* operation), the value field for variables is not filled by the sender of the message.

The Trap PDU has slightly different message format (see Fig. 13.4 (b)). It contains the following fields:

- **Enterprise:** This field identifies the type of object generating the trap.
- **Agent address:** This field provides the address of the object generating the trap.
- **Generic trap type:** This field specifies the generic trap type, permitted values of which include coldStart, warmStart, linkDown, linkUp, authenticationFailure, egpNeighbour Loss and enterpriseSpecific.
- **Specific trap code:** This field provides the specific trap code.
- **Time stamp:** This field carries the value of time elapsed between the last re-initialization of the network entity and the generation of the trap.
- **Variable bindings:** This provides additional information about the trap.

In essence, SNMP PDUs are categorized into two broad categories: those involving Get/Set operations and those indicating a Trap. The difference in the functionality between the two categories results in the different message formats.

13.5 EXAMPLE: TELECOMMUNICATIONS MANAGEMENT NETWORK (TMN)

Telecommunication Management Network (TMN) is a framework defined by the ITU-T to manage telecommunication networks and services. TMN provides the means for transporting, storing, and processing the information necessary to support the management of telecommunication network and services. TMN is applicable to all types of telecommunication networks and network elements that includes analog and digital networks, private and public networks, switching systems, transmission systems, telecommunication software, and logical resources of the network (e.g. circuits or paths). The principal concepts of

TMN are provided in ITU-T specification [ITU-T M.3010]. The complete TMN specification covers a number of recommendations (in the M.3xxxx series).

[ITU-TM.3010 defines TMN as "conceptually a separate network that interfaces a telecommunications network at several different points to send/receive information to/from it and to control its operations."

Figure 13.5 depicts the general relationship of a TMN with respect to a telecommunication network. TMN performs management functions by interfacing with the exchanges and transmission systems. The exchanges and transmission systems, also called network elements, lie partly within the TMN domain and partly within the telecommunication network, because they perform both the telecommunication functions and management functions. The datacommunication network provides a transport mechanism for the exchange of management information between network elements and operation systems. Operation systems form the core of TMN, performing most of

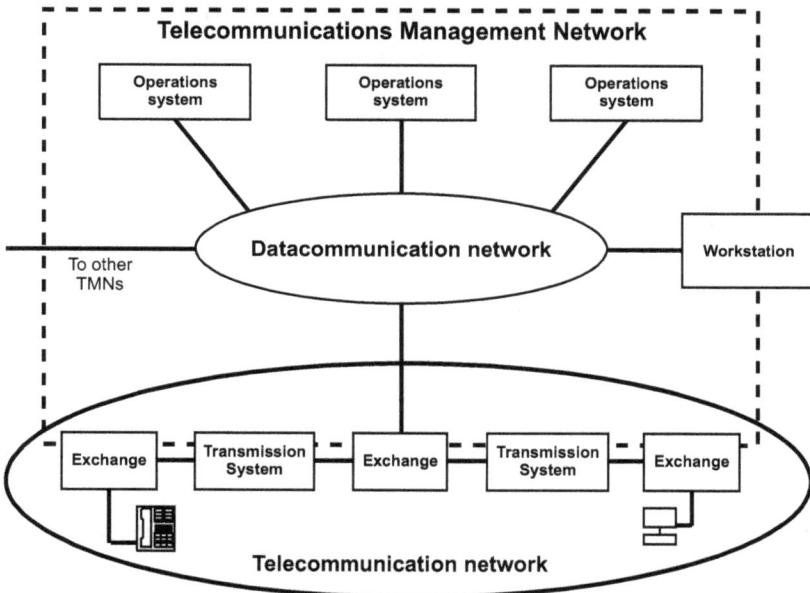

Fig. 13.5 *General relationship of a TMN with a telecommunication network*

the management functions. Workstations are required to interpret TMN information for human user and vice-versa.

TMN architecture is defined by using four types of abstraction. These abstractions are as follows:

- **TMN functional architecture:** This defines the distribution of functionality within the TMN.
- **TMN physical architecture:** This describes how a TMN can be made by using physical components and interfaces.
- **TMN information architecture:** This provides an object-oriented approach for applying the OSI management principles to the TMN principles.
- **TMN logical-layered architecture:** This defines a layered approach for the distribution of management functionality.

The next four sub-sections detail each of these abstractions.

13.5.1 TMN Functional Architecture

The *functional architecture* of TMN is specified in terms of functional blocks and reference points. A functional block provides TMN with a set of functions that enables the TMN to perform management functions. Pairs of TMN functional blocks, which exchange management information, are separated by reference points. Reference points define the service boundaries between two management functional blocks. The concept of functional block and reference point are explained below.

13.5.1.1 TMN Functional Block

The TMN functional architecture defines five different functional blocks. These functional blocks are depicted in Fig. 13.6. In the figure, the TMN functional boundary divides some of the functional blocks [e.g., Q Adaptor Function (QAF) and Network Element Function NEF)] into two regions. This indicates that the functionality of these functional blocks is not completely specified by TMN recommendations. In contrast, those functional blocks which are completely within the TMN functional boundary [e.g. Operations System Function (OSF) and Mediation Function (MF)] have their functionality completely specified by the TMN recommendations.

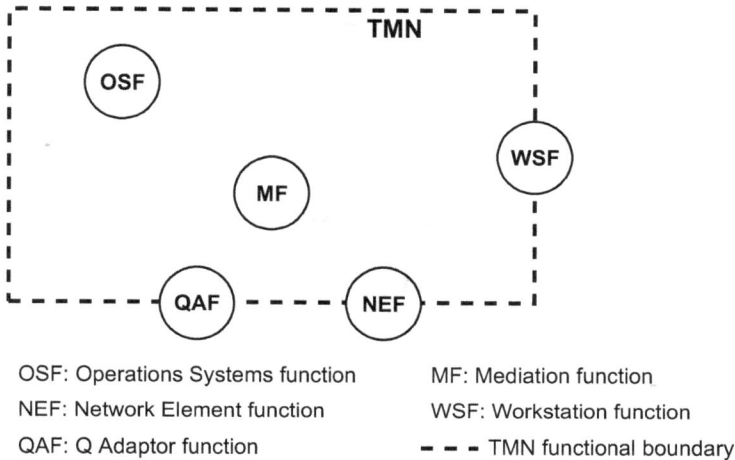

OSF: Operations Systems function   MF: Mediation function

NEF: Network Element function   WSF: Workstation function

QAF: Q Adaptor function   – – – TMN functional boundary

Fig. 13.6 *TMN functional blocks*

The five functional blocks are as follows:

* **Operations Systems Function (OSF) block:** This block is responsible for initiating management operations and receiving notifications. This block also monitors, co-ordinates and controls the telecommunication functions and management functions (including the TMN itself).
* **Network Element Function (NEF) block:** This block is comparable to the concept of agent used in the SNMP management model. NEF essentially performs two sets of functions. One set of functions, referred to as telecommunication functions, provides the means for data exchange between the users of telecommunication networks. Although these functions are a subject of management, they are not directly a part of the TMN, but are represented to the TMN by the NEF. The other set, referred to as management functions, provides the means to the NEF for managing telecommunication functions. These functions of the NEF, that provides management support, are within the purview of TMN.
* **Workstation Function (WF) block:** The WSF block provides the means to interpret TMN information for the human user and vice versa. Since WSF also translates between a TMN reference point and non-TMN reference

point, some portion of WSF is kept outside the TMN boundary.

- **Mediation Function (MF) block:** The MF block is responsible for passing information between the OSF and NEF or between OSF and QAF. Using an MF ensures that the functional blocks attached to the MF get information in the way they want. Thus, MF may store, adapt, filter or condense information.

- **Q Adaptor Function (QAF) block:** This block is used for connecting the TMN to those non-TMN entities, which are NEF-like or OSF-like. Since QAF interfaces with a non-TMN entity, some part of it lies outside the functional boundary.

TMN functional blocks exchange management information using another block called the Data Communication Function (DCF). DCF provides information transport mechanisms including routing, relaying and interworking. Thus, a DCF is equivalent to the OSI layers 1 to 3.

13.5.1.2 TMN Reference Point

A *TMN reference point* defines service boundaries between two management functional blocks. Putting it differently, reference points provide the means to delineate functional blocks and help in defining the interaction between various functional blocks. Note that a reference point is not a protocol in itself. It is merely a conceptual notion that makes the task of defining TMN interfaces easier.

Reference points are classified as TMN reference points and non-TMN reference points. The first category includes three classes of reference points, namely *q*, *f* and *x*. The second category includes *g* and *m* class of reference points. All five classes of reference points are depicted in Fig. 13.7. The different classes of reference points are discussed below:

- **q reference points:** The *q* reference points are located between two TMN-conformant functional blocks that are within the same TMN domain.

- **f reference points:** The *f* reference points are located between WSF and OSF functional blocks and/or between WSF and MF functional blocks.

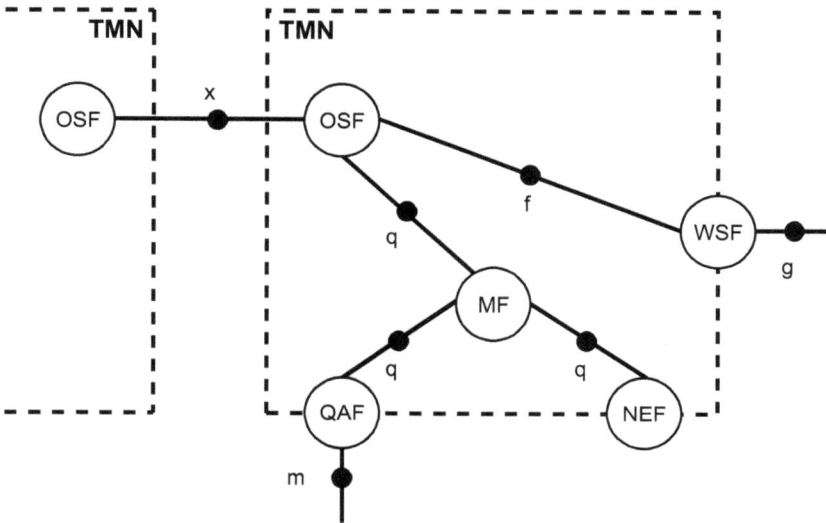

Fig. 13.7   *TMN reference points*

- ***x* reference points:** The *x* reference points are located between the OSF functional blocks belonging to different TMNs.
- ***g* reference points:** The *g* reference points are located between WSF and human users. Although *g* reference points help in conveying TMN information, their detailed definition is outside the scope of TMN recommendations.
- ***m* reference points:** The *m* reference points are located between the QAF functional block and non-TMN managed entities (i.e. those entities that do not conform to TMN recommendations).

Now that the different functional blocks and classes of reference points are defined, consider how the functional blocks are paired. Table 13.4 provides the relationship between functional blocks and reference points. A non-blank entry in the table gives the reference point over which a functional block at the top of a column can exchange management information with a functional block at the left of a row. For example, functional block OSF exchanges management information, with WSF using reference point f. A blank entry in the table indicates that the corresponding functional blocks cannot

Table 13.4 *Relationships between Functional Blocks and Reference Points [ITU-T M. 3010]*

	NEF	OSF	MF	QAF (q3)	QAF (qx)	WSF	non-TMN
NEF		q_3	q_x				
OSF	q_3	x^*, q_3	q_3	q_3		f	
MF	q_x	q_3	q_x		q_x	f	
QAF (q_3)		q_3					m
QAF (q_x)			q_x				m
WSF	f	f					g^ϕ
non-TMN				m	m	g^ϕ	

* = x reference point only applies when each OSF is in a different TMN.
$^\phi$ = The g reference point lies between the WSF and the human user.

exchange management information between each other. For example, pairing between NEF-NEF and WSF-NEF is not allowed. The implication of subscripts is explained in the next subsection.

13.5.2 TMN Physical Architecture

In the previous section, the functional blocks required for building a TMN were discussed. In this section, the *physical architecture* of TMN is discussed. The physical architecture of TMN describes how, by using physical components and interfaces, a TMN can be made. In this sense, physical components and interfaces of the physical architecture are analogous to the functional blocks and reference points of the functional architecture. In fact, the physical architecture of a TMN provides the means to map a functional block to a physical component, and a reference point to an interface.

13.5.2.1 TMN Building Blocks

Recall from the previous section that the functional architecture of TMN defines five functional blocks. Accordingly, the TMN physical architecture defines five building blocks. These building blocks are:

- Operations Systems (OS)
- Network Element (NE)
- Workstation (WS)
- Mediation Device (MD)
- Q Adaptor (QA)

The functions of each of these building blocks can be derived from the corresponding functional block. For example, Mediation Device performs the Mediation Function while the Operations Systems perform operations systems functions. Apart from these five building blocks, [ITU-T M.3010] defines another building block called the Data Communication Network (DCN). This block performs the Data Communication Function (DCF) mentioned earlier.

The TMN physical architecture permits a building block to contain more than one functional block. This is required because it is not always possible to clearly delineate the functional blocks physically. Table 13.5 shows the relationship between TMN building blocks and TMN functional blocks. Specifically, the table states the mandatory and optional functional blocks contained within a building block. Note that all entries in the diagonal (top-left- to right-bottom) are 'Mandatory'. This should be obvious because a building block must at least perform the functions defined by its corresponding functional block. The 'optional' fields in the table indicate that the building block can contain the functional block (but there is no requirement to do the same). The decision as to how to name a building block that contains multiple functional blocks depends on the predominant usage of that block.

Table 13.5 *Relationships between TMN Building Blocks and TMN Functional Blocks*

	NEF	*OSF*	*MF*	*QAF*	*WSF*
NE	Mandatory	Optional	Optional	Optional	Optional*
MD		Mandatory	Optional	Optional	Optional
QA			Mandatory		
OS		Optional	Optional	Mandatory	Optional
WS					Mandatory

* MSF can be present only if OSF or MF is also present.

13.5.2.2 TMN Interfaces

Besides the building blocks, TMN physical architecture also defines three types of interfaces, viz. Q (this is further classified as Qx and Q_3), F and X. The relationship between interfaces and reference points is similar to that between building blocks and functional blocks. It follows from the above that Q interface is applied to q reference point, F interface is to f reference point, and X interface to x reference point. Note that no interfaces exist for g and m reference points, because these interfaces do not fall within the purview of TMN recommendations. The Q_3 interface is q interfaces where one of functional blocks is OSF. All q interfaces not involving OSF are Q_x interface.

13.5.3 TMN Information Architecture

The *TMN information architecture* provides an object-oriented approach for information exchanges. This object-oriented approach is based on manager/agent/objects relationships; an approach similar to that used in the OSI and SNMP standards. Although the concepts are similar and have been detailed, the manager/agent/objects relationships, as presented by [ITU-T M. 3010], are again explained.

For the purpose of information exchange, TMN information architecture introduces the following concepts:

- **Manager:** A manager is an entity that issues management directives and receives notifications.
- **Agent:** An agent processes directives received from the manager and also sends notifications to the manager, stating the current status of the managed object.
- **Managed object:** A managed object is a conceptual notion of modelling resources. Putting it differently, a managed object is the abstraction of such a resource that represents its properties as seen by (and for the purpose of) management. Managed objects are defined by the following:
  — Attributes or characteristics of the object that are visible at the boundary of the object
  — Operations that can be performed on the object
  — Behaviour exhibited by the object in response to the operations
  — Notifications emitted by the object

Figure 13.8 depicts the interaction between manager, agent and managed objects. As can be seen from the figure, both agents and managed objects reside within the managed systems. Agents receive directives from the managers, which they use to

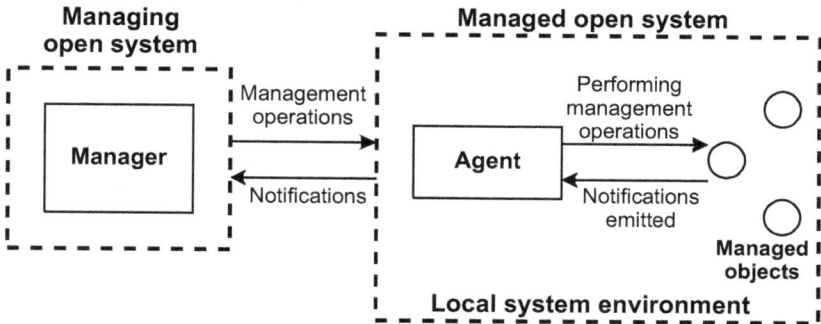

Fig. 13.8 *Interaction between manager, agent and managed objects*

perform certain management operations on the managed objects. A single agent can act on behalf of several managers. Similarly, a single manager can issue management directives to several agents. Our brief discussion concludes with the note that TMN information architecture (like OSI and SNMP) uses the concept of MIBs for modelling management information.

13.5.4 TMN Logical-layered Architecture

It is well-known fact that in most practical situations, layered structures, as compared to flat structures, are preferable. In fact, layered structures are so advantageous that they form the basis of all protocol software. Taking a cue from this fact, TMN defines a layered architecture, called the *Logical Layered Architecture (LLA)*, to distribute management functions. [ITU-T M. 3010] defines LLA as 'a concept for structuring of management functionality which organizes functions into groupings called logical layers, and describes the relationship between layers.' Further, a logical layer is defined as 'the clustering of management information supporting those aspects of management which the logical layer reflects.' Indeed, LLA is one of the most important concepts of the TMN architecture.

The distribution of management functionality into logical layers leads to grouping of Operations Systems Functions (OSF) into layers. (Recall from the discussion on functional architecture that the OSF is the heart of TMN; it monitors, co-ordinates and controls all management functions). The LLA groups the OSF into the following layers:

- **Element management layer:** This layer, consisting of one or more OSFs and/or MFs, is responsible for managing network elements, either individually or in groups. The principle functions of the element management layer includes the following:

  — Controlling a sub-set of network elements on an individual basis, thereby facilitating interaction between network management layer and network element layer.

  — Controlling a sub-set of network elements on a collective basis, thereby managing the relationships between the NEFs and providing a singe-entity view of a number of NEFs.

  — Maintaining statistical information about the elements.

- **Network management layer:** Unlike the element management layer, this layer is not concerned with the internal working of the network elements. Rather, the network management layer only deals with the interaction between individual network elements. In essence, this layer has an overview of the network and manages it at a macro level. The principle functions of the network management layer are:

  — Controlling the network view of all the network elements within the domain of the layer

  — Provisioning and modifying network capabilities for supporting different customer services

  — Maintaining statistical information about the network

- **Service management layer:** This layer is responsible for the contractual aspects of services provided to the users (both current users and potential users). For this, the service management layer uses the information provided by the network management layer. The service management however cannot control the internal working of the network, a function that is performed by the network management layer. The principle functions of the service management layer are:

  — Accounting

— Provisioning and terminating of services
— Maintaining statistical information (e.g. QoS)
— Interaction with service providers.

- **Business management layer:** The business management layer is responsible for the overall working of the enterprise. In essence, the task of this layer is to set goals rather than to achieve goals. The latter is left for other management layers. The principle functions of business management layer are:

   — Decision-making on investments
   — Budgeting
   — Maintaining information about the whole enterprise

Figure 13.9 depicts the four layers of OSF. Note that the lowest layer (i.e. Network Element Layer) is not explicitly

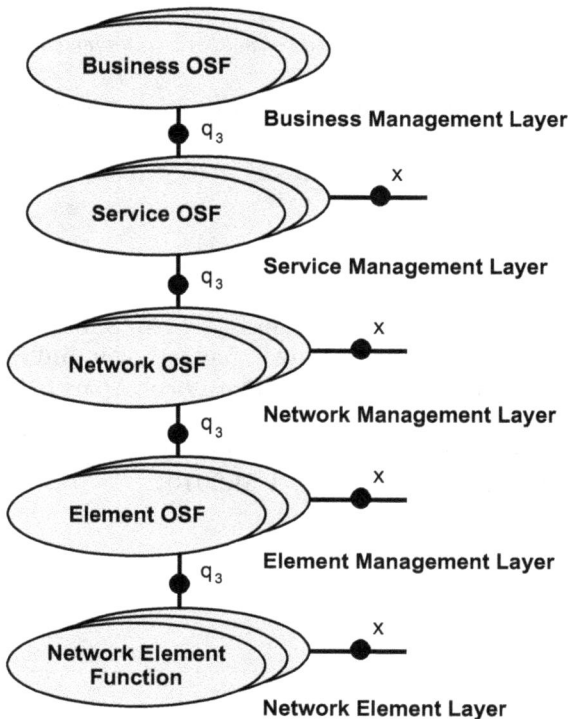

Fig. 13.9 *Layering of OSF*

discussed. This is because it is not a management layer *per se*. It merely comprises the network elements, which have very little management functionality.

⚡ 13.6 CONCLUSION

This chapter provided an overview of network management. The essential concept here is the FACPS, which defines the five functional areas of network management. Each of the functional areas was discussed in this chapter. As examples, the network management protocol for IP-based networks (SNMP) and for telecommunication networks (TMN) were discussed.

REVIEW QUESTIONS

Q 1. *What is the need for network management? What are the essential objectives of network management?*

Q 2. *What are the functional areas of network management? Explain with reference to the FCAPS.*

Q 3. *Differentiate between the TMN functional architecture and the TMN physical architecture.*

Q 4. *What is the importance of logical layers in TMN?*

FURTHER READING

A good text for network management is [NM M. Subramanian] which provides a good overview of network management and in particular, network management based on SNMP protocol. Apart from this, there are a number of RFCs for SNMP, in particular [RFC 1157] and [RFC 1441]. For TMN too, there are a number of ITU-T specifications in the M.3xxxx series, in particular [ITU-T M.3010].

14

SECURITY MANAGEMENT

14.1 INTRODUCTION

In the previous chapters, some of the important concepts related to communication networks were covered. These concepts were related mostly to the services that are offered by the network. The focus of this chapter, however, is not on the services provided by the network, but on how these services can be provided securely. This will lead us to the definition of security, and some of the key terms associated with security.

Today, computers have become a means to store and process all sorts of information. In older days, the same information was kept in lockers to safeguard it from falling into wrong hands. Similarly, when information is stored in computers, security of confidential information is required. The means and tools used to provide this security come under the purview of computer security. While computer security mainly deals with how information can be stored securely in a computer, network security is related to the secure transfer of this information from one node in the network to another. In common parlance, however, the terms network security and computer security is normally used to mean the same, and the terms are hence used interchangeably in this chapter. Together, computer security and network security form the basis for security management, and are the focus areas of this chapter. However, before looking at the concepts of security management, it is essential to formally define the term *security management*.

⚓ 14.2 WHAT IS SECURITY MANAGEMENT?

Security management in any given network is limited to a few key concepts. These include *data integrity*, which means that the contents of data packets are not altered in an unauthorized manner. Next is *origin authentication*, which refers to the corroboration that the source of data is as claimed. Typically, data integrity and origin authentication is achieved using a signature or a message digest of the complete or partial message. The message digest is also known as the *Message Authentication Code (MAC)*. It is customary to club data integrity and origin authentication together, and simply refer to it as *authentication*. Another important concept for security management refers to the *confidentiality* of data. Confidentiality refers to the property that information in a data packet is not made available or disclosed to unauthorized entities. Confidentiality is achieved by encrypting the data by using ciphering techniques. Authentication and confidentiality are two key concepts with regards to security management.

Apart from these main concepts, a few other security aspects exist, which include *replay-protection*. In order to understand this concept, consider a case wherein the attacker can neither breach the contents of the packet, nor impersonate someone else. In such a scenario, it may still be possible for the attacker to simply capture a valid packet (or a series of valid packets), while they are being transmitted, and then replay the same (series of) packets after sometime. This replay of stale packet(s) may cause limited or significant disruption of service. At the very minimum, replay of stale packets can be used to make a *denial-of-service attack*. A denial-of-service attack can be made by loading a server with invalid (or stale) requests, as a result of replay of stale packets. Hence, the server becomes so loaded with invalid requests that it is not able to service other valid requests, coming from genuine clients. A protocol is replay-protected if a captured packet loses its relevance when stored over a pre-defined period of time. Replay-protection is ensured through the use of sequence numbers and/or time-varying parameters.

Here, it is important to clarify the difference between the terms security management, the heading of this chapter, and the term *cryptography*. Cryptography is the science of using mathematics to encrypt and decrypt data. Thus, some view cryptography as the art of encryption and decryption. By this definition, cryptography forms only a part of security management. However, some others refer to cryptography as the process of converting plaintext into ciphertext at the transmitting end and then deriving plaintext from ciphertext at the receiving end, while fulfilling four broad objectives: confidential transfer (preventing unauthorized read/access), data integrity (detecting/preventing unauthorized alteration), non-repudiation (the inability of sender to deny that he/she had not sent the data) and authentication (source of data received is as claimed). Going by this definition, there is little that separates the elements of cryptography and security management and the two terms may thus be used interchangeably.

Irrespective of the vagaries in the terminology followed in the industry, this chapter looks at some of the key aspects of security management; in particular it focuses on *encryption* and *authentication*. Encryption techniques fall in two broad categories, namely *symmetric encryption* technique elaborated in section 14.3 and *asymmetric/public key encryption* technique elaborated in section 14.4. Apart from these, key management techniques, hashing, digital signatures and firewall concepts are also discussed in this chapter. Algorithms that are used for security management are also discussed . However, since most of the algorithms are quite complex, the chapter mainly tries to introduce the fundamentals that are used to make the algorithm. Details of complex algorithms are omitted, and the reader is asked to refer to other literature for reference, wherever required.

14.3 SYMMETRIC (SECRET KEY) ENCRYPTION TECHNIQUES

Symmetric encryption technique, also referred to as secret key encryption, is the oldest and most commonly used technique for encryption. This technique works on the principle that the *key* used for encryption of the information is kept confidential (hence the name secret key encryption). The encryption process consists of an algorithm each for encryption and decryption, and a secret key, which is independent of the information that is to be encrypted. The process of encryption using symmetric encryption technique is pictorially depicted in Fig. 14.1.

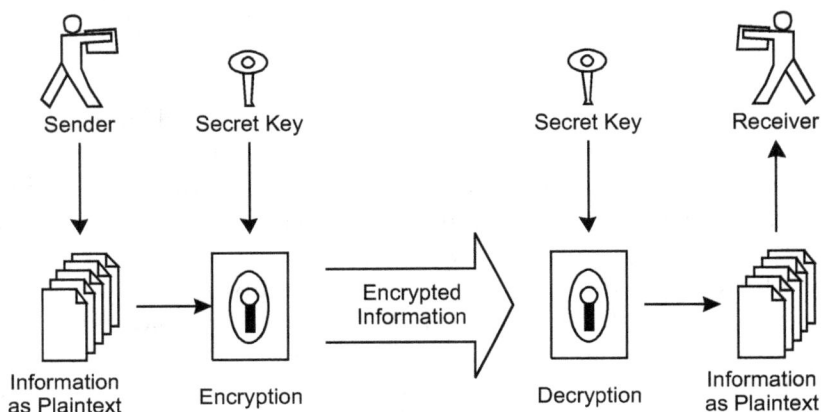

Fig. 14.1 *Symmetric encryption technique*

As depicted in the figure, the sender of information provides the encryption algorithm with the information to be transmitted as Plaintext (meaning information that can be read by anybody without and special means). The encryption algorithm encrypts the plaintext information, using a key that is known only to the sender and the receiver. Information transfer over the network is then done using the encrypted information. At the receiving end, the decryption algorithm takes the encrypted information, and decrypts it back to plaintext using the same key that was used at the sending side to encrypt. The decrypted plaintext is then provided to the user at the receiving end.

The security of the symmetric encryption technique is dependent upon the following two factors:

- **The complexity of the encryption algorithm:** The encryption algorithm must be such that the (output) encrypted information is impractical to decrypt without knowing the key used for encryption.

- **The secrecy of the key used for encryption:** The key used for encryption and decryption must be kept secret between the sender and the receiver. Hence, while the encryption/decryption algorithm itself is not necessary confidential (i.e. the algorithm can be public-knowledge!), it is the secrecy of the key that prevents an unauthorized person to decrypt the information, even though the decryption algorithm is known.

The latter property of symmetric encryption technique is what makes it easy for implementation. Since the algorithm is not secret, the algorithm itself can be implemented and installed at each network entity *a priori*. When any two network entities wish to establish a secure communication channel, the only thing required is to be in possession of a secret key to be used for encryption/decryption. Thus, the symmetric encryption techniques require a means using which two communicating nodes can possess a secret key, which is not known to any other third person. The secret key can be attained through the following means:

- The secret key could be generated at the sending node. The key is then transmitted to the receiving end by using another secure channel.
- The secret key is generated by a reliable third party, and securely delivered to the sender and receiver.

The process of key exchange is hence an important aspect of symmetric encryption and is discussed in detail in the section on Key Management for Symmetric Encryption Techniques (section 14.5.1). The following sections discuss two of the well-known algorithms for symmetric encryption, namely *Data Encryption Standard (DES)* and *Advanced Encryption Standard (AES)*.

14.3.1 Data Encryption Standard (DES)

The Data Encryption Standard (DES) algorithm is designed to encrypt and decrypt blocks of data consisting of 64 bits using a 64-bit key. If the size of the data to be encrypted is greater than 64 bits, then the data is first chopped into multiple blocks of 64 bits each, and the DES algorithm is applied to each block in sequence, to produce the encrypted data. Hence, DES algorithm falls into a category of encryption algorithms that are sometimes also referred to as *Block Encryption Algorithms*.

Figure 14.2 depicts the encryption algorithm followed in DES for encryption of a 64-bit input block. The following steps are involved in the encryption process:

1. The input block is first subject to an *Initial Permutation (IP)*, wherein bits in the input data are permuted as per a defined order. In other words, the bits in the input block are shuffled such that, for example, the permuted input has bit 58 of the input as its first bit, bit 50 as its second bit, and so on. The Initial Permutation is a pre-defined and documented procedure in the DES Algorithm. This initial permutation is applied to remove any patterns that might be visible in the original input. This makes it hard to understand encrypted data and look for patterns in the

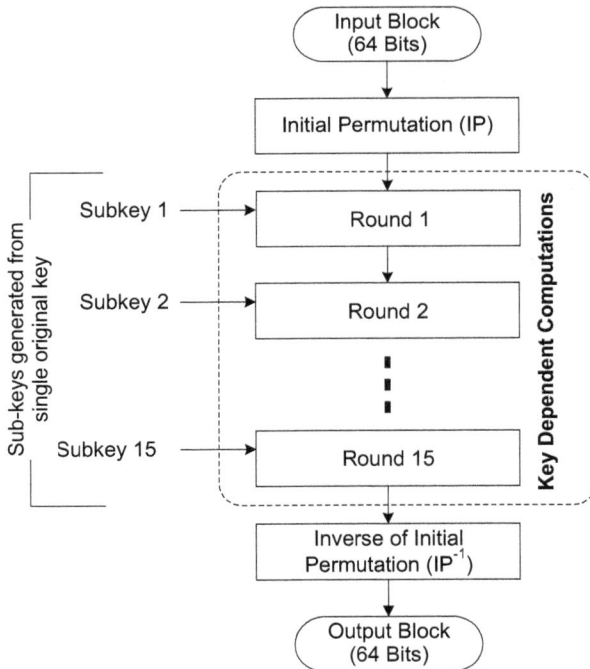

Fig. 14.2 *DES encryption algorithm*

encrypted data, thus making it extremely hard to decrypt the data without knowing the shared secret key.

2. The permuted input is then subject to 16 rounds of a Key-*Dependent Computation*. For each round, a 48-bit Sub Key is used, which is derived from the 64-bit shared secret key, by using a documented procedure. This key-dependent computation forms the core of the DES encryption algorithm. Each round of the key dependent computation can be described as follows. Consider the 64-bit input block to each round of the key dependent computation to consist of two blocks of 32-bit each. Let these be L (left 32-bits) and R (right 32-bits) such that the input block is LR. Further, let $SK(n)$ denote the 48-bit sub-key that is used in round n of the key-dependent computation. Then, the output $L'R'$ of each round of computation can be defined as:

$$L' = R \text{ and } R' = L \ (+) \ f[R, SK(n)],$$

where $(+)$ denotes bit-wise addition modulo 2, and f is a function that takes as inputs a 32-bit block R and a 48-bit block $SK(n)$, and produces a 32-bit block as output. The function f is also documented as part of the DES algorithm.

3. After the permuted input block is subject to 16 rounds of key-dependent computation, a second round of permutation is applied again. This permutation is such that it can be considered as an inverse of the initial permutation . In other words, if an input block is subject to an initial permutation, and then immediately, the inverse of IP is applied, then the same original input block can be recovered. In the DES algorithm, after application of the inverse of initial permutation, the output so produced is the encrypted output block.

The decryption process of DES is similar to the encryption process. The same algorithm as depicted in Fig. 14.2 is applied for decryption. However, the only thing that is different is the order of use of the sub-keys for the key-dependent computation. The sub-keys, at the time of decryption, are used in the reverse order of their use during the encryption process. For a more detailed reading of the DES algorithm, the reader is referred to [Secu W. Stallings].

14.3.2 Advanced Encryption Standard (AES)

The Advanced Encryption Standard (AES) is a symmetric key encryption technique that is expected to replace the commonly used DES standard. The Advanced Encryption Standard was the outcome of a worldwide call for submission of encryption algorithms, issued by NIST in 1997. The winning algorithm, Rijndael, forms the basis of the AES. Two Belgian cryptologists, Vincent Rijmen and Joan Daemen, developed the Rijndael algorithm. AES provides for stronger encryption mechanisms that can be used in various environments, which includes various standard software platforms, limited-space environments, and various hardware implementations.

AES can encrypt 128, 192 or 256 bit data blocks, using 128, 192 and 256 bit keys. The algorithm is more complex than any of the other known encryption algorithms. The number of rounds of key-dependent computations is not kept fixed in AES, and this is one major difference from the DES. A detailed discussion of the AES algorithm is beyond the scope of this chapter. Interested readers are referred to [FIPS 197] for more details.

14.4 ASYMMETRIC (PUBLIC KEY) ENCRYPTION TECHNIQUES

One of the most difficult problems associated with symmetric encryption techniques is the problem of key exchange. As mentioned in section 14.3, symmetric encryption technique relies on the availability of a secret key known only to the sender and the receiver. This requires that the key is somehow generated and securely shared between the sender and receiver. The problem of sharing the secret key securely is not a trivial problem and various mechanisms are used for key exchange, some of which are discussed in section 14.5.2.

A newer approach to encryption is the asymmetric encryption technique, also referred to as *public key encryption*. This technique for encryption is so named because the key used for encryption is not secret, but known to all (known to the public!). Hence, using public key encryption techniques can do away with the overheads for exchange of a secret key for encryption. Public key encryption is based on the principle that each user who wishes to establish a secure communication channel needs to have a pair of keys. One of these keys, known as the *public key* is made public and is known to all. The other key, known as the *private key*, is known only to the user, and is kept secret from the other users.

Public key encryption techniques can be used to solve the problem of confidential transfer of information, the problem of user/source authentication, and the problem of data integrity. The use of public key encryption to achieve confidentiality, authentication and data integrity is discussed in the following sections.

14.4.1 Confidentiality Using Public Key Encryption

Information transfer between two users in a network can be made confidential by using public key encryption (see Fig. 14.3). As depicted in the figure, the encryption process/ algorithm takes as input the information in plaintext, and encrypts it using the public key of the receiver. Since everyone knows the public key of the receiver, there is no need to exchange the key used for encryption prior to the start of communication (unlike symmetric encryption techniques).

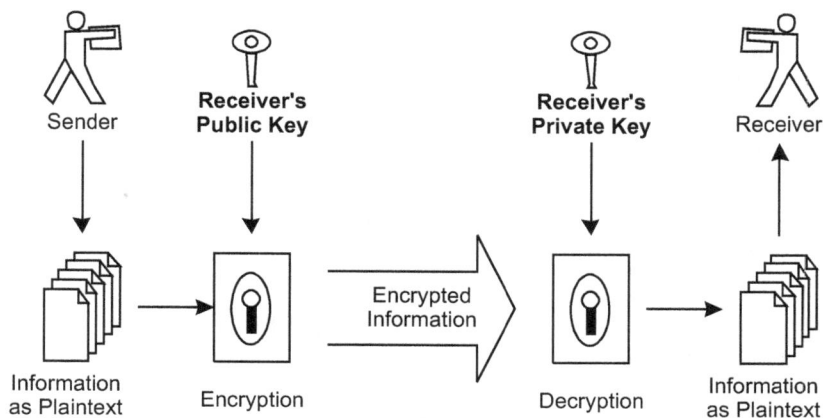

Fig. 14.3 *Confidentiality using public key encryption*

Further, the public key encryption technique is such that to decrypt the encrypted information, the private key of the receiver is required, along with the decryption algorithm. This ensures that none other than the intended receiver can decrypt the information. In this manner, confidentiality can be achieved in communication by using the public key encryption technique.

As discussed above, the only requirement for having a confidential exchange of information using public key encryption is that each party involved in the communication knows the public key of the other party. The public key of each user can be made public using many different schemes. Some of these key exchange schemes for public key encryption techniques are discussed in section 14.5.2. The next section discusses how source authentication is achieved using public key encryption.

14.4.2 Authentication Using Public Key Encryption

While confidentiality seems to be the most important component of network security, it might not always be required. Consider for example, a banking application, which can be used by a user to remotely perform banking operations. In such an application, it is more important to ascertain the identity of the person performing the banking operation (user authentication), rather than protecting the information itself. The information itself may or may not be required to be made confidential, depending upon the nature of the transaction. Public key encryption techniques can be used for user/source authentication as depicted in Fig. 14.4.

As depicted in the figure, the sender uses his/her own private key to encrypt the plaintext information, prior to transfer. The receiver, on the other hand, uses the public key of the sender to decrypt the received information. As is evident, such a scheme for encryption provides no confidentiality to the information transferred. Since everyone knows the public key of the sender, anyone, even other than the intended receiver, can decrypt the transmitted information and obtain the plaintext information. However, the mechanism depicted in Fig. 14.4 facilitates user/ source authentication. Since the private key of the sender is

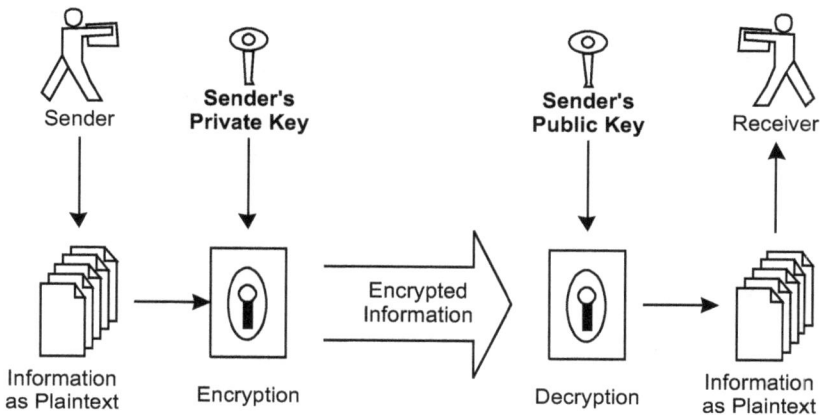

Fig. 14.4 *Authentication using public key encryption*

known to none other than the sender himself/herself, the
receiver can be assured that the received information is coming
from the intended sender only and not from any other source.
Hence, when used in this manner, public key encryption can be
used for user authentication.

14.4.3 Confidentiality and Authentication Using Public Key Encryption

While some secure applications require confidential transfer of
information, for some user authentication is of prime concern.
For some others, however, both confidentiality and user authentication may be required for a secure communication. The
public key encryption algorithm can be used to provide both
confidentiality and user authentication, as depicted in Fig. 14.5.
As depicted in the figure, the means to provide confidentiality
(as depicted in Fig. 14.3) and the means to provide authentication (as depicted in Fig. 14.4) have been merged together to
provide both confidentiality and authentication.

14.4.4 Data Integrity Using Public Key Encryption

The previous sections described how public key encryption
could be used for confidentiality and authentication purposes.
Encryption of information is a time-consuming, processor-
intensive activity. Hence, plaintext information must only be

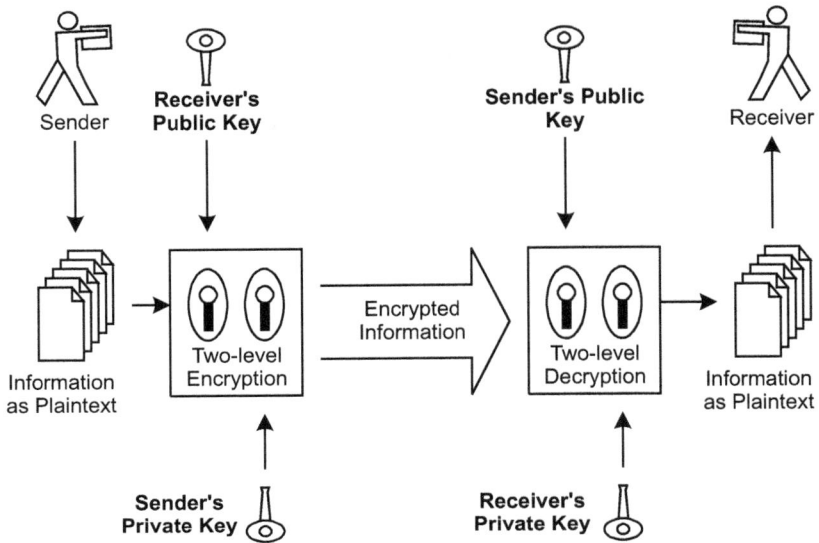

Fig. 14.5 *Confidentiality and authentication using public key encryption*

encrypted if confidentiality of information is of concern. At times, it is only required that information be transferred unaltered via the network, while confidentiality is not of much concern. In other words, data integrity is required instead of confidentiality. Public key encryption technique can be used to provide data integrity as well, as depicted in Fig. 14.6.

While Fig. 14.6 is self-explanatory, the following important points are worth mentioning:

- For data integrity, instead of encrypting the entire message using the sender's private key, only a message hash is encrypted. The message hash is obtained by applying a publicly known *hash function* on the plaintext information. The benefit of this hashing process is that while the plaintext information might be of a huge size, message hash is normally of a fixed smaller size. Hence, the amount of information to be encrypted is significantly reduced. Hash functions used for this purpose are described in more detail in section 14.6. The message hash, after encryption, also known as the *Message Authentication Code (MAC)*, is used for data integrity.

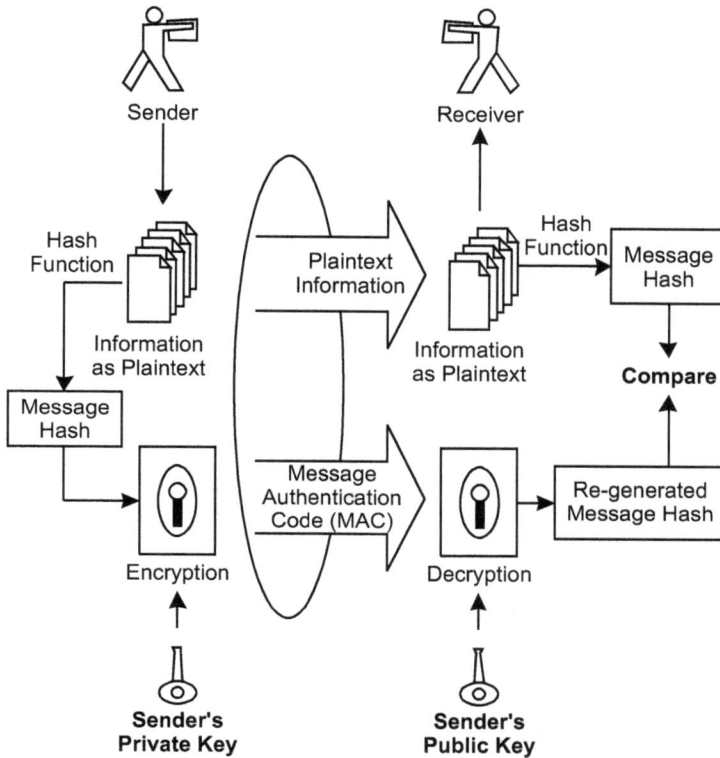

Fig. 14.6 *Data Integrity using public key encryption*

- Both the plaintext information, as well as its MAC is transmitted together as a bundle via the communication channel. At the receiving end, the receiver uses the same hash function to calculate the message hash of the received plaintext information. This message hash is compared against the message hash received in the MAC to verify integrity of the received plaintext information.

- If any intruder in between has altered the plaintext information, then the receiver can figure out that the information has been altered by using the process described above. Note that while the intruder can alter the plaintext information, it cannot adjust the MAC of the message in accordance with the alteration in the plaintext information. This is because the MAC is actually an encrypted message-hash, wherein the encryption was done by using a key known only to the sender (the sender's private key).

Public key encryption can be used to implement any combination of security features, namely, confidentiality, authentication and data integrity. Section 14.4.3 discussed how public key encryption could be used for both confidentiality and user authentication. Other valid combinations of security mechanisms using public key encryption are left for the reader to identify and define.

14.4.5 RSA Algorithm

The RSA algorithm is one of the most popular public key encryption algorithms. The letters (R, S and A) in the name of the algorithm stand for the names of the inventors: Rivest, Shamir and Adleman. The algorithm is based purely on mathematical fundamentals, and describes not only the process for encryption and decryption, but also the mechanism for the generation of the public and the private keys. The algorithm is simple to understand, and can be described in the following steps:

1. Select two large prime numbers p and q, and calculate their product n, i.e. $n = p*q$.
2. Calculate $\Phi(n)$, the Euler totient function, as a product of $(p-1)$ and $(q-1)$, i.e. $\Phi(n) = (p-1) * (q-1)$.
3. Next, select an integer e, such that the greatest common divisor (gcd) of e and $\Phi(n)$ is 1, i.e. e and $\Phi(n)$ are relatively prime to each other.
4. Calculate another integer d, such that $e*d = 1 \bmod \Phi(n)$, i.e. $d = e^{-1} \bmod \Phi(n)$.
5. The public key for the encryption process can then be defined as the set (e,n), and the private key as the set (d,n).

The process of encryption and decryption can next be defined as follows:

1. Choose a block size k for encryption, such that $2^k < n <= 2^{k+1}$. Once the value of k is computed, the encryption and decryption process is applied on blocks of k-bit each.
2. If P is the plaintext input to the encryption process, and C is the ciphertext (encrypted text), then the relation

between P and C (Encryption process) can be defined as $C = P^e \bmod n$.

3. Similarly, the decryption process is also based on the calculation of an exponential, modulo n, and is defined as $P = C^d \bmod n$.

The strength of the RSA algorithm is based on the mathematical fundamental that it is extremely difficult to factor n into its two prime factors (p and q). Thus, even though n is known as part of the public key, factoring it into its two prime factors is a huge computational effort. Hence, if brute force mechanisms were to be applied to factor n, and to thus compute the private key, it would require huge computational complexity, and a large amount of time, making the process useless. As a practical example, calculations have shown that if a 428-bit value of n is chosen, it would take 5000 MIPS-years to factor n using brute force mechanisms (where MIPS-years is defined as a million-instructions-per-second processor running for one year). Thus, RSA algorithm is one of the most simple and robust public key algorithms known, and is commonly used in many commercial implementations.

14.4.6 Elliptic Curve Cryptography

Elliptic Curve Cryptography (ECC), similar to RSA algorithm, is based purely on mathematical principles. As the name suggests, the cryptographic principles in this scheme are based on the concepts related to elliptic curves. An elliptic curve can be defined mathematically by using the following equation:

$$y^2 + xy = x^3 + ax^2 + b$$

where x and y are variables, and a and b are constants. The values of x, y, a and b need not necessarily be from the real number space. Instead, these numbers can belong to any number-space, which includes real numbers, complex numbers, integers, integers modulo a prime number and rational numbers. An important principle of Elliptic curves that is used in elliptic curve cryptography is that for two points on the Elliptic curve, it is possible to define a rule for addition such that the result of addition is another third point that lies on the Elliptic curve itself. This rule for addition fulfils the normal mathematical rules for addition, namely Associative Rules and Commutative Rules.

With this definition of elliptic curves, the steps involved in Elliptic Curve Cryptography can thus be defined as follows:

1. Choose an elliptic curve, which is known to all (i.e. the elliptic curve itself is not secret). Also, choose a point F on the elliptic curve that is made public to all the parties that will use ECC for communication amongst them.

2. Next, define a rule for addition of points that lie on the elliptic curve, which fulfils the requirements of addition as noted above.

3. Each party (A and B) then chooses a secret random number (Pvt(A) and Pvt(B)) on the Elliptic curve as the private key.

4. Each party next computes the public key by multiplication of the private key with the number F. Multiplication, in elliptic curves, is similar to multiplication in any other number-space, and involves multiple addition operations (i.e. Pvt(A) $* F$ = Pvt(A) + Pvt(A) + ... Pvt(A), where the number of times Pvt(A) occurs in the addition is F times). Thus, for parties A and B, the public keys are defined as Pub(A) = (Pvt(A) $*$ F) and Pub(B) = (Pvt(B) $*$ F).

5. Public keys are next exchanged between the parties that wish to communicate by using ECC.

6. Each party then computes the session key, which is a shared secret key that can subsequently be used for the communication session, by using symmetric encryption algorithms (for e.g. DES). The session key is computed as:

   Session Key = [Pvt(A) $*$ Pub(B)] = [Pvt(B) $*$ Pub(A)]

   Thus, ECC is used to calculate public and private keys, which are subsequently used for the calculation of a shared secret key that can be used for communication using symmetric encryption schemes.

The strength of Elliptic Curve Cryptography arises from mathematical principles. Mathematically, it can be seen for example, that even though Pub(A) and F are publicly known, computation of Pvt(A) from the product (Pvt(A) $*$ F) is a processing intensive operation. Hence, it is assumed that the computation of Pvt(A) using brute-force methods shall at least take more time than the duration of the communication session, and by then, the attacker would have no benefit of knowing Pvt(A).

ⵏ14.5 KEY MANAGEMENT

Sections 14.3 and 14.4 discussed two varieties of security mechanisms that are used for security management. However, as mentioned in these sections, both approaches for security management rely on a mechanism for sharing keys between the sender and the receiver. For the symmetric encryption techniques described in section 14.3, a shared secret key between the sender and the receiver is the prime basis for the success of the technique. While the same is not true for public-key encryption techniques, some mechanism for exchange of public keys is still required, as was discussed in section 14.4.1. This section discusses some of the key management techniques that can be used for exchange of keys.

14.5.1 Key Management for Symmetric Encryption Techniques

As mentioned above, symmetric encryption techniques require the availability of a shared secret key between each pair of communicating entities. This secret key can be shared between the pair of communicating entities using any of the following mechanisms:

- **Physical delivery of the key:** In this approach, one of the communicating parties chooses a secret key, and physically delivers the key to the other party. Alternatively, a third party can choose this secret key and physically deliver the key to both the parties that wish to engage in a secure communication. While this approach is the easiest of all approaches, its implementation in large networks is infeasible. In other words, this approach can work for link-level encryption between pair of adjacent nodes, but is hard to implement for end-to-end encryption.
- **Remote delivery of keys via a secure channel:** In this approach, the communicating parties can share the keys to be used for a communication session over another secure channel. There are two alternatives for this approach. The first alternative assumes that the first key between any pair of communicating nodes was delivered physically, as mentioned in the previous approach. Subsequently, keys

for subsequent communication sessions can be exchanged in an encrypted way by using the old key used for communication. However, this alternative has the drawback that if an attacker succeeds to gain access to even one of the keys, he/she can then obtain all subsequent keys being exchanged. Hence, this alternative is rarely used.

The second alternative, which is more popular, and is being used in a lot of communication networks, is based on the concept of third-party distributing keys to each pair of communicating nodes, over an existing secure channel. In this approach, each node in the secure network is designed to have a secure communication channel with a third party *Key Distribution Center (KDC)*. This secure communication channel is implemented by using a shared secret key, known only to the KDC and the node in the network. This key is also sometimes called the *Master key*. Using this secure communication channel, a pair of nodes that wish to participate in a secure communication can obtain *session keys* from the KDC. The session keys are shared secret keys between the communicating nodes, which are used only for a single communication session, and then discarded. Thus, each time two nodes wish to securely communicate with each other, they would first obtain a session-key for that communication session from the KDC. A practical usage of this alternative is seen in the Kerberos authentication application that is discussed in the following section.

14.5.1.1 Kerberos

Kerberos is a system for security management that was developed at MIT as part of project Athena. It implements two broad functions of security management, namely user authentication, and key management, for the exchange of shared secret keys. The system can be used in any network, where users request for service from different types of servers. In such a network, Kerberos can be used to restrict access to the servers to only authorized users. Also, Kerberos can be used to exchange secret keys that can be used for confidential communication between the users and the servers.

The functioning of the Kerberos system is depicted in Fig. 14.7. The Kerberos system consists of two servers, an *authentication server* and a *ticketing server*. The authentication server in the Kerberos system is responsible for restricting access to only legitimate and authorized users. Thus, it authenticates users before they can be granted any service from any of the other servers on the network (including the ticketing server). Once authenticated, users are issued a ticket, which can only be used to obtain service from the ticketing server. Using this ticket issued by the authentication server, users then contact the ticketing server, which issues another ticket that can be used by the user to obtain service from any other server on the network. Thus, the entire process of authentication and service grant can be seen as a three-step process, as follows:

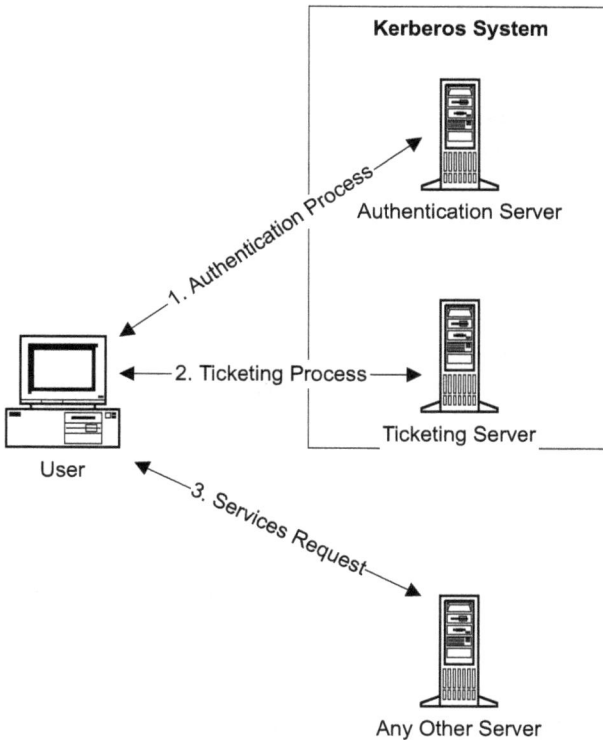

Fig. 14.7 *Kerberos system*

- **Authentication process:** In this step, the user first requests the authentication server to authenticate it, and grant access to the ticketing server. The user sends its identity in the request to the authentication server. On receipt of this request, the authentication server verifies if the user (as known by the identity) is a legitimate user. If so, the authentication server sends back two pieces of information back to the user. This includes the shared secret key (shared between the user and the ticketing server) to be used for communication with the ticketing server, and a ticket that is to be sent to the ticketing server to obtain service from the ticketing server. Both these pieces of information are encrypted using the master key, which is the shared secret key between the user and the authentication server. This ensures that only the legitimate user can read the response from the authentication server. Any attacker masquerading as a legitimate user will not be able to read this response, since the attacker will not know the master key.

- **Ticketing process:** Using the response received from the authentication server, the user requests the ticketing server to grant it access to a particular server on the network. The user sends the ticket received from the authentication server to the ticketing server, along with the identity of the server it wishes to obtain service from. The ticket issued by the authentication server is itself encrypted by using the master key shared between the authentication server and the ticketing server. This prevents the user from tampering with the ticket, and ensures that only the ticketing server can read the information within the ticket. Inside the ticket, the information includes the shared secret key (shared between the user and the ticketing server). Thus, using this ticket, the ticketing server gets to know the secret key to be used for confidential communication with the user. Since both the user and the ticketing server have now securely obtained the shared secret key, the ticketing server uses this secret key to encrypt its response back to the user. In its response back to the user, the ticketing server sends back another shared secret key, which is to be shared between the user and the other server on the network, which the user wishes to obtain service from. Also, the ticketing server sends back another ticket, which is to be used by the user to get service from the other server on the network.

- **Service request:** After the user has been successfully authenticated, and the ticketing server has issued it a secret key and ticket for communication with the other server on the network, the user sends the service request to the server. In the request to the server, the user sends the ticket issued to it by the ticketing server. Similar to the ticket issued by the authentication server, this ticket issued by the ticketing server is also encrypted using the master key shared between the ticketing server and the other server on the network. Hence, the user cannot tamper with this ticket as well, and only the legitimate server can read the information within the ticket. Within the ticket, it contains the shared secret key to be used for communication between the server and the user. Using this method, the user and the Other Server have now securely obtained the shared secret key for communication, and can now proceed to have a confidential dialogue.

To prevent replay attacks, the tickets issued by the authentication server and the ticketing server can additionally contain timestamps, and the validity time for the ticket. This can prevent attackers from capturing legitimate tickets, and then replaying them multiple times later. For more details on the Kerberos system, the reader should refer to [Secu W. Stallings].

14.5.2 Key Management for Asymmetric Encryption Techniques

As discussed in section 14.4, the only requirement for having a confidential exchange of information using public key encryption is that each party involved in the communication knows the public key of the other party. The public key of each user can be made public using many different schemes. Some of the possible schemes are as follows:

- **Announcement of public keys:** In this approach, users can broadcast their public keys via different communication applications. As an example, users can send emails to all parties with which they might ever be interested in communication, and include the public key as part of the email text. Alternatively, public keys might be distributed into multiple multicast groups and/or mailing lists. While this approach is simple to implement, its major weakness is its inability to authenticate the sender of the public key. For example, it is possible for an attacker to impersonate another user and yet distribute its own public key. Other parties involved in communication with this user will never know that in reality, they are communicating with the attacker. The attacker, on the other hand, can simply decrypt all packets being sent to the user, by using his/her own private key. Because of this major weakness, announcement of public keys is not considered a secure means for key distribution.

- **Directory of public keys:** Like a telephone directory, a directory of public keys can be published, which holds the public keys of all known users. New users can be added to the directory just like new telephone subscribers are added to the telephone directory. Anyone wishing to participate in a secure communication with a user can look up this directory to fetch the public key of the user. A certified authority can be given the responsibility to maintain the directory of public keys.

- **Exchange of public keys prior to the start of communication:** Using this scheme, two parties wishing to engage in a secure communication can exchange their public keys with each other prior to start of secure communication. Note that the exchange of the public keys does not require a secure communication channel, since the public key is anyways meant to be known to all. One of the well-known mechanisms for key exchange that is based on this principle is the Diffie-Hellman key exchange, discussed later in section 14.5.2.1.

However, as mentioned earlier, while the public key itself can be exchanged prior to communication start over a non-secure channel, there is no guarantee of the user sending the public key. As discussed in the first approach for key exchange (using public announcement), an attacker can send its own public key and yet impersonate someone else. Hence, even with this approach, user authentication is not possible.

- **Use of public-key certificates:** To solve the above problem, another scheme commonly used for public key exchange is by using public key certificates. Certified authorities issue these certificates, which contain, at the bare minimum, the user identity and public key of a user. Each user can thus obtain its public key certificate from a certifying authority. This certificate can be exchanged between the two parties involved in communication, prior to communication start, to exchange public keys. The concept of public key certificates is discussed in more detail in section 14.7.
- **A hybrid approach:** A hybrid approach for public key exchange can be used. In this approach, the public key directory can be first referred to obtain the other parties public key. Incase an entry is not found for a user in the directory, then any of the other schemes can be used in which exchange of the public keys happens formally prior to start of secure communication.

The next section discusses the Diffie-Hellman approach for key exchange, which has become popular due to its simplicity in generating public-private keys, and also session keys. While the Diffie-Hellman approach might not have the same security as public-key certificates (discussed in more detail in section 14.7), its simplicity makes it a popular approach, and is a good case study.

14.5.2.1 Diffie-Hellman Key Exchange

The Diffie-Hellman key exchange algorithm is a simple and elegant approach for exchange of session keys using public-private keys, which are themselves generated during the time of key exchange. The algorithm is based on purely mathematic fundamentals. The effectiveness of the algorithm depends on the mathematical understanding that it is extremely difficult to compute *discrete logarithms*. Before progressing on to discuss the key exchange algorithm, a brief introduction to discrete logarithms is in order.

Consider a prime number p. A primitive root of this prime number can be defined as any number n, such that powers of n when divided by p gives as remainder all numbers between 1 to $(p-1)$. In other words,

Set $(n \bmod p, n^2 \bmod p \ldots n^{p-1} \bmod p)$ = Set $(1, 2 \ldots p{-}1)$

Alternatively defined, if x is any number in the range from 1 to p-1, then, it is possible to find a unique exponent y such that:

$$x = n^y \bmod p, \ 1 <= y <= (p{-}1)$$

Using the above definition, we can define the exponent y as the discrete logarithm of the number x for the base $n \bmod p$. For discrete logarithms, we shall use the following notation:

$$y = dlog_{np}(x)$$

Mathematical experience has shown that while it is extremely easy to calculate exponentials of a number modulo a prime number, it is extremely hard to calculate discrete algorithms. In other words, in the notation defined for discrete logarithms, even if the numbers n, p and x are known, it is extremely hard to calculate the value of y.

Using the above brief discussion of discrete logarithms, we can now proceed to understand the Diffie-Hellman key exchange algorithm. The Diffie-Hellman approach for key exchange is as follows:

1. Define two publicly known values: a prime number p, and a primitive root of p, namely n. These two values form the basis of the algorithm, and must be made public to all parties that wish to use this algorithm for key exchange.

2. Consider next that the two parties, A and B, wish to exchange shared session keys for use in secure communication. User A can independently select a random number $Pvt(A)$ as its private key, such that $1 <= Pvt(A) <= (p{-}1)$. A then computes its public key $Pub(A)$ as:

$$Pub(A) = n^{Pvt(A)} \bmod p$$

3. Similarly, party B can calculate its public and private keys $Pub(B)$ and $Pvt(B)$ respectively.

4. Next, both parties exchange their public keys with each other.

5. With the public key of party B, party A calculates the session key as:

$$Ks\ 2 = [Pub\ (B)]^{Pvt(A)} \bmod p$$

6. Similarly, party B calculates the session key as:

$$Ks = [Pub\ (A)]^{Pvt(B)} \bmod p$$

Using the above steps, both parties A and B have exchanged the session key Ks that can be used for secure communication. Mathematically, it is easy to see that both the session keys are identical in value, and the calculation is left for the reader.

The strength of the Diffie-Hellman algorithm is based on the difficulty in computing discrete logarithms. While the numbers n and p are publicly known, an attacker can intercept the exchange of public keys, and also gain access to $Pub(A)$ and/or $Pub(B)$. However, with our understanding of discrete logarithms, we know that even with these information in hand, it is extremely hard for the attacker to calculate $Pvt(A)$ and $Pvt(B)$, which are nothing but discrete logarithms. This becomes even more difficult for the attacker if the prime number p is chosen to be an extremely large number. This gives Diffie-Hellman key exchange algorithm its strength.

📡 14.6 HASH FUNCTIONS

As introduced in section 14.4.4, hash functions are commonly used for providing data integrity (or message authentication), which means that the message is authentic and is not tampered with. Hash functions can also be used for providing user authentication, which means that it is possible to determine that the message is coming from an authentic user.

For providing data integrity, instead of encrypting the entire message using the sender's private key, normally, only a message hash is encrypted. The message hash is obtained by applying a publicly known hash function on the plaintext information. The benefit of this hashing process is that while the plaintext information might be of a huge size, message hash is normally of a fixed smaller size. Hence, the amount of information to be encrypted is significantly reduced.

The message hash, after encryption, is also known as the Message Authentication Code (MAC). Since only the original sender can encrypt the message hash (only the sender has the private key), no attacker in between can tamper the message and adjust the MAC accordingly. Further, the MAC, when sent along with the message, also solves the purpose of user authentication. This is possible because only the original sender of the message could have encrypted the message hash to form the MAC, and no one else. The following sections discuss two most commonly used hashing algorithms, namely MD5 and SHA-1.

14.6.1 Message Digest 5

The Message Digest 5 (MD5) algorithm takes as input a message of any arbitrary length and produces a 128-bit message hash as output. In MD5 terminology, this message hash is also called the *message digest* of the input message. While the MD5 algorithm is computationally complex to understand, a simple description of the algorithm is provided in this section. The MD5 algorithm can be defined as a sequence of the following steps:

1. **Append padding bits:** The input message is padded/extended such that its length (in bits) modulo 512 is 448. In other words, the message is extended in a manner so as to make it just 64 bits less than being a multiple of 512 bits.

2. **Append length:** Next, a 64-bit representation of the length of the message is appended to the end of the message. The length of the message is the length before the padding bits were added. In case the length of the message is greater than 2^{64}, then only the low-order 64 bits of the representation of length (in bits) are used. After appending the length to the message, the resulting message now has a length that is an exact multiple of 512 bits.

3. **Divide into blocks:** After the message is made a multiple of 512 bits, it is divided into blocks of 512 bits each. Each block of 512 bits is itself considered as a sequence of sixteen 32-bit words. The MD5 algorithm is then applied on each of these blocks in a sequence, as described in the next step.

4. **Apply MD5 Algorithm on each block:** The MD5 algorithm is a *block-chained* hashing algorithm. In MD5, a hash function is first applied to the first block in the message. An initial seed value is provided as a parameter to the hash function, along with the 512-bit message block. The output (hash) of the first block is then added with the seed itself, and this sum then becomes the seed for the next block. When the hash of the last block is computed, its value becomes the hash for the entire message. Thus, the seed for each block depends upon the hash of the previous

block and hence, blocks cannot be hashed in parallel, but are hashed in a sequential fashion.

At the heart of the MD5 algorithm are four auxiliary functions, each of which take as input three 32-bit words and produce as output one 32-bit word. These auxiliary functions are used to perform the hashing for each of the blocks in the entire message.

5. **Output the hash:** The message digest produced for the entire message is the output of the hash function when it is applied to the last block in the message. This hash is a 128-bit sequence, and works as the Message Digest for the entire message.

The strength of the MD5 algorithm arises from the fact that it is computationally infeasible to produce two messages having the same message digest, or to produce any message having a pre-defined message digest. Since the MD5 algorithm is designed to work with 32-bit words, it is quite fast when run on 32-bit machines. For more details of the MD5 algorithm, the reader should refer to [RFC 1321].

14.6.2 Secure Hash Algorithm 1

The Secure Hash Algorithm 1 (SHA1) is a hashing algorithm similar to the MD5 algorithm. SHA1 takes as input a message of length less than 2^{64} bits and produces a 160-bit output. This output, similar to MD5, is called a message digest in SHA1 terminology. Two different methods for the computation of the message digest are described in the SHA1 algorithm. While these methods differ from the method used in MD5, the computation of the message digest is based on similar principles as in MD5.

The strength of SHA1 lies in the fact that it is computationally infeasible to find a message that corresponds to a given message digest. Similarly, it is computationally infeasible to find two different messages that produce the same message digest. In this regard, SHA1 algorithm characteristics are similar to the characteristics of the MD5 algorithm. However, since the message digest in SHA1 is larger in size than that in MD5, SHA1 is considered more secure against brute-force attacks when compared to MD5. For more details on the SHA1 algorithm, the reader is referred to [RFC 3174] .

⚓14.7 DIGITAL SIGNATURES AND CERTIFICATES

Digital signatures and *digital certificates* are two other popularly used concepts related to security management, and a brief discussion of these concepts is worthwhile to cover. Digital signatures, as the name suggests, solve a function similar to handwritten signatures: that of non-repudiation. Suppose that party A sends a confidential message to party B, by using symmetric encryption schemes, e.g. DES. There are two scenarios that can arise. In the first scenario, party B can modify the message, and then re-encrypt it by using the shared secret key between A and B. B can then later claim the message to have come from A. Thus, B can forge the contents of the message from A. In the second scenario, it is possible that A refuses to accept that it sent a message to B, even though B is in possession of the legitimate message. In both the scenarios, what is required is something similar to handwritten signatures. In other words, all messages from party A to party B must be signed digitally, so that neither B can forge the message, nor can A refute that it sent the message. This is the concept behind Digital Signatures.

Digital Signatures are implemented in a similar fashion to Message Authentication Codes (MAC), as discussed in section 14.6. A digital signature augments any message that is transmitted, similar to MAC. To form a digital signature, the sender first computes a hash of the original message, using the hash functions as discussed in section 14.6. This message hash can be encrypted by the private key of the sender, and appended to the original message, as a digital signature. Since MAC can be used to solve the user authentication problem (as discussed in section 14.6), hence, using the same concepts in digital signatures solves the problems described above.

While digital signatures seem to solve the problems listed in this section, it is still not a foolproof mechanism. For digital signatures to be successful, the receiver is required to have the public key of the legitimate sender. Since it is impractical for the receiver to maintain the public keys of all the potential senders, normally, as a practice, the sender himself/herself sends the public key along with the digitally signed message. However, this makes it very simple for an attacker to masquerade as the legitimate sender, and send his/her own public key along with the digitally signed message. The digital signature in this case is easy to generate, since using the attackers own private key would generate the signature. In this scenario, there is no way for the receiver to know that the sender is an attacker and not a legitimate sender. Hence, a secure means for exchange of public keys is required, using which, the receiver can authenticate that the public key is coming from a legitimate sender and not from an attacker. It is here that the concept of digital certificates finds its significance.

Digital certificates are issued by certifying authorities, which are well-known and trustworthy agencies. A digital certificate, at the bare minimum, contains the identity of a user, and the user's public key. The certificate itself is digitally signed by the certifying authority, using digital signatures. The use of digital certificates solves the problem of user authentication discussed above. When party A sends a digitally signed message to party B, it also sends across the digital certificate that it has received from the certifying agency. Party B now performs the following steps. Firstly, it uses the well-known public key of the certifying agency to verify that the digital certificate is indeed signed by the legitimate certifying authority. Once party B determines that the digital certificate is authentic, it uses the information in the digital certificate to crosscheck the identity of the sender, and to recover the public key of the sender. Thus, by using digital certificates, two parties involved in a communication session can exchange their public keys with each other in a secure manner.

14.8 FIREWALLS

While concepts related to confidentiality, data integrity and user authentication form the crux of security management, the discussion on security management principles is incomplete without a mention of firewalls. As depicted in Fig. 14.8, a firewall is a network entity that resides between a private network (for example, the local area network of an organization) and a public network (for example, the Internet). The firewall implements one or more of the security mechanisms discussed

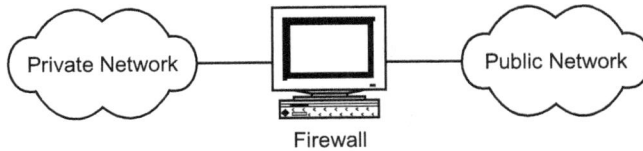

Fig. 14.8 *Location of firewall in a private network*

in this chapter. This removes the burden for implementation of the security mechanisms from the other nodes within the private network.

As depicted in Fig. 14.8 , all traffic coming in and going out from the private network is necessarily made to pass through the firewall. The firewall ensures that only authorized traffic is allowed to pass through it. The network administrator can configure local security policies at the firewall to control the traffic that is allowed to pass through the firewall. For example, the network administrator could define service-level controls to control the services (telnet, http, ftp, etc.) for which traffic can pass through the firewall. This can be done, for example, by defining restrictions on the TCP port addresses in the IP packets. The network administrator may also define user-level controls at the firewall, to control the users who have access to some specific services. For example, the network administrator can control which all users are allowed to access the Internet.

14.9 EXAMPLE: SECURITY MANAGEMENT IN THIRD GENERATION UMTS NETWORK

In the earlier sections, this chapter has discussed the elements of security management. This section looks at the security management in the context of third generation UMTS network. In general, wireless networks pose greater security challenges as compared to wireline networks. The reason for this is the shared air interface that is vulnerable to external attacks. While the notion of security applies end-to-end, the focus has always been the access over the air interface. In the context of mobile cellular networks, there has been steady increase in security capabilities of the network.

In Second Generation (2G) mobile networks, there was *limited support* for the security aspects. One of the major limitations of 2G networks was the user's inability to check whether the network providing the service was authorized to do so or not. Apart from this limitation, there was no support for data integrity in 2G. The cipher keys and authentication data were transmitted in cleartext within the network. Moreover, the security features between two network elements were also nonexistent.

When the Third Generation (3G) networks were designed, it was decided that the security architecture would be built upon the security architecture of 2G networks. Those aspects that were found to be robust would be retained. The areas requiring improvement would be looked into. Thus, the use of Subscriber Identity Module (SIM), as a removable hardware was retained. The subscriber identity confidentiality on the radio interface was also retained. The use of shared secret key between the user and the network was also continued. As is explained later, the shared secret key forms the basis for all authentication procedures.

To strengthen security features, the notion of *mutual authentication* was introduced. This allowed the user to authenticate the network. Provision for the network to authenticate the user already existed in 2G. By allowing the user to authenticate the network, mutual authentication was possible. Then, the possibility of brute force attack was also reduced by increasing the length of ciphering key from 8 bytes to 16 bytes. Apart from this, network domain security was introduced to provide security features between different nodes of a network.

Figure 14.9 shows the *four basic functional blocks* of UMTS security, explained in the following four sections. The first block—*user domain security*—provides users a secure access to mobile stations. This is achieved through the use of PIN codes at the interface between the user and USIM.

The next functional block—*network access security*—encompasses security features to facilitate secure access to UMTS services. Network access security is provided through *user identity confidentiality, mutual authentication, confidentiality* and *data integrity*.

Network Access Security: Set of security features that provide users with secure access to UMTS services and protect against attacks on the air interface. Network Access Security includes:
- User identity confidentiality
- Mutual authentication
- Confidentiality
- Data Integrity

Components of UMTS Security

User Domain Security: Set of security features that enable users secure access to mobile stations.

Network Domain Security (network layer): Set of security features that enable nodes in the provider domain to securely exchange signalling data, and protect against attacks on the wireline network. These security features are applied at network level (e.g. at IP layer).

Network Domain Security (application layer): Set of security features that enable nodes in the provider domain to securely exchange signalling data, and protect against attacks on the wireline network. These security features are applied at application level (e.g. at MAP layer).

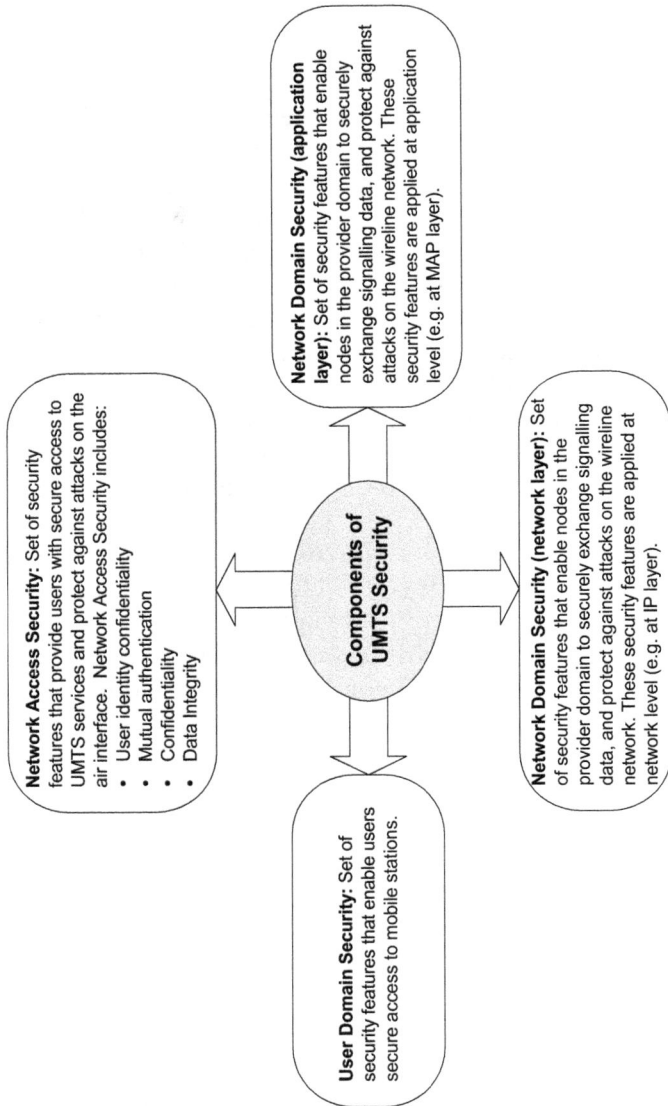

Fig. 14.9 *Functional Blocks of UMTS Security*

The third functional block is of *network domain security*. This security feature enables the network nodes to securely exchange signalling data. Network domain security can be provided at two levels: *application layer* and *network layer*. Application layer security in UMTS networks is provided through MAP security protocol. *MAPsec* is used between two entities (e.g. between HLR and VLR) that use MAP protocol to securely communicate with each other. Here, note that the term 'application' is seen in relation to the application layer signaling protocol and not in the context of user applications.

At the network layer, the network domain security is provided through *IPSec* or IP Security protocol. The IPSec protocol suite includes a key distribution protocol, e.g. Internet Key Exchange (IKE) protocol, and a protocol that provides authentication and confidentiality, e.g. Encapsulating Security Payload (ESP) protocol. Various architectures are possible to deploy IPSec; in the case of UMTS, the IPSec is used to provide secured communication between two signalling gateways. Thus, two entities of different domains use the services of signalling gateways to ensure secured communication. Within a domain, the exchange may or may not be secure. Further, network domain security applies only to control plane, not for user plane. In other words, network domain security seeks to protect signalling data and not user data.

14.9.1 User Domain Security

The user domain security function ensures that only an authenticated user can gain access to a mobile station. This functionality is achieved by restricting access to the Universal Subscriber Identity Module (USIM) until the user has been authenticated. For this, a secret password (called the PIN) is shared between the user and the USIM. The password resides in the USIM and the user has the option to modify it. Access to USIM is granted only when the correct password is provided by the user. In order to ensure that brute force is not applied to obtain the pin, the USIM is rendered unusable after a given number of unsuccessful attempts. Among the various security features, the user domain security is simplest.

14.9.2 Network Access Security

In the UMTS, the phrase 'access security' encompasses a number of security aspects. The term access refers to accessing or utilizing the services provided by the network. Access security principles are applied to ensure that only legitimate users are allowed access to the network. In this regard, two concepts are defined. The aspect, whereby the serving network corroborates the identity of the user, is referred to as *user authentication*. The second aspect, whereby the user corroborates that the serving network is authorized by the user's home network to provide service, is known as *network authentication*. The user authentication and network authentication are collectively referred to as *mutual authentication*, which is one of the most important components of access security.

A scheme called the *UMTS Authentication and Key Agreement (UMTS AKA)* provides the means to achieve mutual authentication. UMTS AKA also provides the means whereby security keys are generated as part of the authentication procedure. These security keys are of two types: *Integrity Key (IK)* and *Cipher Key (CK)*. The integrity key, IK, is used for maintaining the integrity of signalling data between the mobile terminal and the access network. The integrity protection principles are not applied for the user plane due to performance reasons. The ciphering key, CK, is used to encrypt the user and signalling data. The encryption provides confidentiality over the air interface.

The mutual authentication is accomplished by using a secret key K that is shared by the Authentication Center (AuC) of the Home Environment (HE) and the USIM of the Mobile Station (MS). This secret key is never exchanged between the user and the network. Rather, it is a part of the USIM at the time it is made. The AuC keeps a mapping between the permanent user identity IMSI and the secret key K. Thus, AuC knows the secret key K of a given user identified by the *International Mobile Subscriber Identity (IMSI)*. The knowledge of secret key K and that of security algorithm used for mutual authentication is proof of an authentic user. Thus, authentication in UMTS is based on private key cryptography technique (as against public key cryptography technique).

Apart from the secret key K, the USIM and AuC also share a sequence counter called Sequence Number (SQN). The SQN maintained at AuC is called the SQN$_{HE}$ (referring to SQN maintained by HE). The SQN$_{HE}$ is essentially an individual counter that AuC maintains for each user identified by its permanent identity IMSI. At the MS, the USIM also maintains a sequence number SQN$_{MS}$. The SQN$_{MS}$ denotes the highest sequence number accepted by the USIM. The use of SQN$_{MS}$ and SQN$_{HE}$ is to ensure the presence of a dynamic variable for authentication (apart from the static key and algorithm).

Using the secret key K and sequence counter SQN, the UMTS Authentication and Key Agreement (UMTS AKA) works as follows (see Fig. 14.10): In order to authenticate a user, the Visitor Location Register (VLR) or Serving GPRS Support Node (SGSN) (depending upon whether the domain is circuit-switched or packet-switched) requests AuC to send an array of authentication vectors. The authentication vectors are a set of five elements (generated using the secret key K and sequence counter SQN). The elements of authentication vectors are as follows:

- **RAND:** 16 byte field, refers to a random number computed using pseudo-random-number generators
- **XRES:** 4 to 16 byte field, refers to the expected response from USIM
- **CK:** 16 byte field, refers to cipher key, which is used for ciphering purposes
- **IK:** 16 byte field, refers to integrity key, which is used for integrity protection
- **AUTN:** 16 byte field, refers to authentication token, which is used by the serving network to authenticate itself to the MS. The SQN used to generate an authentication vector forms a part of the AUTN.

Each authentication vector is used for one authentication and key agreement between the VLR/SGSN and the USIM. Each set of vectors achieves the same function. The vectors are used by the VLR/SGSN in the order in which they are received. The VLR/SGNS pre-fetches an array of vectors from the AuC to expedite authentication when it takes place again. From a given authentication vector (which is the first in the available set of

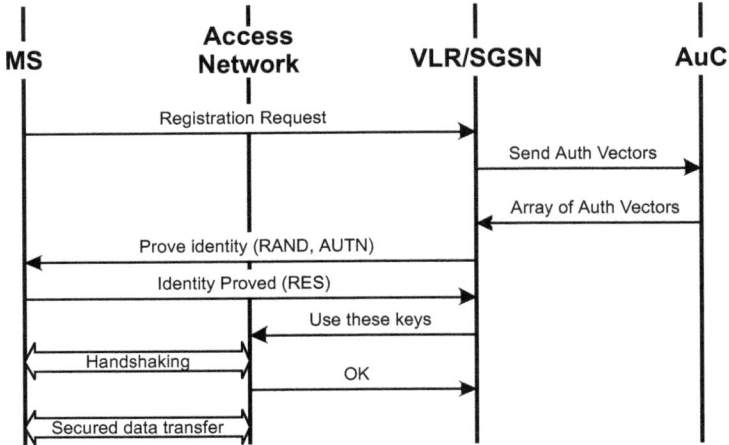

Fig. 14.10 *Mutual authentication in UMTS network*

vectors), the VLR/SGSN only picks the elements RAND and AUTN and sends them to USIM.

On receiving these fields, USIM runs the same algorithm that was earlier used by AuC to generate the authentication vectors. During the execution, of the authentication algorithm the USIM uses the RAND and a part of AUTN as input. The output includes RES, XMAC, CK and IK. The XMAC is compared with the last 8 bytes of AUTN (which is called the MAC). In case XMAC is equal to MAC, the network is said to be authenticated by the user. The RES is sent by the USIM to VLR/SGSN. The CK and IK generated during the process are identical to the CK and IK received by VLR/SGSN from AuC. This CK and IK are later used for encryption and data integrity. When the VLR/SGSN receives the RES, it checks whether the latter is equal to the XRES (belonging to the authentication vector used for authentication). In case RES is equal to XRES, the user is said to be authenticated by the network. This completes the process of mutual authentication. The CK and IK are now ready to be used at the USIM and Access Network (AN) for encryption and data integrity.

In summary, the essence of the whole algorithm is based on a random number RAND. Only this number is sent over the air interface along with a Message Authentication Code (MAC). Exchange of RAND and MAC do not pose any security threat.

Note that MAC is a one-way function and knowledge of MAC and RAND is not adequate to break the secret key. Thus, network sends MAC to authenticate itself, while user sends RES (a type of authentication code) to authenticate itself. This forms the mutual authentication. The same RAND is also useful for generation of Cipher Key (CK) and Integrity Key (IK). This is also important because no separate key management protocol is required and the random number along with shared key K is adequate for generation of CK and IK. Since random number changes with each authentication, the CK and IK also change, thereby ensuring frequent changes in the security keys.

Apart from mutual authentication, data integrity, and confidentiality, another important feature of access security is user identity confidentiality, which refers to the property that the permanent identity, IMSI, of a user is not eavesdropped on along the radio access link. In order to achieve this objective, the user is allocated—and is identified by—a temporary identity called the *Temporary Mobile Subscriber Identity (TMSI)*. However, in some rare cases, it is required that the user provides its permanent identity *International Mobile Subscriber Identity (IMSI)*. Such scenarios occur when the core network does not have knowledge of this permanent identity, nor can it obtain this information from any other means. Apart from such specific scenarios, a user is always identified by the TMSI.

The mobile equipment identification function is also considered to be part of access security. In this, the serving network seeks the *International Mobile Equipment Identity (IMEI)*, from the MS to verify its legitimacy. The Equipment Identity Register (EIR) is used to verify the IMEI. The EIR maintains three lists of IMEI, namely the white list, black list and the gray list. These lists are used to check the legitimacy of the mobile equipment.

14.9.3 Network Domain Security Using MAPSec

In the 2G networks, the absence of security features in SS7 was viewed as an important security weakness. However, this weakness was justified in view of the fact that the SS7 networks were controlled by a small number of large companies. Thus, security threats arising out of the interoperability between networks were not seen as a prime threat. With the large-scale deployment of UMTS networks, and the burgeoning demand for inter-operability and global roaming, the security requirements have drastically altered. Thus, in addition to the access security mechanism (using AKA), recent specification of Third Generation Partnership Project (3GPP) mandates the use of network domain security mechanisms for control plane signalling within and between the core networks. Since Mobile Application Part (MAP) protocol is the most important protocol used in the core network, the focus has been to incorporate security features in MAP protocol.

At the MAP application level, the security protocol used to protect MAP messages is referred to as the *MAPsec protocol*. Formally [3GPP TS 33.200] specifies that 'the complete set of enhancements and extensions to facilitate security protection for the MAP protocol is termed MAPsec and it covers transport security in the MAP protocol itself and the security management procedures.'

MAPsec provides four security services. First, it provides data integrity. The integrity of the MAP message is protected by using a 32-bit Message Authentication Code (MAC). The MAC is also used for origin authentication. MAPsec also provides anti-replay protection. This is done by including a 32-bit time stamp in the MAP message header. The presence of a time-stamp ensures that the MAP messages cannot be stored and replayed (if the MAP messages are actually stored and replayed, the presence of a stale time-stamp indicates a possible violation). Lastly, MAPsec optionally provides confidentiality. Confidentiality is achieved by encrypting the original MAP message. To provide these four security services, MAP peers establish a *Security Association (SA)* between them. The SA contains the parameters (i.e. security keys) necessary for the operations of MAPsec.

Defining a mechanism for obtaining security keys in MAPSec has been a challenging task for specification writers. During the early phases of implementation, manual key exchange was provided (implying that the nodes engaged in MAPSec did not have means to obtain or change security keys automatically). Even during the phases where automatic key exchange was suppose to be incorporated, the standards were incomplete. This only proved the difficulty in designing a good key management protocol. This also explains the beauty of UMTS AKA that not only provides authentication but also provides means for key agreement.

14.9.4 Network Domain Security Using IPSec

In the UMTS Core Network, the Circuit Switched (CS) domain and Packet Switched (PS) domain primarily use MAP protocol for exchange of signaling information. Thus, security provisions have been added in MAP. This security enabled MAP, called MAPsec, has been discussed in the previous section.

Another development in UMTS has been the introduction of the *IP Multimedia Sub-system (IMS)*. The IMS domain uses the IP protocol at the network layer, the Stream Control Transmission Protocol (SCTP) protocol at the transport layer, and Session Initiation Protocol (SIP) at the application layer. Among these protocols, SCTP, SIP protocols do not provide any security features. Thus, the onus of secured exchange is with the IP layer using the *IP Security (IPSec) protocol*. The IPSec, as defined in [RFC 2401], provides the security architecture for IP-based transport.

The applicability of IPSec for IMS domain and for other domains using IP-based transport (e.g. Sigtran for CS/PS domain) is defined in [3GPP TS 33.210]. This specification is also referred to as NDS/IP (meaning Network Domain Security for IP-based protocols). Network Domain Security for IP-based protocols (NDS/IP), as defined in [3GPP TS 33.210], is not a new protocol specification. Rather, it merely refers to the IPSec protocol suite specified in RFC 2401 to RFC 2412 and defines how this protocol suite is applicable to UMTS. For example, while IPSec defines use to two security protocols, namely Authentication Header (AH) and Encapsulating Security Payload (ESP), [3GPP TS 33.210] specifies that only ESP is adequate to satisfy the requirements of the Network Domain Security (NDS/IP).

The security features provided by NDS/IP are similar to those of MAPsec. The features include data integrity, data origin authentication, anti-replay protection and confidentiality (optional). NDS/IP also provides limited protection against traffic flow analysis when confidentiality is applied.

14.10 CONCLUSION

In this chapter, concepts related to security management were discussed. The term 'security' is an overloaded term, and involves multiple concepts, namely authentication, confidentiality, data integrity, anti-replay and non-repudiation. Security algorithms are divided broadly into two categories, namely symmetric (or secret-key) algorithms and asymmetric (or public-key) algorithms. Examples of symmetric algorithms include DES and AES, while those for asymmetric algorithms include RSA and elliptic curve cryptography. For both categories of algorithms, key management forms an important aspect. Some of the well-known mechanisms for key management include the Diffie-Hellman mechanism, and the Kerberos implementation.

Another important aspect related to security management is the concept of hash algorithms. Hash algorithms are used to compute a message digest, which can be used to solve the dual purpose of user authentication and data integrity. Use of message digest for user authentication is the important concept behind digital signatures. The use of digital signatures, combined with digital certificates, is one of the commonly used, practical mechanisms for the key exchange for asymmetric algorithms.

Wireless networks pose greater security challenges as compared to wireline networks. The reason for this is the shared air interface that is vulnerable to external attacks. Towards the latter part of the chapter, the UMTS wireless network was discussed, which described the practical use of the security management principles discussed in the first part of the chapter.

REVIEW QUESTIONS

Q 1. *What are the major elements of security management? How are these addressed in UMTS networks?*

Q 2. *What is the difference between symmetric and asymmetric encryption techniques? Describe a scenario wherein a combination of both schemes can be used to solve the dual purpose of secure key exchange and confidential information transfer.*

Q 3. *What is the basis behind the RSA algorithm? Describe the conditions under which this algorithm can be rendered useless.*

Q 4. *What are the pre-requisites of a good security management protocol?*

Q 5. What is the beauty of UMTS AKA protocol? Why does it not require a separate key management protocol?

Q 6. In UMTS, which elements are sent in cleartext during authentication? Do these pose any security threat? If no, why? If yes, what threats?

Q 7. What is network domain security? What are the two levels in UMTS at which Network Domain Security applies?

FURTHER READING

[Secu W. Stallings] is an excellent reference for knowing more about cryptography and related topics. [Secu Kahate], [Secu W. Stallings 2] and [Secu R. Bragg] also provide an overview on the subject. For the case study presented in this chapter, the 3GPP specification [3GPP TS 33.102] provides a comprehensive overview of security related procedures in 3G UMTS. Further, the user domain security function is specified in [3GPP TS 31.101], MAP security in [3GPP TS 33.200] and IP based network domain security in [3GPP TS 33.210]. [Wireless S. Kasera] also discusses aspects related to UMTS security.

REFERENCES

[3GPP TR 25.922] 3GPP Technical Report 25.9222, "Radio Resource Management Strategies".

[3GPP TS 45.002] 3GPP Technical Specification 05.02, "Multiplexing and Multiple Access on the Radio Path".

[3GPP TS 25.214] 3GPP Technical Specification 25.214, "Physical Layer Procedures (FDD)".

[3GPP TS 25.401] 3GPP Technical Specification 25.401, "UTRAN Overall Description".

[3GPP TS 31.101] 3GPP Technical Specification 31.101, "UICC-terminal interface; Physical and logical characteristics".

[3GPP TS 33.102] 3GPP Technical Specification 33.102, "3G Security; Security architecture".

[3GPP TS 33.200] 3GPP Technical Specification 33.200, "3G Security; Network Domain Security (NDS); Mobile Application Part (MAP) Application Layer Security".

[3GPP TS 33.210] 3GPP Technical Specification 33.210, " 3G security; Network Domain Security (NDS); IP network layer security".

[Addr 3COM IP Addressing] 3COM White paper, "Understanding IP Addressing: Everything You Ever Wanted To Know", http://www.3com.com/other/pdfs/infra/corpinfo/en_US/501302.pdf.

[ATM D. McDysan] David E. McDysan and Darren L. Spohn, "ATM Theory and Application", McGraw-Hill.

[ATM M. Prycker] Martin De Prycker, "Asynchronous Transfer Mode: Solution for Broadband ISDN", Prentice-Hall.

[ATM S. Kasera] Sumit Kasera, "ATM Networks: Concepts and Protocols", Tata McGraw-Hill.

[ATM U. Black] Uyless D. Black, "ATM Volume I: Foundation for Broadband Networks", Prentice-Hall.

[ATM W. Goralski] Walter J. Goralski, "Introduction to ATM Networking", McGraw-Hill.

[ATMF TM4.1] ATM Forum specification, "Traffic Management Specification 4.1", af-tm-0121.000, March, 1999.

[Bri R.Perlman] R. Perlman "Interconnections: Bridges and Routers", Addison Wesley.

[DLC U. Black] U. Black, "Data Link Protocols", Prentice-Hall.

[FIPS 197] Federal Information Processing Standard (FIPS) 197, "Advanced Encryption Standard".

[Gen A. Tanenbaum] Andrew Tanenbaum. "Computer Networks", Prentice-Hall of India.

[Gen D. Comer] Douglas Comer, "Computer Networks and Internets", Pearson Education India.

[Gen J. Kurose] James F Kurose, "Computer Networking: A Top-down Approach Featuring The Internet", Pearson Education India.

[Gen S. Keshav] Srinivasan Keshav, "An Engineering Approach to Computer Networks", Pearson Education India.

[Gen W. Stallings] William Stallings "Data and Computer Communications", Prentice-Hall of India.

[IEEE 802.1D] IEEE specification 802.1D, "MAC Bridges".

[IEEE 802.2] IEEE specification 802.2, "Logical Link Control".

[IEEE 802.3] IEEE specification 802.3, "CSMA/CD (Ethernet)".

[IEEE 802.5] IEEE specification 802.5, "Token ring access method and physical layer".

[IEEE 802.5x] IEEE specification 802.5x, "Supplement to 802.1Q: Virtual Bridged LANs: Source Routing".

[ITU-T E.164] ITU-T specification E.164, "The international public telecommunication numbering plan".

[ITU-T G.704] ITU-T specification G.704, "Synchronous frame structures used at 1544, 6312, 2048, 8448 and 44 736 kbit/s hierarchical levels".

[ITU-T I.113] ITU-T specification I.113, "Vocabulary of terms for broadband aspects of ISDN".

[ITU-T I.150] ITU-T specification I.150, "B-ISDN Asynchronous Transfer Mode Functional Characteristics".

[ITU-T I.321] ITU-T specification I.321, "B-ISDN Protocal Reference Model and its Application".

[ITU-T I.361] ITU-T specification I.361, "B-ISDN ATM Layer Specification".

[ITU-T I.371] ITU-T specification I.371, "Traffic Control and Congestion Control in B-ISDN."

[ITU-T I.430] ITU-T specification I.430, "Basic User-network Interface-Layer 1 Specification".

[ITU-T I.431] ITU-T specification I.431, "Primary Rate User-network Interface - Layer 1 Specification".

[ITU-T M.3010] ITU-T specification >3010, "Principles for a Telecommunications Management Network".

[ITU-T Q.761] ITU-T specification Q.761, "ISDN User Part Functional Description".

[ITU-T Q.762] ITU-T specification Q.762, "ISDN User Part General Functions of Messages and Signals".

[ITU-T Q.763] ITU-T specification Q.763, "ISDN User Part Formats and Codes".

[ITU-T Q.764] ITU-T specification Q.764, "ISDN User Part Signaling Procedures".

[ITU-T Q.921] ITU-T specification Q.921, "ISDN User-network Interface – Data Link Layer Specification".

[ITU-T Q.931] ITU-T specification Q.931, "ISDN User-network Interface Layer 3 Specification for Basic Call Control".

[ITU-T X.25] ITU-T specification X.25, "Interface between Data Terminal Equipment (DTE) and Data Circuit Terminating Equipment (DCE) for Terminals Operating in the Packet Mode and Connected to Public Data Networks by Dedicated Circuit".

[ITU-T X.700] ITU-T specification X.700, "Management framework for Open Systems Interconnection (OSI) for CCITT Applications".

[NM M. Subramanian] Mani Subramanian, "Network Management: Principles and Practice", Pearson Education Asia.

[RFC 1058] C.L. Hedrick, "Routing Information Protocol", June, 1988.

[RFC 1155] M.T. Rose, K. McCloghrie, "Structure and Identification of Management Information for TCP/IP-based Internets", May, 1990.

[RFC 1157] J. D. Case et. al., "Simple Network Management Protocol (SNMP)", May, 1990.

[RFC 1212] M.T. Rose, K. McCloghrie, "Concise MIB Definitions", May, 1991.

[RFC 1213] M.T. Rose, K. McCloghrie, "Management Information Base for Network Management of TCP/IP-based internets: MIB-II", May, 1991.

[RFC 1321] R. Rivest, "The MD5 Message-Digest Algorithm", April 1992.

[RFC 1441] J. Case, et. al., "Introduction to Version 2 of the Internet-standard Network Management Framework", April, 1993.

[RFC 1483] J. Heinanen, "Multi-protocol Encapsulation over ATM Adaptation Layer 5," July, 1993.

[RFC 1771] Y. Rekhter, T. Li, "Border Gateway Protocol 4", March, 1995.

[RFC 1902] J. Case, *et. al.*, "Structure of Management Information for Version 2 of the Simple Network Management Protocol (SNMPv2)", January, 1996.

[RFC 1903] J. Case, *et. al.*, "Textual Conventions for Version 2 of the Simple Network Management Protocol (SNMPv2)", January, 1996.

[RFC 1904] J. Case, *et. al.*, "Conformance Statements for Version 2 of the Simple Network Management Protocol (SNMPv2)", January, 1996.

[RFC 1907] J. Case, *et. al.*, "Management Information Base for Version 2 of the Simple Network Management Protocol (SNMPv2)", January, 1996.

[RFC 2328] J. Moy, "OSPF Version 2", April, 1998.

[RFC 2453] G. Malkin, "RIP Version 2", November, 1998.

[RFC 2460] S. Deering, R. Hinden, "Internet Protocol, Version 6 (IPv6)", December, 1998.

[RFC 2960] R. Stewart, *et. al.*, "Stream Control Transmission Protocol," October, 2000.

[RFC 3174] D. Eastlake 3rd, P. Jones, "US Secure Hash Algorithm 1", September, 2001.

[RFC 768] J. Postel, "User Datagram Protocol", August, 1980.

[RFC 791] J. Postel, "Internet Protocl", September, 1981.

[RFC 793] J. Postel, "Transmission Control Protocol", September, 1981.

[RFC 826] D.C. Plummer, "Ethernet Address Resolution Protocol: Or Converting Network Protocol Addresses to 48.bit Ethernet Address for Transmission on Ethernet Hardware," November, 1982.

[Rou M. Sportack] Mark Sportack, "IP Routing Fundamentals", Cisco Press.

[Secu Kahate] Kahate, "Cryptography and Network Security", Tata McGraw-Hill.

[Secu R. Bragg] Roberta Bragg, *et. al.*, "Network Security (the Complete Reference)", Tata McGraw-Hill.

[Secu W. Stallings 2] William Stallings, "Network Security Essentials", Pearson Education India.

[Secu W. Stallings] William Stallings, "Cryptography and Network Security: Principles and Practice", Pearson Education Asia.

[Sig R. Onvural] Raif. O. Onvural and Rao Cherakuri, "Signalling in ATM Networks", Artech House.

[Swi Fouad A. Tobagi] Fouad A. Tobagi, "Fast Packet Switch Architectures for B-ISDN," Proceedings of the IEEE, Vol. 78, No. 1, Jan 90, pp 133–167.

[Swi H. Ahmadi] H. Ahmadi and W. E. Denzel, "A Survey of Modern High-performance Switching Techniques," IEEE Journal on Selected Areas in Communication, Vol 7, No. 7, September, 1989, pp. 1091–1102.

[Swi J.S. Turner] J.S. Turner "Design of a Broadcast Packet Network," INFOCOM '86, Miami, 1986.

[Swi M. G. Hluchyj] M. G. Hluchyj and Mark Karol, "Queueing in High-performance Packet Switching," IEEE *Journal on Selected Areas in Communication*, Volume 6, Number 9, Dec, 1988, pp. 1587–1597.

[Swi Nick McKeown] Nick McKeown and Thomas E. Anderson, "A Quantitative Comparison of Scheduling Algorithms for Input-queued Switch", Computer Networks, Volume 30, Number 24, Dec, 1998.

[Swi Y. S. Yeh] Y. S. Yeh, *et al.*, "The Knockout Switch: A Simple, Modular Architecture for High Performance Packet Switching," IEEE Journal on Selected Areas in Communication, Volume 5, Number 8, Oct, 1987, pp. 1274–1283.

[TCP/IP D. Comer] Douglas Comer, "Internetworking with TCP/IP, Volume I", Prentice-Hall.

[TCP/IP R. Stevens] Richard Stevens and Gary R Wright, "TCP/IP Illustrated", Volumes 1 and 2, Pearson Education India.

[TM H. G. Peros] H. G. Peros and K. M. Elsayed, "Call Admission Control Schemes: A Review," IEEE Communication Magazine, Vol. 34, No. 11, pp. 82–91, November, 1996.

[TM J. Roberts] J. Roberts, U. Mocci and J. Virtamo, "Broadband Network Teletraffic," Heidelberg: Springer.

[TM K. Shiomoto] K. Shiomoto et. al., "Overview of Measurement-based Connection Admission Control Methods in ATM Networks," IEEE communication surveys, http://www.comsoc.org/pubs/surveys, First Quarter, 1999.

[TM Raj Jain] Raj Jain, Congestion Control and Traffic Management in ATM Networks: Recent Advances and A Survey, Computers Networks and ISDN Systems, Volume 28, 1998, pp. 1723–1738.

[TM V. Jacobson] V. Jacobson, Congestion Avoidance and Control, Proceedings of SIGGCOMM, 1988.

[Wireless A. Viterbi] A. J. Viterbi, "Principles of Spread Spectrum Communication", Addison Wesley.

[Wireless B. Walke] B. Walke, "Mobile Radio Networks", John Wiley & Sons.

[Wireless C. Bettstetter] C. Bettstetter, H.J. Vogel, J. Eberspacher, "General Packet Radio Service GPRS: Architecture, Protocols, and Air Interface", IEEE Communications Survey, Third Quarter 1999, volume 2, Number 3.

[Wireless H. Holma] Harri Holma and Antti Toskala, "WCDMA for UMTS: Radio Access for Third Generation Mobile Communications", John Wiley & Sons.

[Wireless J. Korhonen] Juha Korhonen, "Introduction to 3G Mobile Communications", Artech House.

[Wireless M. Mouly] M. Mouly and M. B. Pautet, "The GSM System for Mobile Communications", Published by the Authors.

[Wireless S. Kasera] Sumit Kasera and Nishit Narang, "3G Networks: Architecture, Protocol and Procedures", Tata McGraw-Hill.

INDEX
